MICROBIAL GEOCHEMISTRY

Microbial Geochemistry

EDITED BY

W. E. KRUMBEIN, PhD

Professor of Geomicrobiology
University of Oldenburg, West Germany

BLACKWELL SCIENTIFIC PUBLICATIONS
OXFORD LONDON EDINBURGH BOSTON MELBOURNE

Editorial offices:
Osney Mead, Oxford, OX2 0EL
8 John Street, London, WC1N 2ES
9 Forrest Road, Edinburgh, EH1 2QH
52 Beacon Street, Boston
Massachusetts 02108, USA
99 Barry Street, Carlton
Victoria 3053, Australia

First published 1983

Printed at the Alden Press, Oxford

Microbial geochemistry
1. Microbial ecology
I. Krumbein, W. E. ~~II. Series~~
576′.15 QR100

ISBN 0-632-00683-8

DISTRIBUTORS

USA
Blackwell Mosby Book Distributors, 11830 Westline Industrial Drive, St Louis, Missouri 63141

Canada
Blackwell Mosby Book Distributors, 120 Melford Drive, Scarborough, Ontario, M1B 2X4

Australia
Blackwell Scientific Book Distributors, 31 Advantage Road, Highett, Victoria 3190

CONTENTS

LIST OF CONTRIBUTORS

STANLEY M. AWRAMIK Department of Geological Sciences, University of California at Santa Barbara, Santa Barbara, California 93106, USA.

JACQUES BERTHELIN Centre de Pédologie Biologique du C.N.R.S., B.P. 5, 54501 Vandoeuvre-les-Nancy cédex, France.

T. HENRY BLACKBURN Institute of Ecology and Genetics, University of Aarhus, Ny Munkegade, DK-8000 Aarhus C, Denmark.

CHARLES D. CURTIS Department of Geology, University of Sheffield, Mappin Street, Sheffield S1 3JD, UK.

BO B. JØRGENSEN Institute of Ecology and Genetics, University of Aarhus, Ny Munkegade, DK-8000 Aarhus C, Denmark.

ANDREW H. KNOLL The Biological Laboratories, Harvard University, Massachusetts 02138 USA

WOLFGANG E. KRUMBEIN Geomicrobiology Division, University of Oldenburg, D-2900 Oldenburg, West Germany.

KENNETH H. NEALSON University of California in San Diego, Scripps Institution of Oceanography, MBRD, A-002, La Jolla, California 92093, USA.

PETER K. SWART Fisher Island, Rosenstiel School of Marine and Atmospheric Sciences, University of Miami, Miami Beach, Florida 33139, USA.

DIETRICH WERNER Fachbereich Biologie, Philipps-Universität, Marburg-L, West Germany.

PREFACE

Truth is a daughter of time

Two colleagues, my seniors in science and experience, have encouraged me to edit this book and to invite my fellow scientists to contribute. They proposed the title of *Microbial Geochemistry*, but two other, almost synonymous terms are also related to this title and to my field of work: these are 'Geomicrobiology' and 'Biogeochemistry'.

Geomicrobiology has only relatively recently been recognized as an academic subject. When I first became interested in studying it in 1957 it was not possible. However, some 22 years after I was informed that there was no such science as geomicrobiology, I was nominated a geomicrobiologist at the University of Oldenburg; apparently the first university position in the subject.

I owe much of this achievement to Georg Knetsch who had enough intuition to stimulate and encourage study in both these fields. Vernadsky and Kuznetsov, Winogradski, Hutchinson and others have also had a great impact on the areas covered in this book and on myself. However, at the time when I was a student neither the microbiologists nor the geologists in Germany could envisage the importance of these new interdisciplinary studies and of work in these fields. Science had lost its unitarian views and was dominated by specialization in many new disciplines and methodologies. It is only recently that the reintegration of the study of nature has progressed and attention has been directed towards unifying concepts such as the 'Gaia-hypothesis' of Jim Lovelock and Lynn Margulis.

Many friends and colleagues not mentioned here have contributed to this book through their work and discussions with me. I should like to thank them and my thanks go also to Heia, Franziska, Daniel and Friederike for their indulgence while this book was being compiled.

The scientific data presented in my chapters and the exchange of information with colleagues was made possible by grants from the following institutions: Deutsche Forschungsgemeinschaft, Volkswagenstiftung, Lotto-Fonds of Niedersachsen, Stifterverband der Deutschen Industrie, NASA, Minerva-Stiftung, and by invitations from many scientific societies and universities. I am also grateful to Gisela Gerdes, Gisela Koch, Christine Giele, Klaus Liebig, Immo Raether, Peter Rongen and Brigitte and Jürgen Steinfeld who helped with drawing, typing and work on the SEM, and also for their encouragement. In addition I would like to thank my students for their patience during the writing and editing of this book.

Work in this field has barely begun and I hope that this book will encourage young biologists and geoscientists to attempt the seeming impossible, to try to reintegrate the study of nature and natural history at all levels as Giordano Bruno and Baruch Spinoza did before them. When we have achieved an intuitive understanding of the structure and function of nature and of life, then we may be able to understand the survival of life on this planet and the impact of life on geological cycles. We may also understand and believe that man, although it seems untenable today, has little impact on the structure and function of life on earth.

CHAPTER 1
INTRODUCTION

WOLFGANG E. KRUMBEIN

The sum of all matter filling the infinite universe is constant.

Antoine de Lavoisier (1789)

The sum of all energy, active in the infinite universe and causing all phenomena is unalterable.

Robert Mayer (1842)

The sum of all energy and matter filling the universe is constant but can be transferred in one another.

Albert Einstein (1921)

The reciprocal effects between mass, space and time to form 'substance' are not yet studied in detail but it is evident that the corpuscular nature of 'aether' can be envisaged as another 'Wechselwirkung' of light and mass.

Ernst Haeckel (1900)

Wie wundervoll sind diese Wesen,
Die, was nicht deutbar, dennoch deuten,
Was nie geschrieben wurde, lesen,
Verworrenes beherrschend binden
Und Wege noch im ewig Dunkeln finden.

Hugo von Hoffmannsthal (1894)

A textbook on microbial geochemistry and geomicrobiology should perhaps not try to solve the critical questions of what 'substance' is as compared to 'matter' and how 'aether' can be defined. One is, however, tempted to look into the 'Wechselwirkungen' (interplay) between life and inanimate material in terms comparable to those that try to define the universe as infinite and non-infinite in one; wherein black holes, created by the collapse of matter, consume matter and create new masses and light to balance the expansion of mass and space at other places. Giordano Bruno's heroic death still sends its message of life through the centuries. Giordano (Filippo) Bruno (1548–1600) had his most fruitful years in London writing and working out the three major books: *De la causa, principio e uno*, *De l'infinito universo e mondi* and *Lo spaccio della bestia trionfante*. The dualistic view of the world partitioned in matter and energy, finite and infinite, world and God is fused into a pantheistic view of the world, man and God. God is nature. The creation concept is replaced by the theory of interactions in eternal change.

Death is transferred into division and composition, redivision and recomposition.

The mineral and microbial mineral cycles as we view them today on the basis of experimental work have been envisaged as the unifying concept of world and universe, creating the principle of the one living nature of Bruno and Spinoza. Their monad theory was later reappraised in a rather non-critical way by Ernst Haeckel. The basic approach of Bruno, however, is still alive and is evidenced in scientific and mathematic terms by non-Euclidian geometry, by the modern field theories and Einstein's relativity and gravity theories, as well as by Lovelock's 'Gaia-hypothesis'. Giordano Bruno has deeply influenced Baruch Spinoza, Leibnitz, Kant, Goethe and Schelling. He still influences unitarian thought in science and philosophy.

It is evident, therefore, that a textbook on microbial geochemistry, which will try to define the intimate relations between inorganic material and living matter, between the energy input and storage of energy in living organisms, its release and cybernetic control on all major mineral cycles on earth, must come back to Bruno's original thoughts of 'cyclic developments' rather than 'creation and destiny' as revealed in the clerical Christian thoughts of his time which have so severely inhibited the development of science.

Microbial geochemical cycles have been going on since life on earth began. The idea of life on earth as an entity composed of myriads of monads all contributing to one single purpose, i.e. the continuation and stabilization of a 'steady state' for that entity via constant input of energy and transformation of material would have fascinated Giordano Bruno. This is what Lovelock's Gaia-hypothesis states and what he expresses as Gaia: 'The physical and chemical condition of the surface of the earth, of the atmosphere, and of the oceans has been and is actively made fit and comfortable by the presence of life itself. This is in contrast to the conventional wisdom which held that life adapted itself to the planetary conditions as it and they evolved their separate ways'. He continues: 'Gaia is a complex entity involving the earth's biosphere, atmosphere, oceans, and soil; the totality constituting a feedback or kybernetic system which seeks an optimal physical and chemical environment for life on this planet. The maintenance of relatively constant conditions by active control may be conveniently described by the term homeostasis.' The unitarian view of nature which is filled with scientific data by the Gaia-hypothesis (Lovelock 1979) can also be seen in some modern physical approaches to space, matter and energy as derived from thoughts of Giordano Bruno and later Baruch Spinoza. Bruno, Spinoza, Haeckel and Lovelock have one thought in common which separates them from all deterministic views of the world and

its structure; they claim, in close relationship to the basic laws of matter and energy, that 'the sum of life once established on the earth (or in the universe) will tend to keep itself constant.'

This last provocative statement is based mainly on our studies of geochemical, mineral and rock cycles as they are influenced, modified, altered and cybernetically controlled by the sum of life on this planet. After more than 150 years of geomicrobiological studies and searching for the traces of life in the fossil record; after all studies on the manifestations of life in rocks, sediments, water and atmosphere from the present day and far back into the Precambrian, we may be tempted to conclude from Heisenberg's and Prigogine's statements and calculations on the interaction of masses of material and amounts of energy, once it is proven to be a cyclic interaction, that the mass of life is constant to the same degree and underlies the same physical principles as matter, energy and space. The general relativity theory may, therefore, also be applied to living systems. Evolution, which in principle could be regarded as increasing organization and thus decreasing entropy (Riedl 1973, 1979) on the basis of statistical physics, should control itself by the equilibration of increase *and* decrease of order to yield a balance to changing external energy flows. When external energy pressure increases, more information should be stored, when it decreases, life should continue with less storage of energy and information.

The implication of these assumptions is that life forms an integral part of the basic principles of the universe as we understand it today. This implies a 'field theory' of life which means that any kind or quality of life which is extinguished will produce new kinds and qualities of life to continue the basic principle. If one then wonders how life emerged and covered the planet during the early Precambrian, one must ask of the physicists, How is matter destroyed and turned into energy? How is energy transferred into matter and how are the relative points of view to be respected? In other words, the question of the reality of space and time in an objective sense is the same as the question about the reality of life in space and time.

Masses and concentrations of matter can be regarded as transformable into energy and vice versa and they are related to velocity. The mass of a certain body is naturally much more important than its weight which can be altered with its degree of acceleration. Life consists of acceleration and retardation of chemical reactions and of the embodying of continua of space and time into a special reference system, in non-linear reaction networks. Natural selection and evolution can be regarded as natural consequences of self-reproducing entities. Eigen and Schuster (1979) with their hypercycle theory combine many of the thoughts of modern physics of the space–time system with the line of thoughts derived from Darwinian evolution. The realistic hypercycle model is a target for experimental testing whereby the

target itself can be changed as postulated by Einstein in his approaches. They also cite Einstein, 'Only theory can tell us which experiments are to be meaningful.' They end the introduction by the 'Gaian' statement that life is a 'regularity of nature'.

Geomicrobiology may help us to understand the regulation principles which rule the cybernetics of the planetary biogeochemical system. Self-organizing reaction networks are familiar to life, to space and matter in space, as well as to biogeochemical cycles when these are cybernetically controlled by microbes and microbial material cycles and hypercycles. This book deals with many, though not all aspects of microbial geochemical cycles in the past and present. Its stress is on the present environment but it tries to look back into other periods of time and life and its spatial distribution.

References

Eigen M. & Schuster P. (1979) *The hypercycle. A principle of natural self-organization.* Springer Verlag, Berlin.

Lovelock J.E. (1979) *Gaia. A new look at life on earth.* Oxford University Press, Oxford.

Riedl R. (1973) Energie, Information und Negentropie in der Biosphäre. *Natürwiss. Rundschau* **26,** 413–20.

Riedl R. (1979) *Biologie der Erkenntnis.* Parey, Berlin.

CHAPTER 2
THE MICROBIAL CARBON CYCLE

W. E. KRUMBEIN AND P. K. SWART

2.1 INTRODUCTION

Carbon is the central atom of life. The capacity of carbon to make covalent bonds as well as to share electron pairs with oxygen, hydrogen, nitrogen and sulphur is a necessary feature of life. The cycles of all other important elements tie in with that of carbon. There is a seemingly endless number of possible organic compounds, which can be produced or degraded by microorganisms (e.g. 10^6 different compounds in crude oil). Microbial pathways, however, can be reduced to a relatively small and surprisingly logical number of molecular building blocks and reaction chains essential for the maintenance of life.

The three essential ingredients of life: matter, energy and information, rule the basic biological building blocks in a way that leads to a diversified multitude of compounds, variations in energy transformation, molecular and cellular construction and transfer and exchange of information. The amount of information within the cell or the cell's nucleic acids can possibly be exceeded by man-made information carriers or storage units. However, the astonishing recombination and variation principle which is also embedded within the informational molecules of the cell, namely nucleic acids and their building and transfer machinery has never been matched by man's industrial and communicational systems.

2.2 PRODUCTION OF ORGANIC CARBON COMPOUNDS (PRIMARY PRODUCTION)

Primary production is usually regarded as the process by which cells produce reduced organic carbon compounds from inorganic radicals using energy input from either solar energy or chemical bond energy of inorganic compounds. In order to understand the complexity of the present day organic carbon cycle and to relate it to the fossil remains of former variations of the carbon cycle it is necessary to try to assemble palaeontological, geochemical, genetic, biochemical and physiological data on the evolution of microbial activities. Today it is possible to analyse phylogenetic routes of the organic carbon cycle and to compare them with the scarce palaeontological evidence we have of early life. It is possible to relate data on the

distribution of specific organic molecules in sedimentary rocks to specific groups of microorganisms and their activities in the carbon cycle. Furthermore it seems possible to draw conclusions from the distribution of different inorganic radicals and gases of the present day geochemical reservoirs as they are biologically influenced and to attempt to extrapolate from the chemical status of former sediments and sedimentary cycles the microbial communities of those times. These two sets of data can ultimately be compared and thus be used to verify the two approaches:
(1) the phylogenetic approach based on tRNA and rRNA-sequencing, DNA-hybridization or computer analyses of complex proteins;
(2) the geochemical-palaeontological approach based on the analysis of microfossils, 'fossil marker molecules' and carbon isotope fractionation data.

One has to keep in mind, however, that the approaches of molecular genetics as well as in physiology, geochemistry and palaeontology are still not sufficiently sound to describe accurately the history of different types of microbial carbon cycles which definitely do exist and definitely must have emerged in a logical and temporal sequence. Perhaps we will never be able to decipher the past from the two sets of information we presently use. One set is the palaeontological and geochemical study of often highly metamorphosed sedimentary rocks. These rocks have been altered chemically and physically by diagenetic changes (heat, pressure and solutions circulating through the rock crevices and interstitial pores) and metamorphic changes (pressure and temperature-controlled mineral transformations) which occurred after the hard rocks had been formed. By means of these two processes the original, biologically produced chemical and physical system has been altered to such an extent that it may make firm conclusions difficult (Hayes *et al.* 1983).

The second set of information lies in the cell itself and in the relation of different cells of different organisms to each other. It is difficult to analyse from existing genetic information of present day microorganisms which bacterial genus developmentally preceded the other.

Schidlowski (1980) deduced from hundreds of geochemical analyses of rocks that the ratio between reduced, i.e. biologically fixed, carbon in sediments and oxidized, i.e. biologically excreted and precipitated, carbon compounds in rocks has been stable on this planet for 3.5–3.7 billion years. He and others (Golubic *et al.* 1979) claim that only few minor oscillations have occurred in the relatively stable global system of biological cycling of carbon (Fig. 2.1). Krumbein (1980) concluded that practically all carbonate rocks which are forming now and which have been formed in the past are as 'biogenic' as any of the reduced carbon compounds embedded in sediments and rocks. Por (1980), on the other hand, claims that a tremendous increase

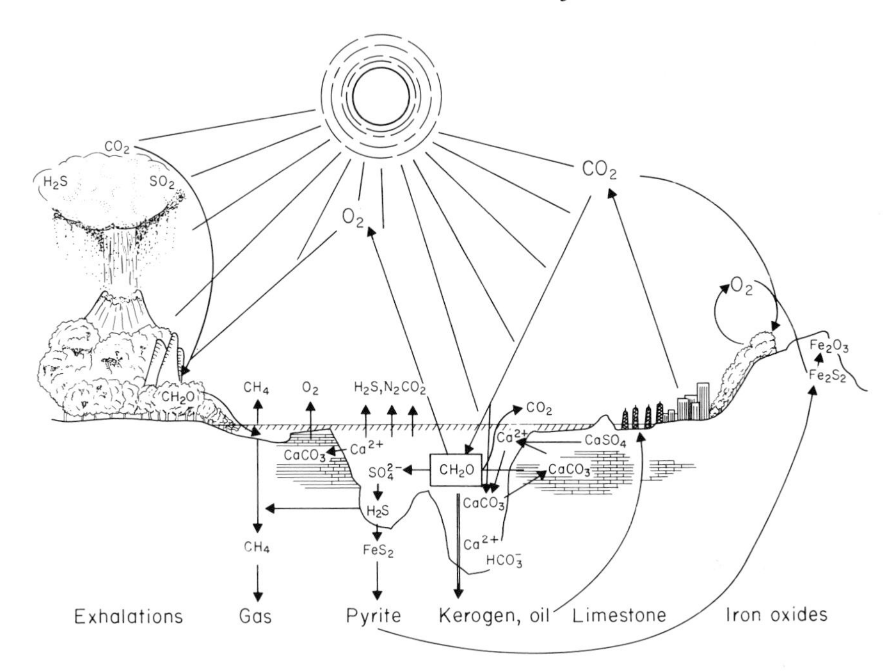

Fig. 2.1. The carbon cycle in relation to pools of oxidized and reduced substances in the exogenic cycle of materials.

of the global biomass can be attributed to the evolution of land plants and faster-accumulating climax ecosystems, leading to a present day maximum.

Before a scenario of the evolution of primary productivity is given, the data which are presently used for the reconstruction of former cycles and phylogenetic trees as well as the methods, and problems of the methods to determine primary productivity and organic carbon pools and reservoirs shall be briefly described.

2.2.1 Methods of study related to the biogeochemical cycles of carbon and primary productivity

A classical example of microbial ecosystems which have the potential of preservation as a whole or in parts are stromatolites, i.e. laminated, microbially made, sedimentary structures usually formed by photosynthetic bacteria of various groups (Krumbein 1983). No precise method has yet been developed to define quantitatively the production of organic material in these systems and to estimate the turnover rates of the organic carbon compounds produced. The methods used in plankton productivity cannot be directly transferred, inasmuch as they usually work with homogeneous suspensions of material. Such an approach was taken by Cohen *et al.* (1977) when laminated microbial mats were homogenized and $^{14}C\text{-}HCO_3^-$ was added so that the amount of radioactivity in comparison to total bicarbonate was a measure of the incorporation of inorganic carbonate. These figures were then expressed as gram carbon fixed per surface area or volume of total carbon. The validity of such data is further limited by determinations of and correlation to total protein, chlorophyll, ATP and even DNA or RNA (Vollenweider 1971, Sorokin & Kadota 1972, Cohen *et al.* 1977, Karl *et al.* 1980, Lorenz *et al.* 1981).

Oxygenic photosynthesis can be measured directly by oxygen measurements in samples exposed to light as compared to the same samples or aliquot samples kept in darkness. The amount of oxygen liberated via oxygenic photophosphorylation is compared to the amount of respiratory oxygen demand and then to surface area, volume or any relatively stable biomass parameter. In the case of mixed populations of oxygenic and anoxygenic photosynthetic microorganisms or when cyanobacteria switch from oxygenic to anoxygenic photosynthesis (Cohen *et al.* 1975) and when sulphate-reducing or sulphur-oxidizing bacteria contribute to primary production, or when turnover is so rapid that the study intervals are too short, the methodology becomes even more complicated and conclusive data cannot be derived from the ^{14}C or O_2 methods. In these cases microelectrode studies of the flux of O_2 and H_2S can be used in order to determine quantitatively the productivity in laminated benthic photosynthetic populations (Revsbach *et al.* 1980). Another possibility of studying carbon fluxes from the inorganic to the organic phase is the study of the stable isotope transfer during these processes.

Carbon has two stable masses, ^{12}C and ^{13}C, as well as various other short-lived radiogenic isotopes of which ^{14}C is the most important. Organic material is depleted in ^{13}C (approx. $-25‰$) relative to atmospheric CO_2 ($-7‰$) and inorganic carbonate ($0‰$; the $\delta^{13}C$ notation expresses the $^{13}C/^{12}C$ ratio as parts ‰ deviation from international standard PDB (Pee-Dee-formation belemnite standard). The standard PDB is carbonate from a belemnite of the Cretaceous Pee Dee formation, South Carolina; it has an absolute carbon-13 to carbon-12 ratio as defined by Craig (1957) of 0.011 2372). This depletion in organic material occurs during the incorporation of CO_2 during photosynthesis and chemolithotrophy or any other autotrophy for carbon and is the most significant process which affects the fractionation of carbon isotopes in the terrestrial environment. Plants with different routes of photosynthetic assimilation, i.e. the C-3, C-4 or CAM (crassulacean acid metabolizing) plants, fractionate their isotopes to differing extents. Therefore, carbon with a depleted isotope ratio is usually regarded as indicative of an organic origin while more positive ratios are taken as inorganically derived or inorganic in nature.

There are, however, exceptions to this general rule: numerous measurements of diamonds reveal ranges from 0‰ to −30‰ (Deines 1980a); gases emanating from volcanic and hydrothermal vents show large amounts of isotopically light abiogenic gases (Craig 1953), and extraterrestrial samples reveal large ranges of $\delta^{13}C$ values (Grady *et al.* 1982 and others).

The data on the fractionation of stable carbon isotopes in modern geochemical studies of ancient biogenic organic compounds are usually supported by the study of other isotopes and their ratios, e.g. ^{16}O, ^{17}O, ^{18}O, ^{14}N and ^{15}N, and ^{32}S and ^{34}S, where basically the same rule of preferential uptake of the lighter isotope is observed. The combination of these studies with laboratory experiments on the ratio of uptake by several different microorganisms under different physiological conditions and with the associated enrichment of the remaining 'heavier' fraction in sediments (e.g. heavy sulphate remaining in a sedimentary basin in which high activities of sulphate-reducing bacteria were taking place) enables teams of microbiologists, geologists and geochemists to decipher partially the fate of organic matter and to deduce ancient primary productivity data from the trace fossils in sedimentary rocks. All these deductions are, of course, heavily biased by the already mentioned diagenetic and metamorphic changes with geological time. Corrections for these changes, however, are perhaps possible (Hayes 1983, Hayes *et al.* 1983).

2.2.2 Methods of molecular biology which enable extrapolations into past ecosystems

The phylogenetic tree derived from the relatedness of nucleotide homologies of oligonucleotides produced from 16 sRNA (Fox *et al.* 1980) gives some indication as to the relative time of evolutionary separation of several groups of bacteria. The analysis of biochemical capacities, e.g. presence or absence, gain or loss of specific enzymes may also be used in this kind of approach. Trüper (1982), for example, has used information of this sort to discuss the question whether the evolution of photosynthesis may have preceded the evolution of sulphate reduction, or vice versa. Also the onset of oxygenic photosynthesis in molecular biology terms may be related to the onset of anoxygenic photosynthesis by analysing and comparing sulphur isotope and carbon isotope fractionation data in Precambrian sediments. These data can, furthermore, be related to the question of whether certain sediments were deposited under anaerobic or aerobic conditions. Working under these premises the set of mineralogical data on the world-wide distribution of the reduced uraninites or the oxidized iron and manganese minerals of the Banded Iron Formations (BIF) of the Precambrian and Cambrian eras (3.5 billion years to 1 billion years before present) will produce additional information (Krumbein *et al.* 1979, Clemmey & Badham 1982, Grosovsky *et al.* in press).

A final method to study the biogeochemical cycles of carbon compounds in the past and present is the ecosystems approach which combines the methods of palaeobiology with microbiological, botanical, and zoological analyses of present day ecosystems and their productivities. It is evident that the study of primary production, standing crop, degradation and preservation of synthesized organic material in contemporary ecosystems as well as the transfer and flux of organic carbon from one system to the other has to be analysed by ecological methods which include productivity, environment definition, population density and diversity, stability of the system and systems dynamics. Palaeobiology has evolved through the same steps from a mere species-describing science to an analytical science which includes palaeoecology, population dynamics of past periods of earth history, even palaeoproductivity studies, and the analysis of the question of whether populations and ecosystems as, for example, the stromatolites of the Precambrian or the huge animals of the Cretaceous were extinguished or reduced in importance by planetary evolution or biological evolution or both. For these reasons palaeontology has progressed rapidly from a descriptive science ruled by taxonomy to complex palaeoecology and studies of population dynamics as described in Chapter 10.

2.2.3 Evolution of primary productivity

It is assumed that the first steps of primary production of organic carbon compounds were completely abiological, creating the so-called primordial soup. The second step towards the evolution of primary productivity is a problem of semantics. If one regards the production of organic compounds via fermentative pathways from other organic compounds and inorganic CO_2 or CH_4 as a species of primary production, then anoxygenic photosynthesis and special fermentations are the true candidates for the first biological primary production systems. Thus the cycling of organic carbon and its regeneration can be initiated under anoxic and reducing conditions via fermentation and anoxygenic photosynthesis using mostly organic precursors. Primary production, however, has, for long periods, been regarded as the process which yields organic compounds exclusively from inorganic radicals (e.g. CO_2, N_2, SO_4^{2-}, PO_4^{2-}, H_2O) via light or inorganically bound chemical energy (photolithotrophy or chemolithotrophy).

Under inclusion of classical pathways of primary production the following production pathways for organic matter may be imagined. They are named in the sequence of their probable evolutionary occurrence. The approximate times of their initiation and thus the deadline for their influence on geochemical cycles would perhaps be derived on the basis of palaeontological and phylogenetic considerations. Primary production here is redefined as any process that leads to a net increase in reduced carbon compounds in the natural geochemical cycle of carbon, via input of light energy or energy released from chemical bonds.

Abiogenic condensation and polymerization via unorganized (accidental) energy transfer

The most primitive samples of organic matter known to man are arguably contained in type I carbonaceous chondrites (CI) (Hayatsu & Anders 1981). Various different 'cosmothermometers' consistently give temperatures of formation for these meteorites of 360 K, an age of approximately 4.55×10^9 years, and a time scale of formation of the order of 0.2 Myr. At the high temperatures and low pressures present in a contracting solar nebula CO is the stable form of carbon; this transforms to CH_4 around 600 K. It is believed that this change proceeds through complex tarry compounds brought about by partial hydrogenation of CO. Such a process which is thought to liken Fischer Tropsch type synthesis (FTT), has been suggested to employ magnetite and phyllosilicates as catalysts. One problem of abiogenic synthesis is that the organic compounds are thermodynamically unstable with respect to their starting elements. In the case of the FTT synthesis, however, CO is the reactant and reactions to form organic compounds are exoergic thus requiring no external energy.

An alternative process is that when UV or electric discharges produce free radicals and/or unstable molecules. These then react further to yield organic compounds; this type of reaction is known as the Miller–Urey reaction. The extent to which either of these reactions is predominant in extraterrestrial environments is unknown, with various groups of workers supporting one or the other (Hayatsu & Anders 1981).

Primitive primary production (experimental productivity)

Aggregations of organic compounds either absorbed on particles having highly reactive surfaces; or condensed in crevices and interstitial spaces of early sediments, or floating in the primitive ocean combined into 'coacervates' and, when large enough, dispersed into smaller ones via mechanical energy. Special molecules (of enzyme character) catalysed these processes (mostly large abiogenic molecules with central metal cations). Many 'wild-type' enzymes were probably producing substances which no longer exist, or yielded one and the same substance under different energy input and thermodynamic reactions. Later organized membranes became stable, leading, together with probably very large (cell) units in areas of low mechanical disturbance, to the establishment of molecules which were ultimately capable of spontaneous partial copying of energetically probable proteins. Some cell wall peptides and enzyme peptides are still generated without the aid of ribosomal RNA. These may represent a preserved example of an early evolutionary phenomenon. During the prebiotic period membrane structures, primitive RNA and DNA molecules and copy-ready double helices were organizing themselves, probably in great numbers. Some of the mathematical models for this 'pre-proterozoic' experimentation time have been developed by Eigen and Schuster (1979) involving coexistence, competition and chirality (selection of one-handed, biologically significant molecules). tRNAs seem to be the most logical messenger–adapter molecules for the evolution of precise copy procedures. Many geological time periods must have passed before competition and selection ruled out the coexistence of a great variety of primary products and later of 'primary or primitive nucleic acids and organisms'. During this time several 'experimental' ways of CO_2-fixation must have been evolved and may have prevailed for some time by chance. Some of them may have been under some kind of 'genetic' control, which must not have been as efficient as today.

Modern plants exhibit an isotopic range from -10 to $-30‰$ (Deines 1980b) and generally fractionate carbon isotopes to differing extents depending on the initial CO_2 acceptor in the Calvin cycle and environmental conditions. C-4 plants, for example, utilize PEP-carboxylase rather than RuBP carboxylase and consequently exhibit a smaller isotopic effect. CAM plants, on the other hand, show a range of intermediary isotopic values and can employ either system (O'Leary 1981). It is generally believed that C-4 plants are evolutionarily more advanced than C-3 plants as they do not show the phenomena of photorespiration. This is essentially a malfunctioning of the photosynthetic cycle occurring in the presence of oxygen. As C-3s are not only isotopically lighter than C-4s, but also show a greater discrimination against ^{13}C with increasing O_2 concentrations (Troughton 1972), one may speculate that similar variations would be present in the fossil organic record, not only as C-4 plants evolved from C-3 species, but also perhaps as atmospheric O_2 levels increased during the Precambrian and fractionations changed during early evolutionary 'experimental periods'.

Although recent work by Hayes *et al.* (1981) has cast doubts on the geochemical integrity of many of the analyses of Precambrian 'biological' material, they concluded that even after correlation for dehydrogenation carbon isotope abundances were much more variable in Precambrian than Phanerozoic times.

The steadily decreasing range of carbon isotope fractionation data can be perhaps interpreted as a gradual replacement of random CO_2-fixations and energetically less efficient 'experimental CO_2-fixation pathways' by the more efficient and genetically better controlled present pathways.

The most primitive organisms may have been fermenting bacteria which exhibited only partial CO_2-fixation, besides mixed fermentations of abiogenic organic materials, and photoorganotrophic heterotrophs with the capability of fermentation and/or photosynthesis using either CO_2 or abiogenic organic carbon compounds as a carbon source, and either solar energy, reduced carbon compounds, H_2S or reduced metals as an energy source.

Primary production via anoxygenic photosynthesis methane oxidation and iron oxidation

Coccoid and filamentous bacteria had developed by 3.5 billion years ago as witnessed by the Warrawoona formation, West Australia, and in Zimbabwe greenstones, Gnenya formation of the Sebakwian group (Walter *et al.* 1980, Awramik 1981, Orpen & Wilson 1981). These may not have been cyanobacteria although stromatolites have been found. Earlier stromatolites may have been formed by more primitive filamentous photosynthetic or non-photosynthetic bacteria, for example by iron- or methane-oxidizing or heterotrophic bacteria. The interpretation of stromatolites as being of cyanobacterial origin is derived from studies on recent analogues, an interpretation which is neither justified by the morphology of the organisms, nor by organic geochemical analyses, nor by the morphologies of the sedimentary laminated structures (Krumbein 1983). There are also some indications that methane oxidation may occur under anaerobic conditions using terminal electron acceptors other than oxygen. Photosynthesis in the initial biota was probably restricted to reactions involving photosystem I and bacteriochlorophyll *a*. CO_2 was probably reduced via a reductive Kreb's cycle (Fuchs & Stupperich 1981). Both photoorganotrophs and chemolithotrophs would have produced enough organic matter to explain the concentrations reported for the Warrawoona formation. Fig. 2.2 is intended to throw a critical light on the over-interpretation of microfossils (see also Awramik *et al.* 1983). Fe^{++} may also have served as electron donor for photosynthesis with Photosystem I.

*Anoxygenic chlorophyll-*a*-based photosynthesis*

The next step in the evolution of primary production may have been the invention of chlorophyll *a* by ancestral cyanobacteria. Their precursors may

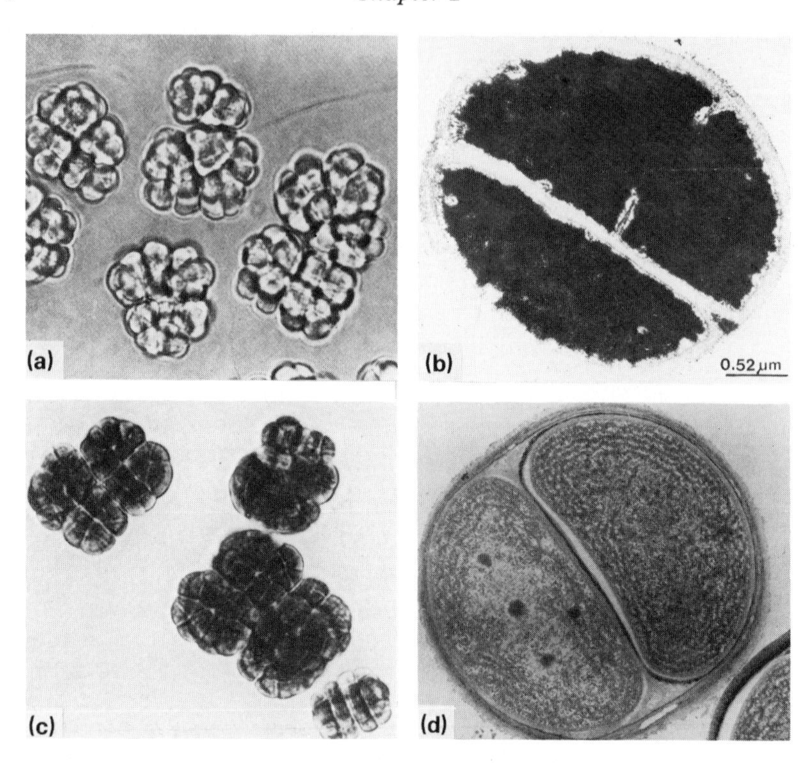

Fig. 2.2. Transmittant light and electron photomicrograph of *Methanococcus* sp. (a,b) and a pleurocapsalean cyanobacterium (c,d). Both organism types belong to very ancient, but physiologically very different groups. Morphological evidence even in such a complete form as in this figure will not be sufficient to define to which fossil genus they belong. Palaeontologists, however, claim repeatedly that the cocci or filaments they see in very old sediments are cyanobacteria because they resemble recent cyanobacteria. In microbial palaeontology, therefore, many other factors and parameters need to be studied (e.g. minerals, isotopes, ecology, trace chemical fossils, etc.) in order to verify the taxonomic position of a given microfossil. (Reproduced with permission from Waterbury & Stanier 1978 and Kandler, personal communication.)

have had chlorophyll *a*, a cyclic photophosphorylation system and exclusively anoxygenic photosynthesis. Later the new pigments (phycobiliproteins) and structures (phycobilisomes) could have been added to the photosynthetic apparatus. The primitive cyanobacteria may already have had two different light-harvesting systems but it is not known whether these were exerting linear photophosphorylation or oxygenic photosynthesis from the beginning. The new hypothesis of Arnon et al. (1981) also assumes independence of photosystems I (cyclic photophosphorylation, anoxygenic) of PS II (linear photophosphorylation, oxygenic). Good palaeontological evidence of cyanobacteria coincides with the world-wide rise of larger and more numerous stromatolite systems and the occurrence of plankton communities contributing different allochthonous morphological forms of organisms to autochthonous benthic stromatolite communities (see Chapter 10; Knoll 1979, Knoll & Golubic 1979). Even then most cyanobacteria may not yet have developed the second light-harvesting system using phycobiliproteins and photolysis of water as electron donor. Cyanobacteria and possibly the Chlorobiaceae previously, are the candidates among the photolithotrophs which may have evolved or incorporated by genetic exchange with other organisms the presently prevailing RuBP-carboxylase-based system of CO_2-fixation.

In terms of CO_2-fixation it is important, furthermore, that the primitive cyanobacteria, as well as any other photosynthetic bacteria, were dependent on sulphide and other quantity-limited electron donors and thus could not increase the primary productivity beyond certain limits. The global primary productivity may still have been rather small relative to later periods.

Bacterial oxygenic photosynthesis

Until a few years ago (and some botanists still believe it is so) a wide gap could be recognized between 'bacterial' and 'plant type' photosynthesis. Bacterial photosynthesis was restricted to some peculiar phototrophic anaerobic groups of bacteria which lived in niches which were anaerobic and bad smelling from hydrogen sulphide—the so-called sulphureta. 'Plant-type' photosynthesis was thought to be the only oxygenic way of photosynthesis based on special organelles, the chloroplasts, which had a special pigment combination of chlorophyll *a* and chlorophyll *b*, existing only in plants and enabling the catalytic hydrolysis of water, thus generating reducing power and oxygen. The cyanobacteria were regarded as plants regardless of their prokaryotic, that is to say bacterial, nature.

Today we know that any type of photosynthesis is basically bacterial or prokaryotic. We further know that cyanobacteria are capable of anoxygenic and oxygenic photosynthesis (linear and cyclic oxygenic and/or anoxygenic photophosphorylation); and it has been shown (e.g. Fisher & Trench 1980, Stackebrand, pers. comm.) that cyanobacteria do exist which have chlorophyll *a* and chlorophyll *b* and no phycobiliproteins, also differing in other structural and biochemical characteristics from the majority of the cyanobacteria but still being phylogenetically, morphologically and biochemically closer to the bulk of cyanobacterial genera than to any other group of bacteria. These researchers present evidence that the photosynthetic bacterium *Prochloron* is actually a cyanobacterium of the Synechocystis type. From this they deduce that practically all eukaryotes exercising photosynthesis are derived from endosymbiotic phototrophic bacteria and that the chloroplast of green plants is probably derived from bacteria very closely related to *Prochloron*. 16 sRNA analyses show, however, no direct relation of *Prochloron* to plastids of plants. The photosynthetic bacteria can thus be divided into the group of obligate anoxygenic photosynthetic bacteria with bacterial chlorophylls and the obligate or facultative oxygenic photosynthetic bacteria with chlorophyll *a* and phycobiliproteins or chlorophyll *a* and *b*.

It is obvious that light energy, which must have played a major role already in the evolution of non-biogenic macromolecules is an unlimited, renewable energy source. All primitive or 'early' sources of reducing power (and perhaps of electron acceptors for fermentation and respiratory chain electron acceptors) were limited. Therefore, it is clear that water, though it is energetically costly to hydrolyse, is an excellent unlimited electron donor and that oxygen in turn is the ideal electron acceptor for the stabilization of living systems far away from thermodynamic equilibrium.

The first problem arising from the water-splitting type of photosynthesis is the fast production and possibly accumulation of a very effective cell poison, namely free oxygen. Free oxygen under natural conditions and in the respiratory chain based on oxygen as terminal electron acceptor produces either the superoxide radical, i.e. O_2^- when oxygen is reduced by a single electron, or the similarly poisonous hydrogen peroxide, when reduced by two electrons. Therefore, mechanisms and enzymes had to be evolved or had to have existed, which transformed this poisonous compound. One enzyme-type essential in ecosystems with high primary productivity is superoxide dismutase, an enzyme which occurs in all aerobic cells and transforms superoxide into molecular oxygen and hydrogen peroxide, which is transformed by catalase. In addition, some simple amino acids and electron scavengers such as ascorbic acid may help to avoid damage by the production of oxygen and its intermediates. Three steps were, therefore, necessary to initiate the expansion of the new principle of oxygenic photosynthesis by bacteria: (1) the new combination of old and new pigments to enable linear photophosphorylation; (2) the development of the complex Mn-containing enzyme catalysing water hydrolysis; and (3) the development of enzymes which detoxify the poisonous side-products of oxygen. These developments permitted the expansion of life all over the globe and into an almost unlimited cycling of carbon from the inorganic to the organic form, and vice versa. The new limiting factors were the associated elements, e.g. phosphorus and nitrogen. The initiation of oxygenic photosynthesis may initially have had little influence on the physiology and biochemistry of the microbial ecosystems and the populations living within them at that time. Photosynthetic oxygen may have come into contact with large amounts of reduced compounds such as the reduced heavy metals, iron, manganese, uranium, etc., and many metabolic products of fermentative and anoxygenic photosynthetic pathways which were then gradually oxidized. Only after this process had reached equilibrium would free oxygen possibly have dissolved in the water and later have diffused into the atmosphere. It may well be that higher oxygen levels were only created with the occurrence of photosynthetic eukaryotes because only these possess the peroxisomes which detoxify oxygen most efficiently. The banded iron formations (BIFs) mentioned by Cloud and Gibor (1974) may represent trapped oxygen of the first oxygenic communities which perhaps consisted of laminated benthic or floating microbial mats. Banded iron formations are the huge iron deposits of the Precambrian which today mined in Brazil and Canada. Also it is conceivable that, for long periods, photosynthetic oxygen was mainly kept and cycled in such complicated laminated systems without diffusing into the surroundings (Krumbein *et al.* 1979). Recent stromatolites do actually exude large amounts of oxygen, but a high percentage of the photosynthetic oxygen remains in the mats and is recycled within them.

Another problem arising with the establishment of oxygenic photosynthesis by cyanobacteria is the question why an equilibrated state was reached with 20% atmospheric oxygen, at least after oxygen-respiring organisms and pathways evolved. Immediately after the firm establishment of oxygenic photosynthesis on the earth there may, however, have been a period of higher concentrations of atmospheric oxygen until the oxygen-consuming heterotrophs brought the system into equilibrium.

Cloud and Gibor (1974) and Golubic *et al.* (1979) follow the original ideas of Rubey who calculated the pools of organic sedimentary matter, carbonates, sulphates, sulphides, iron and other metal oxides as being equilibrated by oxygen in the atmosphere. The problems still remain of the large amounts of water and whether biological regulatory effects on the equilibrium of H_2O, O_2 and (CH_2O) are important or not. These problems will be discussed later. Palaeontological evidence for the evolution of cyanobacteria is scarce (Knoll 1979). Towe (1978) discussed influences other than the biological factors in the rise of oxygen and Schidlowski *et al.* (1979) also came to the conclusion that we have at present no satisfactory answer to the question of the rise of oxygen levels in the atmosphere. The biochemical steps for the evolution of oxygenic photosynthesis are also not yet analysed. Research will have to focus on the relations between RuBP-carboxylase/oxygenase, nitrogenase and ATPases, i.e. the coupling of oxidation-reduction chains within the cell with phosphorylation-dephosphorylation steps on one hand and the coupling of protonmotive forces across membranes with electronmotive forces across photosynthetic membranes. Altogether it is not yet clear which apoproteins and enzymes are the most reasonable candidates in the search for transitional enzymes marking the step from an anoxic to an oxygenated world. Accessory electron sinks outside the cell's membrane system have played an important role as buffer systems for the evolutionary span necessary for adaptation. Besides inorganic acceptors, the extracellular products of some archaebacteria and anoxygenic photosynthetic bacteria may have produced large amounts of such buffering sinks during that transitional period.

In addition, our knowledge is insufficient concerning both the biochemical implications of anoxygenic and oxygenic photosynthetic potential in recent cyanobacteria (Padan 1979) and the evolutionary links between oxygenic photosynthetic bacteria and eukaryotes as indicated by the *Prochloron* problem. It was established, however, by the work of Cohen (Krumbein & Cohen 1974, 1977, Cohen *et al.* 1975, Belkin & Padan 1978) that cyanobacteria are the most versatile organisms with respect to anoxygenic and oxygenic photosynthesis, although some eukaryotes can be forced into very ineffective anoxygenic photosynthesis.

Sulphur and sulphate-reducing bacteria and primary productivity

Sulphate reduction may have originally been exclusively a primary production process. Dissimilatory sulphate reduction or sulphate respiration is also called sulphide fermentation because energy yields are considerable and it is not necessary for Desulfovibrio to use organic carbon compounds as energy *and* cell carbon sources. Sulphate reduction probably split off the stem of Chromatiaceae relatively late. It may have coincided with the evolution of oxygenic cyanobacteria but may well have happened before the advent of cyanobacterial oxygenic photosynthesis.

RNA-analyses and sulphur isotope data (Monster *et al.* 1979, Fox *et al.* 1980, Trüper 1982) imply that sulphate reduction did not occur as early as one might think. Probably it was established later than oxygenic photosynthesis, but clear evidence is not given by isotope palaeochemistry (Holland & Schidlowski 1982).

Chemolithotrophic productivity

Most chemolithotrophs are obligate, though microaerophilic, aerobes. They need large amounts of oxygen as electron acceptor for their respiratory pathways. There is much evidence that these have evolved relatively late and have not been of major importance in the primary production of organic material. There are some chemolithotrophic autotrophs, however, which operate anaerobic respiration based on nitrate or other oxidized substances (e.g. anaerobic CH_4 oxidation) as electron acceptors. The view that primitive chemolithotrophs have had world-wide distribution during the Precambrian and have been restricted to defined niches afterwards cannot be excluded. It may also be that other as yet unrecognized chemolithotrophic pathways exist, but that the organisms have not yet been isolated from the natural environments (e.g. Fuchs & Stupperich 1981). A typical and recently discovered environment for research into such organisms in the present day cycle of carbon is in the deep sea hot vents. These occur in the Atlantic, the Pacific, the Indian Ocean and in the Red Sea and perhaps elsewhere. One can conceive some special productivity patterns around these vents. Chemolithotrophic bacteria of several different physiologies (e.g. sulphide oxidation, iron and manganese oxidation) abound in the absolute darkness of the deep sea, presumably using reduced inorganic compounds exhaled by the volcanic vents. The bacteria use these as energy and electron donors, and build up organic carbon compounds either directly or symbiotically and thus support a large variety of deep-sea animals living in the immediate vicinity of the vents. Thus around the deep sea hot vents 'primary productivity' of chemolithotrophs in total darkness represents oases of the organic carbon cycle in the deep-sea semi-desert (Karl *et al.* 1980, Cavanaugh *et al.* 1981, Ruby *et al.* 1981). These places have even been suggested as possible places for the origin of life on this planet (Corliss *et al.* 1981).

Eukaryotic oxygenic photolithotrophy

Eukaryotic oxygenic photolithotrophic autotrophs (microorganisms and macroorganisms) are genetic chimaeras (Fox *et al.* 1980). All of them seem to have evolved from several different symbiotic systems (see Margulis 1981, for additional references). The concentration of oxygen at the time of their onset must have been raised to modern levels and the flow and transport of oxygen must have enabled respiratory chains using oxygen as terminal electron acceptor to evolve together with complicated patterns of cytoplasmic transfer, specialization of parts of the organisms, etc., to yield larger organisms. All evidence suggests that this endosymbiotic chlorophyll *a* and *b* based type of photosynthesis, that hence has dominated the primary production on the earth, is very young in the evolutionary scale. Multiple endo- and ectosymbioses, genetic exchange and mutation steps must have occurred before the different eukaryotic photosynthetic organisms had evolved and established themselves permanently. Probably, more than one plastid symbiosis has occurred which has given rise to contemporary photosynthetic systems. Doubtless the symbiosis and the resulting chimaera had tremendous advantages and evolutionary potential once established; this definitely was a primary productivity revolution. Simultaneously the broad majority of the aerobic chemoorganotrophs apparently regained an important place by transferring some of the 'primary photosynthetic production' into an immediate secondary productivity level, based on chemoorganotrophic heterotrophy. The only prokaryote operating oxygenic non-cyclic photophosphorylation on the basis of chlorophyll *a* and *b* and having double thylakoids comparable to the plastids of green plants is the cyanobacterium 'Prochloron' which today occurs only as ectosymbiont of marine ascidians (Fisher & Trench 1980, Lewin 1981).

The modern type of primary production, yielding some of the most resistant and refractory organic compounds such as lignins and hydrocarbons (in both photolithotrophs and oxygen-respiring chemoorganotrophs) is evolutionarily recent; it marks the beginning of the present day cycle of organic carbon with its anaerobic and aerobic degradation pathways and its always considerable amounts of buried and non-cycled organic materials which lead to the accumulation of oil, coal and kerogen.

2.3 THE DEGRADATION OF ORGANIC MATTER

In this section the remineralization of reduced carbon compounds of all origins is considered. Polysaccharides, proteins and many other substances are readily remineralized by most microorganisms which use these substances as reserve material, as sources of internal or external energy, and as donors of carbon, other nutrients, electrons and/or protons. The decomposition of these substances under aerobic and anaerobic conditions is straightforward. Problems arise with (1) less degradable complex substances often regarded as metabolic waste products; (2) so-called 'refractory' (non-degradable) substances; and (3) easily degradable substances that are only partially decomposed and thus create accumulations of specific materials which necessitate special organisms or adaptations by organisms for further degradation and complete remineralization. Usually all organic biogenic compounds produced during biological life cycles should ultimately be degradable by microorganisms.

2.3.1 Degradation of 'refractory' organic carbon compounds

Several microbiologists are convinced, without having absolute proof for their statements, that no biogenic reduced carbon compound is actually a refractory non-degradable substance and so has to accumulate in sediments (e.g. Zeikus 1981). Given optimal conditions, biogenic compounds should be recycled within reasonable time. This implies that many, if not all, so-called refractory organic compounds are not really biogenic. One of the most outstanding problems regarding biological and physico-chemical degradability is the breakdown of lignin. However, lignin, the precursor of coal and possibly some hydrocarbons, is a compound which has not been shown to be generated entirely via enzymatic pathways. Many aromatic compounds are produced via enzyme catalysis from cinnamic acids. Lignin, however, is a product of dehydrogenative polymerization of cinnamyl alcohols. The condensation mechanism does not seem to be under strict enzymatic control (Zeikus 1981) and no polymerase has been shown to be active in the reaction. This may be the reason why lignin is such a refractory substance and is not readily remineralized. A peroxidase reaction with H_2O_2 may be involved via pyridine nucleotides of the cytoplasm (Gross 1977).

Thus lignin is a 'biopolymer', but of unique characteristics and 'abiogenic' in the last steps of polymerization. Neish (1968) regarded lignin as a waste excretion product of plants which by their multicellular organization had to find new ways of disposing of by-products. Since, except for the outermost cells of the whole organism, plants cannot excrete these waste products directly they had to evolve ways to transform the waste products into volatile or insoluble non-toxic substances. When plants achieved the step to convert phenylpropanoid amino acids into phenoxyradicals they had practically reached the necessary precursor step for lignin formation. When these phenoxyradicals were condensed and arranged in fibres extracellularly the precursors of lignin were no longer a waste product but useful for the further development of higher plants and trees. The comparison of other secondary products of plants with the lignin precursors shows clearly the importance of such detoxification and purification pathways. All these substances (e.g. the polyphenolic compounds of sphagnum or bacterial phenol-oxidase-catalysed reaction products) are difficult to degrade and undergo postbiotic condensation.

Therefore in considering the question of possibilities of the complete recycling of biogenic reduced carbon compounds one has to consider the possibilities of abiological reactions following initial biological steps; this makes bacterial remineralization processes much more complicated than mere reversed biological reactions.

Another possible mechanism of condensation of organic compounds into refractory material is heat- and pressure-controlled polymerization during early diagenesis (see Chapter 9).

2.3.2 Aerobic and anaerobic degradation pathways

The strongest biologically useful oxidant is oxygen (which was introduced into the global geochemical cycles mainly by a biological activity, i.e. oxygenic photosynthesis). Therefore, it is only natural that oxygen-based respiratory pathways are the most efficient means of degrading organic substances to mineral compounds. Several generalized chemical terms have been developed to summarize the remineralization pathways of complex organic structures in the presence or absence of oxygen. The organic matter (remains of plants and animals) in marine sediments and in soils together with detritus (dissolved and solid organic substances in the sea) are the largest pools of degradable organic matter in the global geochemical cycle. Redfield (1958) has suggested a general sum formula of biologically produced organic matter:

$$(CH_2O)_{106}\,(NH_3)_{16}\,(H_3PO_4)_1. \qquad (2.1)$$

This formula has been modified for some marine sediments by Hartmann *et al.* (1973) into:

$$(CH_2O)_{106}\,(NH_3)_{16}\,(H_3PO_4)_{0.2\text{–}1}. \qquad (2.2)$$

This is because phosphorus is very often already depleted when detritus or 'marine snow' reaches the bottom. These particles of biogenic origin, floating in the water column are scavengers of many chemicals and heavy metals.

It is generally expected (Hartmann *et al.* 1973, Froelich *et al.* 1979; see also Chapter 5) that this material under natural conditions and in the presence of bacteria is oxidized into simple inorganic compounds by pathways using the available oxidant which yields the highest free energy change per mole of organic carbon oxidized; when this oxidant is exhausted, oxidation will continue using the next most efficient oxidant, and so on. In the equations given below it is assumed that all theoretically available oxidants are also used in biological processes even though enzymatic pathways have not always been demonstrated conclusively. It has to be stated, however, that several of the energetically less efficient oxidants may be used simultaneously in sediments in a competitive or co-operative way (Iversen & Blackburn 1981).

Aerobic respiration

Aerobic degradation of organic material usually proceeds according to the generalized formula:

$$(CH_2O)_{106}\,(NH_3)_{16}\,(H_3PO_4)_{(0.2\text{–}1)} + 138\ O_2 \rightarrow 106\ CO_2 + 16\ HNO_3 + 1\ H_3PO_4 + 122\ H_2O. \qquad (2.3)$$

The $\Delta G_0'$ being -3190 kJ (762.4 kcal) mol^{-1} of glucose. This reaction type is responsible for very fast turnover of organic material in oxygenated environments. Most of the oxygen produced via photosynthesis is exhausted by this process in the aquatic and terrestrial aerobic environment.

Garrels *et al.* (1975) have calculated that on a global basis, a total of about 5×10^{15} mol CO_2 annually are fixed of which more than 98% is fixed initially via oxygenic photosynthesis (50% marine, 50% terrestrial primary productivity). The error of determination, however, is more than 5% and may exceed 10%. About 4.7×10^{15} mol of the organic material are remineralized in a year. It is impossible, however, to define with existing methods how much of this is achieved via oxygen respiration. Most probably the bulk of remineralization reactions must operate via oxygen respiration since otherwise within a few years the oxygen budget would be out of balance and the oxygen content of the atmosphere would become enriched. Changes in oxygen respiration and in CO_2-fixation and liberation can be expected to differ in their relative impact on the atmosphere because of the differences in atmospheric O_2 and CO_2 concentrations. The difference is of a factor of over 1000. Alternative oxygen sinks however are the reaction products of anaerobic respiration and physico-chemical and biological weathering. Only the exact budget of all respirations and all (aerobic and anaerobic) weathering processes will give the ultimate percentage of oxygen respiration against total loss of photosynthetic oxygen. In addition it should be emphasized that within the generally oxygenated environment many anaerobic niches occur. The human body provides an excellent illustration of this.

Manganese respiration

Energetically, the next most promising reaction occurs at the expense of oxidized manganese minerals. Trimble and Ehrlich (1970) and Ghiorse and Ehrlich (1976) have provided some evidence for electron transport chains in certain bacteria leading to manganese, but did not isolate an enzyme responsible for the respiratory reduction of manganese oxide or Mn^{IV} as terminal electron acceptor. Nealson (see Chapter 8) also suggests the appealing idea that manganese may serve under anaerobic conditions as electron acceptor for the oxidation of reduced carbon compounds. This fits the thermodynamic situation perfectly. However, he has reservations about the physiological and biochemical evidence for this pathway as a direct process. For general considerations of diagenesis of organic matter in sediments this would not be of absolute importance because other substances may play intermediate roles (e.g. via inorganic processes) in overall redox transactions within a sedimentary system rich in organic substances. The reaction may be represented by the formula:

$$(CH_2O)_{116}\,(NH_3)_{16}\,(H_3PO_4) + 236\ MnO_2 + 472\ H^+ \rightarrow 236\ Mn^{2+} + 106\ CO_2 + 8\ N_2 + H_3PO_4 + 366\ H_2O. \qquad (2.4)$$

The $\Delta G_0' = -3090$ kJ (738.5 kcal) mol^{-1} glucose (birnessite, MnO_2)
-3050 kJ (729 kcal) mol^{-1} glucose (nsutite MnO_2)
-2920 kJ (698 kcal) mol^{-1} glucose (pyrolusite MnO_2)

The mineral birnessite contains variable amounts of Na, K, Ca, Mg and even divalent Mn besides tetravalent Mn.

Nitrate respiration

Logically, the next electron acceptor for anaerobic respiratory chains would be nitrate. In this case two different reactions are possible and have been well studied:

$$(H_2O)_{106}\,(NH_3)_{16}\,(H_3PO_4)_1 + 94.4\ HNO_3 \rightarrow 106\ CO_2 + 55.2\ N_2 + H_3PO_4 + 177.2\ H_2O, \qquad (2.5)$$

with $\Delta G_0'$ of -3030 kJ (724 kcal) mol^{-1} glucose.

$$(CH_2O)_{106}\,(NH_3)_{16}\,(H_3PO_4)_1 + 94.4\ HNO_3 \rightarrow 106\ CO_2 + 42.4\ N_2 + 16\ NH_3 + H_3PO_4 + 148.4\ H_2O. \qquad (2.6)$$

In this case $\Delta G_0'$ is -2750 kJ (657 kcal) mol^{-1} glucose. In this calculation it is assumed that the original organic nitrogen is not totally reduced but liberated as ammonia. Biochemically both reaction types have been shown to take place and the related enzymes have been demonstrated. As can be seen by the free energy yield, nitrate respiration may occur alongside the hypothetical manganese respiration in marine and limnic sediments and anoxic waters, under some circumstances.

Iron respiration

The enzymes of iron reduction have been shown to be active in many microorganisms. It was probably Roberts (1947) who first showed enhanced respiratory activity of *Bacillus polymyxa* on glucose in the presence of ferric iron under anaerobic conditions. Later, Bromfield (1954) and Ottow (1968, 1969, 1970) and Hamann and Ottow (1974) studied the respiratory reduction of iron and found that it is very often associated with nitrate reductase activity (Pichinoty 1973). Ottow and Ottow (1970), on the other hand, have convincingly demonstrated that in soils the oxidation of organic matter under anaerobic conditions via anaerobic respiratory chains seems to be more efficient and the capability of soil microfloras more effective using ferric iron and specific enzymes for the reduction of iron rather than with nitrate reductase acting on ferric iron simultaneously. They have found iron reduction in nitrate reductase negative mutants as well as in wild-type strains not producing nitrate reductases. From these findings it can be concluded that one iron reductase system is more or less identical to nitrate reductase A, and another, more efficient, Fe^{3+} reductase system, yet unknown, would not be capable of nitrate reduction. The overall stoichiometry of iron respiration would be:

$$(CH_2O)_{106}\,(NH_3)_{16}\,(H_3PO_4)_1 + 212\ Fe_2O_3 \text{ or } 424\ FeOOH + 848\ H^+ \rightarrow 106\ CO_2 + 16\ NH_3 + H_3PO_4 + 424\ Fe^{2+} + 530 \text{ respectively } 742\ H_2O. \quad (2.7)$$

The $\Delta G_0'$ being, in the case of hematite (Fe_2O_3) -1410 kJ (-337 kcal) mol^{-1} glucose and, in the case of limonite, goethite (FeOOH) -1330 kJ (-318 kcal) mol^{-1} glucose.

Sulphate respiration

Its abundance means that sulphate is a very important electron acceptor for anaerobic respiration in normal marine sediments, while in the limnic environment and in paddy soils the situation is different. Jørgensen (see Chapter 4), Jørgensen (1978a, b, c), Jørgensen *et al.* (1978) and Jørgensen and Cohen (1977) have shown the importance of sulphate reduction in the oxidation of organic matter for normal marine conditions, for estuaries and for hypersaline highly productive laminated microbial mats. The overall abundance of sulphate makes this compound of key importance in the recycling of products of primary production. As will be shown in Chapter 4, sulphate as a component of sea water in the marine environment links the reservoirs of organic carbon and inorganic carbon, of reduced and oxidized iron and of gypsum and carbonates (Fig. 2.1). Although sulphate reduction is a relatively recently evolved energy-conserving pathway, it must be older than oxygen respiration or at least of comparable evolutionary age and it releases almost as much ATP as aerobic respiration. However, it should be borne in mind that less sulphate is used and, in the absence of oxygen, ammonia is usually not oxidized to nitrate, but remains in its reduced form.

The generalized equation for recycling of reduced carbon via sulphate (and sulphur)-reducing bacteria (see also Chapter 5) is:

$$(CH_2O)_{106}\,(NH_3)_{16}\,(H_3PO_4)_1 + 53\ SO_4^{2-} \rightarrow 106\ CO_2 + 16\ NH_3 + H_3PO_4 + 53\ S^{2-} + 106\ H_2O. \quad (2.8)$$

The Δ is given by Froelich *et al.* (1979) as -380 kJ (-91 kcal) mol^{-1} glucose.

The substrate for sulphate reduction in many cases is not glucose, but several organic acids which are products of mixed acid fermentations prior to sulphate respiration. This indicates that some fermentative pathways lacking a terminal inorganic electron acceptor are energetically much more advantageous.

Carbon respiration

Finally, a reaction occurs within anaerobic and reducing sediments of marine and freshwater environments which has previously been thought to be fermentative rather than respiratory although the terminal electron acceptor is CO_2, an oxidized inorganic compound. There is now evidence of an electron transfer chain in the methanogenic bacteria and it is thus proper to refer to this process as anaerobic respiration. Thauer *et al.* (1977) convincingly summarized the ATP-generating processes involved in CO_2 'respiration'. Keltjens and Vogels (1981) and Whitman and Wolfe (1980) evidence an electron transfer chain which involves the compounds F_{420}, F_{430}, F_{420}, a hydrogenase, a hypothetic CH_3-S-CoM and an unknown electron transport chain from F_{420} to the various intermediates. The amount of ATP generated is unknown. The methanogenic reaction as a fermentation process could be written thus:

$$(CH_2O)_{106}\,(NH_3)_{16}\,(H_3PO_4)_1 \rightarrow 53\ CO_2 + 53\ CH_4 + 16\ NH_3 + H_3PO_4 + 106\ H_2O. \quad (2.9)$$

The $\Delta G_0'$ in this case is -350 kJ (-84 kcal) mol^{-1} glucose as a theoretical value.

This type of methane production has been shown to occur with only 5 substrates; acetate, formate, methanol, carbon dioxide and carbon monoxide. The reactions are listed below (Games & Hayes 1976).

$$CH_3 + CO_2H \rightarrow CO_2 + CH_4$$
$$CO_2 + 4H_2 \rightarrow 2H_2O + CH_4$$
$$4CH_3OH \rightarrow 3CH_4 + CO_2 + 2H_2O$$
$$4HCO_2H \rightarrow CH_4 + 3CO_2 + 2H_2O$$
$$2H_2O + 4CO \rightarrow CH_4 + 3CO_2.$$

It has been suggested that in the case of acetate the CH_3 group is reduced to methane and the carboxyl generates the CO_2. If this were the case, it would be difficult to explain the large carbon isotopic fractionation between CH_4 and CO_2 which is observed in these organisms unless large intramolecular variations were present in the parent organic molecule.

In cases where molecular hydrogen is available as an electron donor, methane and/or other C_1 compounds may be utilized as the sole carbon source. Organisms which are able to undertake the process, the 'C_1-utilizing microorganisms' can be divided into two groups: (1) methylotrophs which assimilate carbon as formaldehyde and CO_2, but by processes distinct from the Calvin cycle; (2) autotrophic types, which oxidize C_1 compounds to CO_2 and then process the CO_2 via various other mechanisms including the Serine cycle, the RuBP cycle and the RMP (Ribulose monophosphate) cycle.

In addition there is now evidence that some bacteria such as *Methylococcus capsulatus* can use both modes simultaneously. A complete review of the processes involved is provided by Colby *et al.* (1979).

Where molecular hydrogen is available as electron donor, methane formation will be a chemolithotrophic process of production of biomass and, in this case, methane would not be regarded as a terminal degradation product but as an organic carbon compound which can be used for methane oxidation via aerobic respiration pathways, for example:

$$3\ CH_4 + 6\ O_2 \rightarrow 3\ CO_2 + 6\ H_2O. \quad (2.10)$$

The complete reaction of methane oxidation in energetic terms would look as follows:

$$CH_4 \rightarrow CH_3OH \rightarrow HCHO \rightarrow$$
$$125\ kJ \rightarrow -192\ kJ \rightarrow -213\ kJ$$
$$\text{methane} \rightarrow \text{methanol} \rightarrow \text{formaldehyde}$$
$$HCOOH \rightarrow HCO_3^- \rightarrow CO_2$$
$$-238\ kJ$$
$$\text{formate} \rightarrow \text{bicarbonate} \rightarrow \text{carbonate.}$$

Until recently it was assumed that methane generated in anaerobic sediments by the above-described carbon dioxide reduction or 'respiration' would have to migrate upwards, or would accumulate within sedimentary rock systems as natural gas because its oxidation under energy gains would only be possible in the presence of oxygen. Recently, however, Reeburgh (1982) has described situations which imply biological anaerobic methane oxidation. Reeburgh (in press) discusses the matter and comes to the conclusion that anaerobic biological methane oxidation, which was predicted from sedimentary models, really takes place. Lidstrom (personal communication) has found methane oxidation in anoxic waters of Framvaren fjord and related some of the biological activity to associated sulphate reduction. However, she has not been able to isolate an organism so far.

The observed layer of anaerobic methane oxidation falls within the limit of the lower end of the sulphate reduction zone which fits very well with all geochemical carbon balances (Reeburgh, in press). Rate measurements in the field as a tool in the balance calculation of biological remineralization have turned out, in most cases, as highly indicative of possible microbial activities and thus it is expected that sulphate-reduction-related anaerobic methane oxidizers will ultimately be isolated. Thus even under anaerobic conditions—given enough time—a complete carbon cycle and total remineralization of carbon compounds is possible. Rates of oxidation measured in the field are usually faster than the mineralization occurring according to pool analyses of chemicals. Therefore, the chances of preservation of organic carbon compounds are low and mostly only very refractile substances will be polymerized or larger amounts of organic material will be buried by fast sedimentation rates to depths which no longer enable microbial processes to exist.

On the other hand, recent findings of the groups working on barophilic microorganisms of the hot vents (e.g. Baross, Oregon and Yayanos, La Jolla) seem to provide good evidence of fast metabolic rates of bacteria under high hydrostatic pressures in temperature ranges of up to 250 °C. It can thus be expected that microbial degradative processes eventually continue under high pressures up to the critical point of water, i.e. 374 °C, under the appropriate pressures enabling water activity. These so far speculative data, which are, however, based on laboratory experiments under high hydrostatic pressures, may explain the many reports of microbial activity in deep drilling cores and in and around deep-sea hot vents. They may, as well, with increasing sediment depth push the borderlines of life further than reported so far.

Although these findings are preliminary, they may throw new light on biological mineralization and transformation rates of organic carbon compounds (for references see Edmond 1982).

We mentioned earlier that the most primitive pathways to emerge on Earth in terms of microbial life may have been fermentation pathways and anoxygenic photosynthesis. It is, therefore, not surprising that methanogens belong to the primitive and very early deviating group of the archaebacteria, which by their 16*s* rRNA affinity values are characterized as very early offspring of biological evolution. Methanogens have a complex internal membrane system, they are characterized by the production of sterols, which play an important role in membrane formation in eukaryotes also. The transformation of one-carbon compounds and their importance in nature as well as in applied microbiology are dealt with in many recent reviews and books (e.g. Mah *et al.* 1977, Schlegel & Barnea 1977, Thauer *et al.* 1977, Zeikus 1977, 1980, Zehnder & Brock 1979). Some of the important reactions involving methane are described in Fig. 2.3.

Fermentation or substrate level phosphorylation

The degradation and complete mineralization of organic compounds produced by any of the previously described pathways has also been demonstrated in several cases to be complete when no external inorganic electron acceptor is available. Carbon dissimilation pathways in which ATP

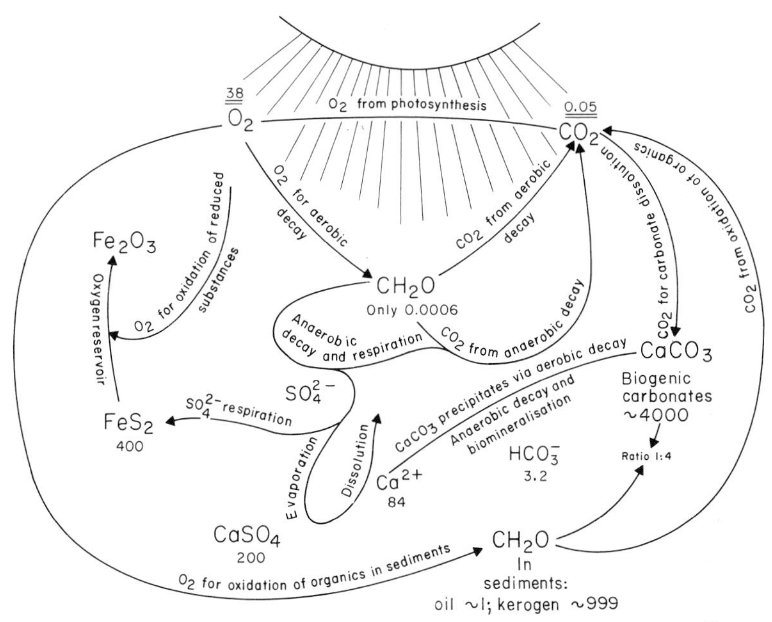

Obviously nothing works without water (the ocean contains 78 000 x 10^{18} moles)

Fig. 2.3. The natural cycle of carbon. The numbers below the indicated compounds in the various reservoirs represent $\times 10^{19}$ moles. Residence times are not given. It is clear, however, that the residence time in sediments is higher than that in the atmosphere by a factor of 10^6 or more.

is formed by substrate level phosphorylation lead to typical 'fermentation products', for example short-chain fatty acids. Hydrogen is formed, for example, via pyruvate- and formate-forming mixed acid fermentations and is recycled to yield methane anaerobically. Since anaerobic methane oxidation has recently been postulated, the cycle could be regarded as also being closed under anaerobic conditions. Usually, however, this is not the case and organic degradation products in many cases accumulate in sediments. On average less than 2% of the annual production is preserved within the sediment in finely distributed form. Krumbein and Cohen (1977) have shown for a very productive area (Solar Lake, Israel) where the sediment itself consists of practically 100% organic material, that the degradation rates were also in the order of 97–99% of the annual production. Fig. 2.4 gives an impression of the possibilities of different productivity and degradation rates in a stromatolitic environment.

Martens and Klump (1980) have analysed carbon turnover in anaerobic shallow estuarine sediments which, in contrast to the stromatolite environ-

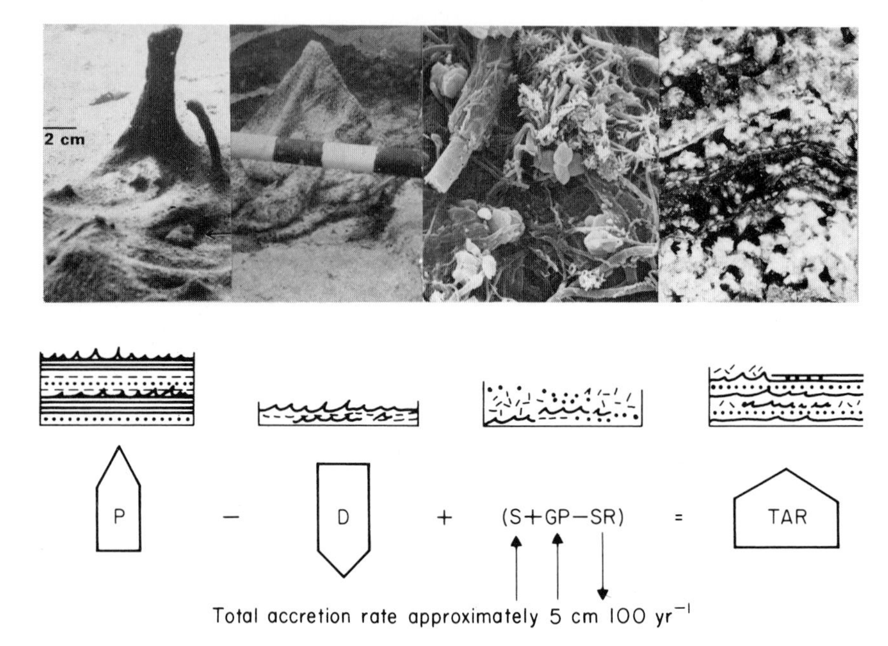

Fig. 2.4. Production and degradation budget or organic carbon compounds in the microbial mats of Solar Lake (Sinai). Laminated mats of filamentous cyanobacteria alternate with coccoid cyanobacteria layers. 99% of the annual production is degraded via bacterial activity. Some eolian and fluviatile (wadi-floods) clastic sedimentation occurs together with gypsum precipitation. Some of the gypsum is redissolved via sulphate-reducing bacteria living on the photosynthates. The final accretion rate is about 5 cm per 100 years. The figures represent oxygen-filled pinnacles emerging from the laminated filamentous surface mat. The white carpet with the 10 cm rod scale is a decaying mat full of myxobacteria, etc. Gypsum precipitation marks the degradation environment with the ensheathed filamentous cyanobacterium. The last figure is a thin section of the ultimate laminated stromatolitic rock.

ment, are constantly fed by organic material from terrestrial sources. In contrast to Solar Lake where methane does not play an important role, huge amounts of methane are produced. It may be that the speed of productivity enables the preservation of polysaccharides and other substances as witnessed by SEM-analysis of recent stromatolites (Figs. 2.7–2.30). The cyanobacterial outer cell wall layers (glycocalyx), however, can be regarded as partially abiologically polymerized substances comparable to lignin in their refractory behaviour in a degradative environment. Organic geochemists are presently analysing which original compounds and which polymerisates of abiogenic diagenetic origin will ultimately be stored in the sedimentary record (Boon *et al.*, in press).

Degradation and bacterial remobilization of organic material in semi-anaerobic sediments has been found to be relatively efficient in shallow, open marine sediments where bacterial numbers, sediment transport and turbulence levels are very high (Krumbein 1971).

The microbiology of anaerobic decay in peat and swamp areas has been studied in some detail but quantitative data are very rare. Küster (1978) gives a summary of peat microbiology. Zeikus (1981) outlines the anaerobic degradation of refractory substances in anaerobic environments including peat and shallow marine sediments. Fenchel and Jørgensen (1977) and Reeburgh (in press) give data tables, but better data are needed really to understand the anaerobic fermentative degradation of organic material in nature. Altogether it seems that fatty acids, steroids, bacterial extracellular polysaccharides and lignocellulosic substances are the most prominent compounds which, after surviving aerobic degradation by bacteria, fungi, and other respirative microorganisms, are preserved within the natural environment and further condensed during diagenesis.

Some additional substances have been added recently by human production of organic compounds which are released into the environment and accumulate in sediments. These, however, cannot be regarded as belonging to the natural cycle of organic carbon compounds. In addition, it has been shown that most of these compounds are also biodegradable. The degradation and mineralization pathways, however, are slower and in cases where these substances are toxic they create major environmental hazards and problems; examples are DDT (2,2-bis (p-chlorophenyl)-1,1,1-trichlor-oethane) with a half-life ($T_{\frac{1}{2}}$) (time until half of the total material released into the environment is degraded) of around 4 years, chlordane with $T_{\frac{1}{2}}$ about 5 years and many other chlorinated compounds used as herbicides or in plastics with $T_{\frac{1}{2}}$ of up to 10 years under unfavourable conditions.

Altogether it seems that geological and sedimentological processes rather than biological ones are responsible for the preservation of organic carbon compounds for periods longer than several decades. Some of the geological aspects of the degradation/preservation equilibria are discussed in Chapter 9.

2.4 CARBON ISOTOPE FRACTIONATION

It is an important fact relevant to the fractionation of stable isotopes within biological systems that the observation of a net isotopic fractionation requires both the existence of an isotopic effect and a division of the carbon flow at a 'branch point' (Monson & Hayes 1980). Consider the following example which is taken from the photorespiration cycle in C-3 plants. Reaction I in Fig. 2.5 is the conversion of glyoxylate to glycine. If this reaction were studied in a closed system, some carbon isotope effect might be shown. However, in an open system in which reactants are added and products withdrawn at similar rates, the isotopic composition of the glyoxylate would match that of the glycine. In the second part of the pathway glycine is converted to serine and CO_2. In this case a kinetic effect causing an enrichment or depletion in the serine would cause a concomitant change in the carbon isotope composition of the released CO_2.

This principal has implications in assessing the effect of the various reactions already described. In most of these there is no splitting of carbon, i.e. all the carbon atoms in the reactants involved end up as CO_2. The CO_2 produced will, therefore, have an isotopic composition similar to the parent organic material; in the case of a C-3 type plant this will be approximately $-26‰$.

However, this is not the case with the methanogenic bacteria in which two carbon species, CH_4 and CO_2 are produced. The difference between these two species represents the most substantial carbon isotope fractionation in the terrestrial environment. Table 2.1 gives the previously reported fractionations between CO_2 and CH_4. From these data Games and Hayes (1976) recognized three separate types of fractionations. Two of these may be separated on the basis of an α value of ≈ 1.07 distinct from those with lower values of ≈ 1.05. In turn, these two can be further subdivided on the relationship between the gases and parent organic matter (see Table 2.1). As yet the precise pathways involved in methane formation remain to be resolved and therefore the point at which fractionation occurs is unknown. Interestingly, the α (CO_2/CH_4) shown in Table 2.1 approximates to that expected from equilibrium thermodynamics at 25 °C (Friedman & O'Neil 1977). However, under standard temperature conditions, such as equilibrium would take a prohibitive time to be established, although the catalytic influence of the organism involved cannot be ignored.

In summary, processes such as aerobic manganese, sulphate and iron respiration do not involve the production of two carbon species. Therefore the $\delta^{13}C$ of the produced CO_2 will be similar in isotopic composition to the parent organic material. The addition of CO_2 from this source results in the CO_2-HCO_3-CO_3^{2-} systematics adopting a more negative composition and consequently any diagenetic carbonates produced will have depleted isotopic ratios.

The anaerobic processes which results in methane and CO_2 production produce highly depleted CH_4, but $\delta^{13}C$-enriched CO_2. This process can produce isotopically heavy carbonates. The very negative values of bacterial methane distinguish it from either hydrocarbon- and/or geothermal-derived methane.

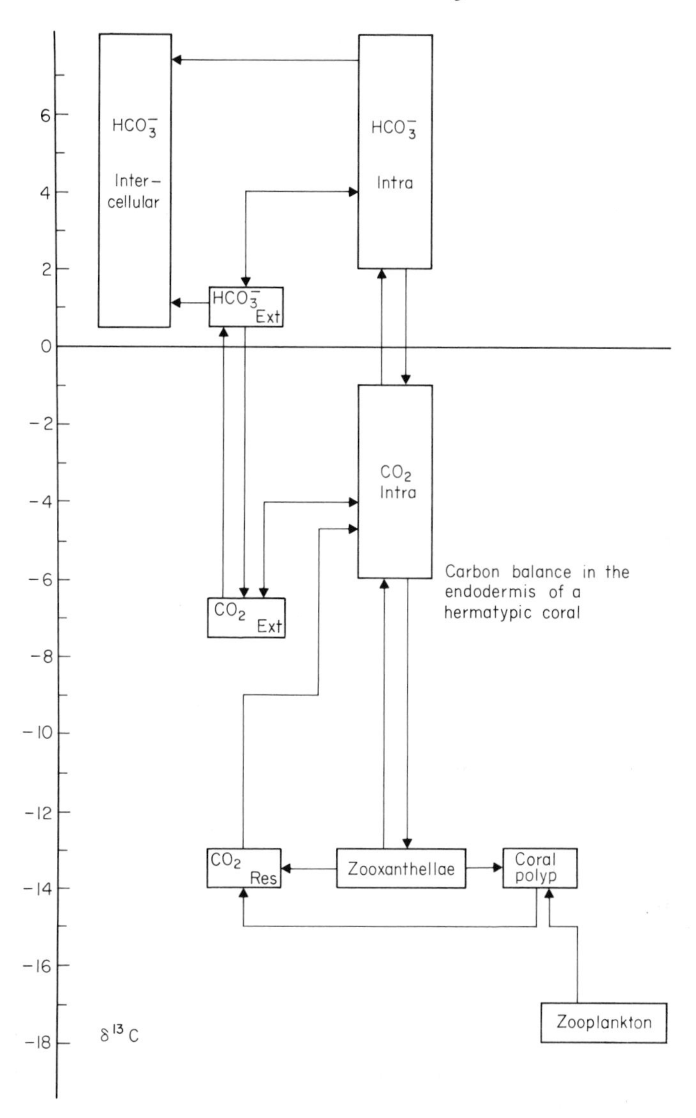

Fig. 2.5. Scheme of carbon flow in a zooxanthellate coral. Two arrows indicate isotopic exchange reactions; a double-sided arrow diffusion in both directions; and a single arrow input only. Vertical scale indicates possible ranges only of ^{13}C. The isotopically light-respired CO_2 ($-13‰$) mixes with inorganic CO_2 ($-7‰$); HCO_3^- in equilibrium with this CO_2 has a range of isotopic compositions between $+8‰$ and $+1‰$. This figure differs from that of Goreau (1977) in that the intracellular carbon pool is shown to be isotopically heavier than the external one. The range of the intracellular CO_2 pool shown is based on a fractionation of CO_2 by RuBP carboxylase of approximately $10‰$. The intracellular pool may, therefore, be very much heavier (after Swart 1983).

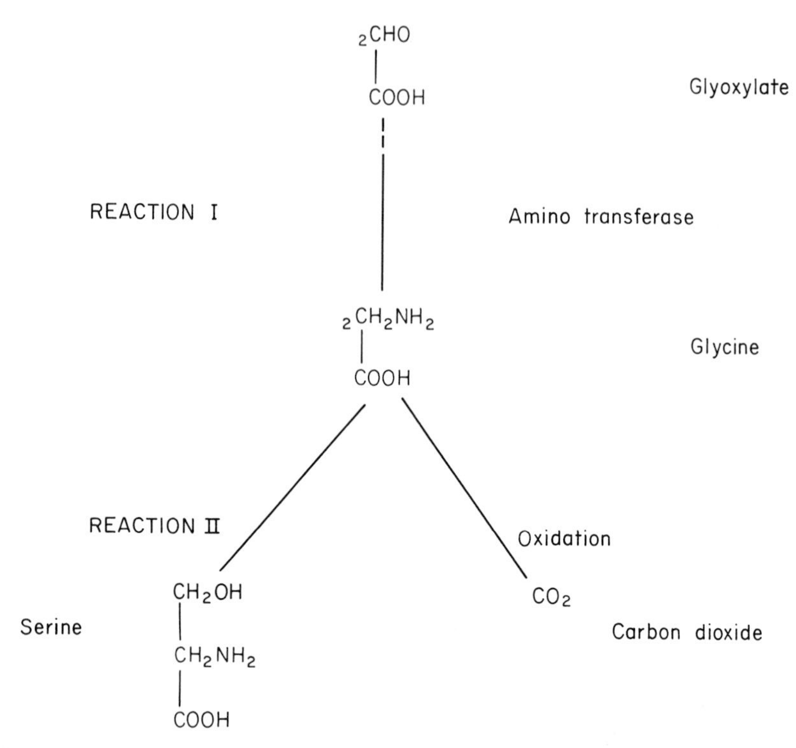

Fig. 2.6. Part of the photorespiration cycle from C-3 plants showing two reactions. Firstly, the conversion of glyoxylate to glycine, through the activity of amino transferase and secondly the conversion of glycine to serine and carbon dioxide. No net isotopic effect would be evident in the first reaction, whereas the second involves a division of the carbon atoms to form two compounds and therefore a possible isotopic effect.

These fractionations, and also the fractionations in products of different primary productivity pathways are heavily biased by closed system effects as has been shown during a study at the Planetary Biology and Microbial Ecology Program (Margulis & Nealson 1983). Deeper parts of laminated microbial mats (potential stromatolites) of Bahia California have fractionation data as low as $-7/‰$, while surface mats of the San Francisco bay yield -18 to $-20/‰$ and a completely open mat exposed to fast-running water at the Alum Rock Park (San Jose) exhibits values as high as $-33/‰$ in some cases (des Marais & Krumbein, unpublished observations). The 'heaviest' biogenic isotope ratios ever found originate from the Solar Lake and Gavish Sabkha (Sinai) with -4.5 and $5.2‰$ respectively in the potential stromatolite mats (Schidlowski & Matzigkeit, in press).

Table 2.1. Summary of previously reported fractionations between CO_2, CH_4 and parent organic material (after Games & Hayes 1976).

	$\delta^{13}C$ (‰) PDB			
	CO_2	CH_4	Organic matter	$\alpha(CO_2/CH_4)$*
Laboratory cultures				
methanol utilizers 30°C	−1.2	−68.6	−16.3	1.072
(Rosenfeld & Silverman 1959)				
Lake mud	−5.0	−77.0	−30.0	1.078
(Oana & Deevy 1960)				
Marine sediment,				
interstitial water	16.7	−55.0	−20.0	1.076
sewage sludge	4.1	−47.1	−22.6	1.054
(Nissenbaum *et al.* 1972)				
Glacial drift gas	−7.9	−75.0	—	1.072
(Wasserburg *et al.* 1963)				
Lake Kivu	−2.0	−45.0	—	1.045
(Deuser *et al.* 1973)				
Sediments from DSDP 11–19	−48.0	−86.0	—	1.039–1.081
(Claypool *et al.* 1973)				
Landfill gases	−20.0	−50.0	−25.0	1.074
(Games & Hayes 1974)				
Holocene sediments } Black Sea	−63.7	−16.5	—	1.050
	−62.4	+9.0	—	1.076
(Alekseev & Lebedev 1975)				
Landfill gases	16.6	−52.1	−25.0	1.072
Sewage sludge tank	−5.5	−49.1	−25.0	1.046
Pure cultures				
Methanosarcina barkerei	−42.3	−82.6	—	1.044
Methanobacterium M.o.H.	−42.0	−65.6	—	1.059
M. thermoautotrophicum	−41.4	−64.9	—	1.025
(Games & Hayes 1976)				
Ruminant Gases C-3	−27.0	−64.5	−27.0	1.039
C-4	−13.0	−49.3	−13.0	1.040
(Rust 1981)				
Holocene sediments } Baltic Sea	−23.0	−64.0	—	1.044
(Lein *et al.* 1981)				

* $\alpha(CO_2/CH_4) = (1000 + \delta^{13}C_{CO_2})/(1000 + \delta^{13}C_{CO_4}$

2.5 CARBON RELATIONSHIPS IN SYMBIOTIC ORGANISMS

The role of microsymbionts in the carbon cycle of numerous organisms has undoubtedly been of major significance, not only during their evolution, but also for their continual survival. Here we shall consider briefly only two

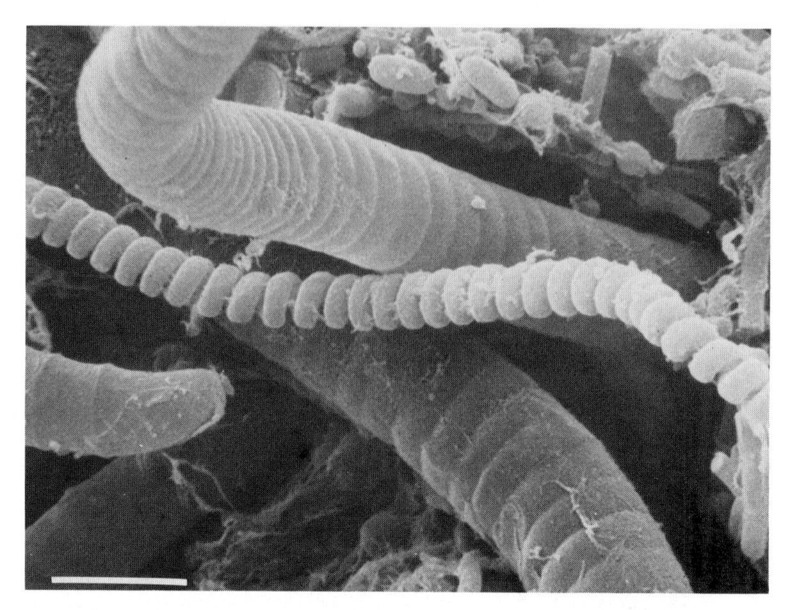

Fig. 2.7. Microbial microcosm of a laminated microbial ecosystem (the 'Farbstreifensandwatt' of Mellum, North Sea). Primary production is high, as is species diversity. *Oscillatoria*, *Spirulina*, *Microcoleus*, *Synechococcus*, *Phormidium*, etc., represent filamentous and coccoid cyanobacteria of a recent stromatolitic environment; bar = 6 μm.

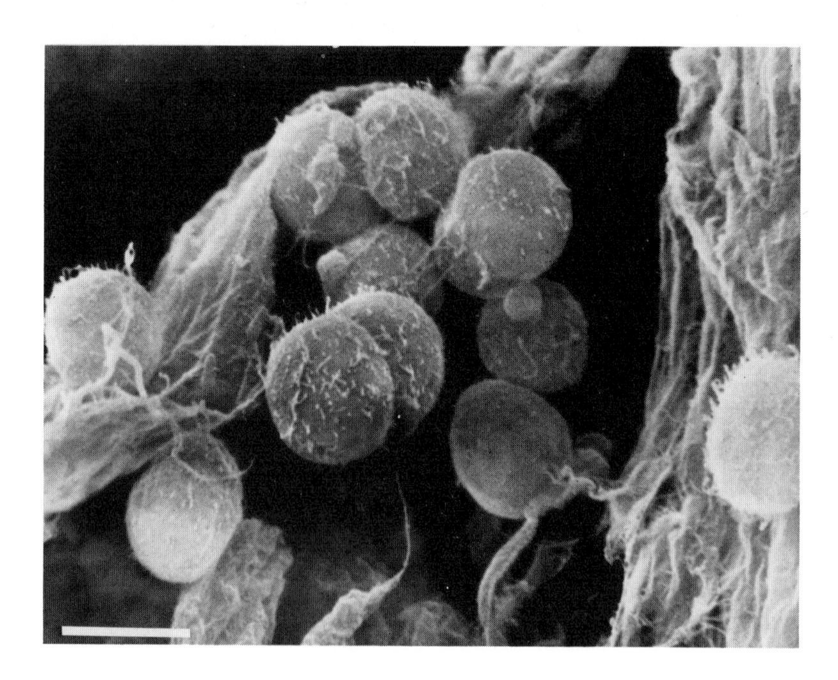

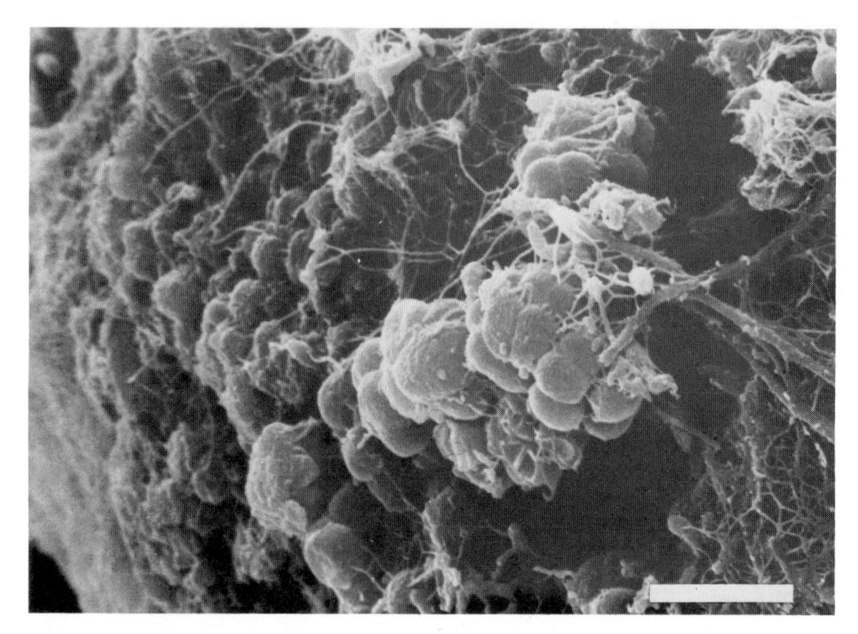

Fig. 2.9. Colony of a coccoid cyanobacterium from the Gavish Sabkha which, in contrast to the coccoid photosynthetic bacterium in Fig. 2.6, is producing organic carbon compounds via oxygenic photosynthesis within the same system. The colony represents a pleurocapsalean cyanobacterium with sheath material in the first steps of decay. Many of the photosynthetic bacteria within recent stromatolitic benthic communities develop common extracellular slime layers of rather complicated structure and composition, the function of which is not yet fully understood. They are called sheath or capsule or glycocalyx; bar = 6 μm.

←

Fig. 2.8. Anoxygenic production of organic carbon compounds in a hypersaline recent stromatolite environment by members of the genus *Thiocapsa*. The samples for this figure and most of the subsequent figures demonstrating decay processes within the sedimentary environment are taken from a hypersaline lagoon environment, the Gavish Sabkha at the shores of the Gulf of Aqaba; bar = 3 μm (Krumbein *et al.* 1979).

Fig. 2.10. Laminated microbial mat (potential stromatolite; Krumbein 1983) of the Solar Lake, Sinai. The bar represents 20 mm. The layers on the left side are from approximately 20–30 cm depth within the mat and represent an age of several hundred years after production of the organic material. The white spots indicate carbonate spherules and voids that generated within the mat during microbial degradation of the cyanobacterial material. SEM-micrographs do not allow any classification at the genus level within these mats. The right side represents the zone between 5 and 15 cm depth, the zone of high activity of sulphate-reducing bacteria. Iron sulphide precipitation causes the black areas which are prevalent in these layers. Later the iron sulphides are transformed into pyrite, and grey and white areas, caused by a mixture of refractory organic material with carbonate and some pyrite, prevail.

At the surface of the mat the potential stromatolite consists of 96% organic material. Towards the bottom, calcium carbonate in several mineral forms increases to 70–80% while organic carbon decreases to 2–4% during microbial degradation of cellular materials and generation of carbonate.

Figs 2.11–2.15. Morphologically documented degradation sequence of a filamentous cyanobacterium of a potential stromatolite (Gavish Sabkha) from surface and production site to 2 m depth within the sediment (see Friedman & Krumbein, in press).

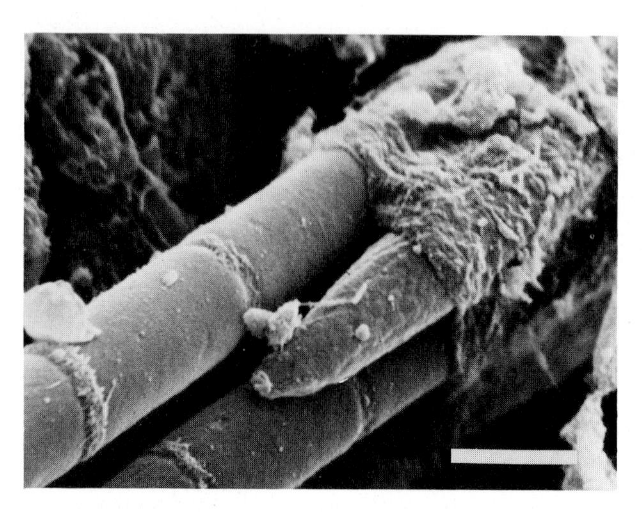

Fig. 2.11. Actively photosynthesizing *Microcoleus* bundle from surface layer in SEM preparation; bar = 3 μm.

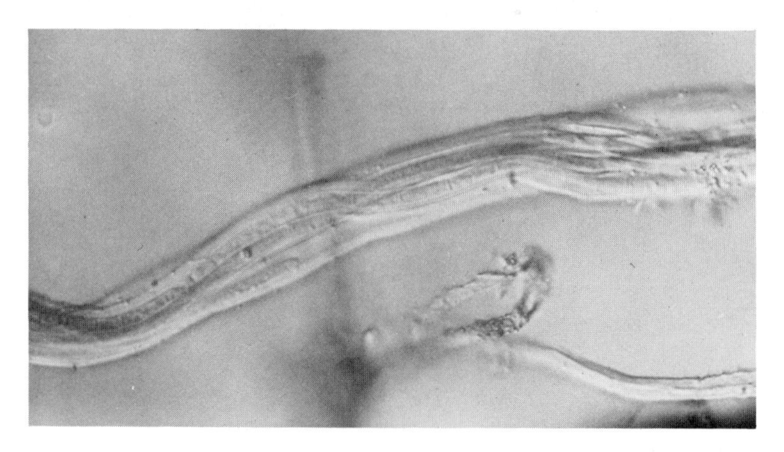

Fig. 2.12. The same species as Fig. 2.11 in transmittant light photomicrograph.

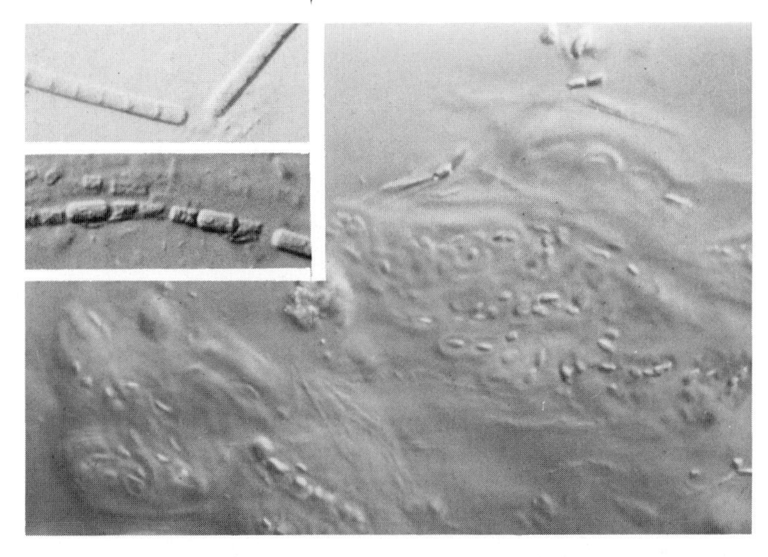

Fig. 2.13. Nomarsky-photomicrograph of *Microcoleus* sp. from the Gavish Sabkha is taken from the top layers. The lower insert shows cells in decay from 24 mm depth within the mat. The major part of the figure is taken from 120 cm depth within the Sabkha sediments from a laminated carbonate system of an approximately 5000-year-old stromatolite. The common sheath of *Microcoleus* bundles is clearly discernible, while only minute remainders of the original trichomes are left after chemoorganotrophic decomposition. Obviously the major part of the fossilized material is derived from the sheath or glycocalyx.

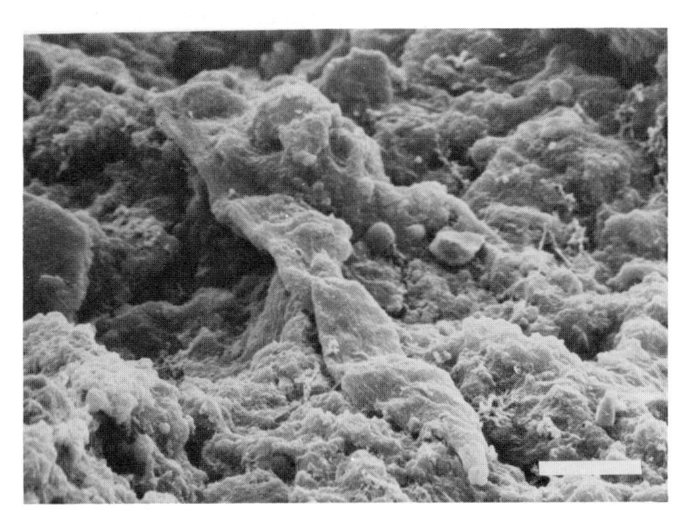

Fig. 2.14. SEM-micrograph of the same layer as in Fig. 2.13. SEM occasionally shows very clear envelopes (glycocalices) of *Microcoleus*, now completely empty and mostly deflated; bar = 20 μm.

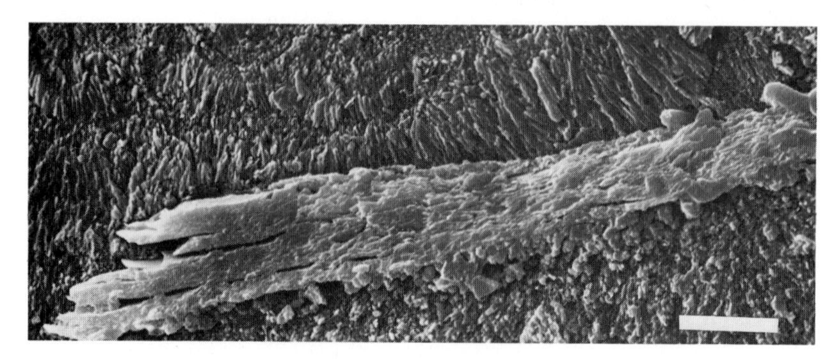

Fig. 2.15. Completely calcified glycocalyx of *Microcoleus* from Solar Lake stromatolites. The structure of the former cyanobacterial layer is now difficult to distinguish; bar = 30 μm.

Figs 2.16–2.26. Degradation and fossilization sequences of coccoid cyanobacteria as documented by light and scanning electron microscopy in gradually older laminated microbial mats of the Gavish Sabhka, Sinai (Friedman & Krumbein, in press).

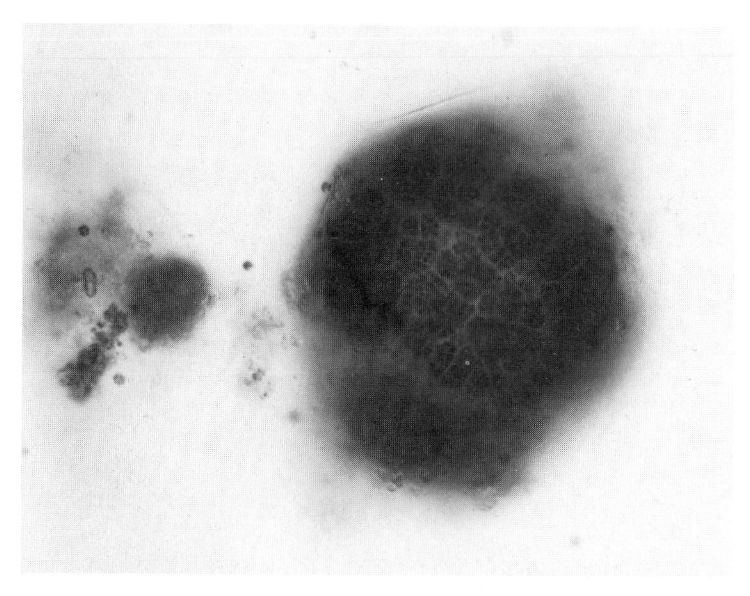

Fig. 2.16. Light micrograph of a densely packed colony of *Gloeothece* sp.

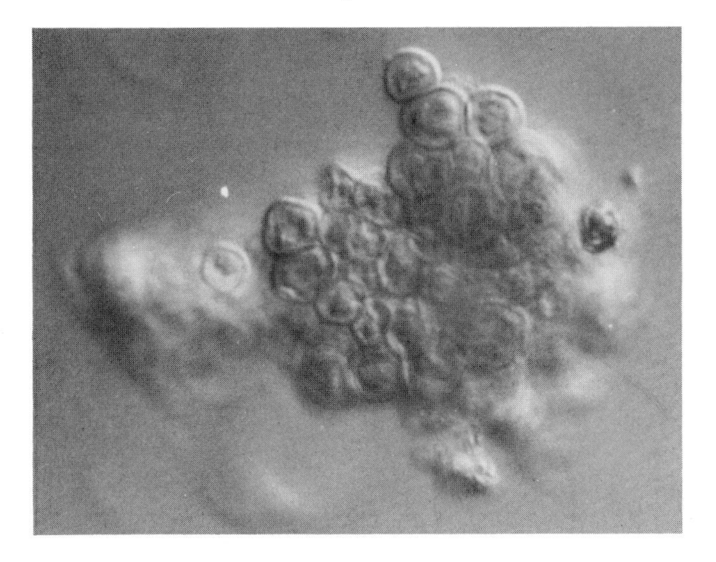

Fig. 2.17. Light micrograph (Nomarsky) of the earliest stage of decay of *Pleurocapsa* sp.

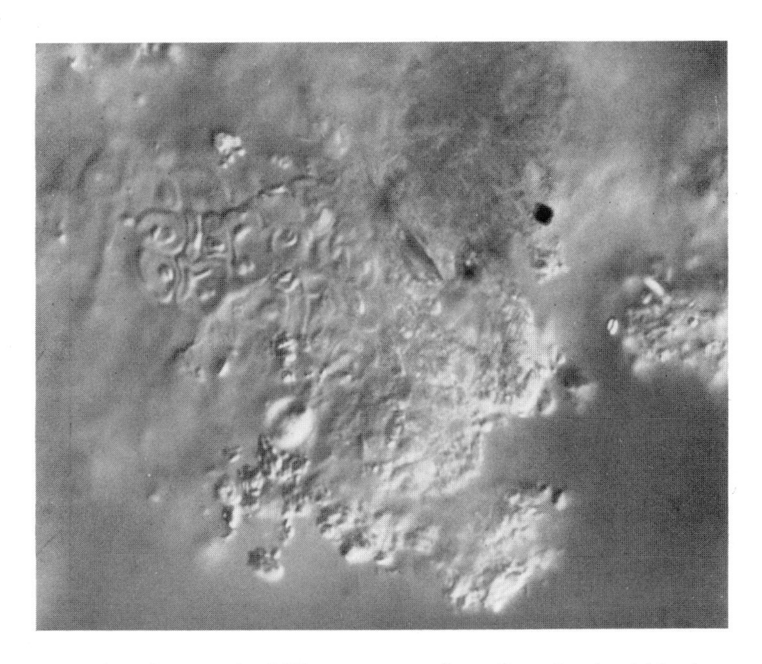

Fig. 2.18. Light micrograph of *Pleurocapsa* sp. from 4 cm depth within the sediment.

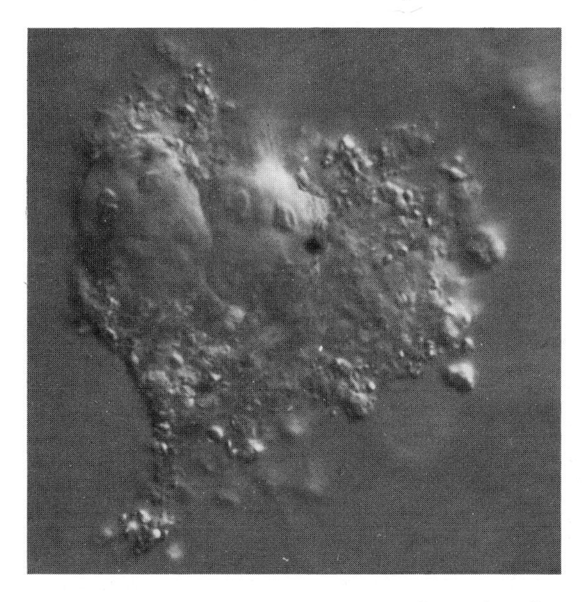

Fig. 2.19. Light micrograph of pleurocapsalean cyanobacterium from 205 cm depth. The majority of preserved organic material again is the capsule or glycocalyx material.

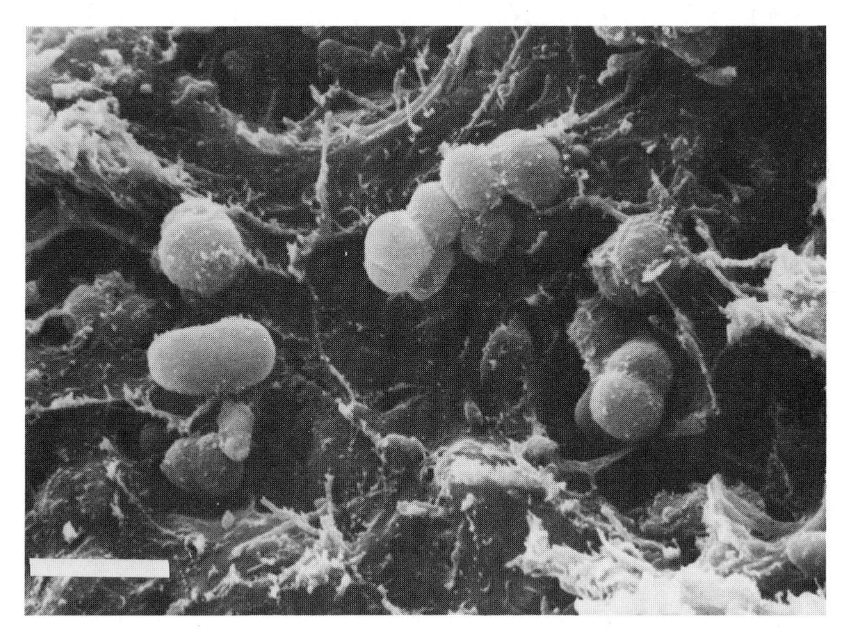

Fig. 2.20. SEM micrograph of *Pleurocapsa* sp. from the Gavish Sabkha; bar = 6 μm.

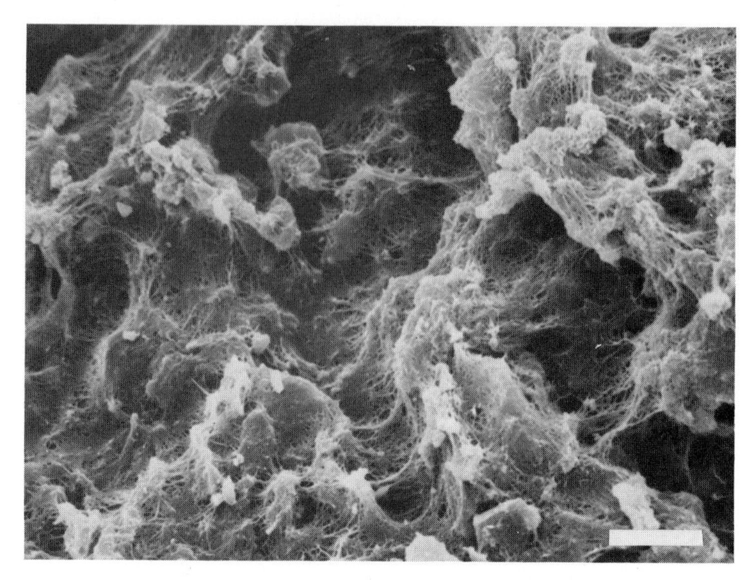

Fig. 2.21. Similar colony in an advanced stage of decay but clearly discernible as the remnants of a pleurocapsalean cyanobacterium; bar = 10 μm.

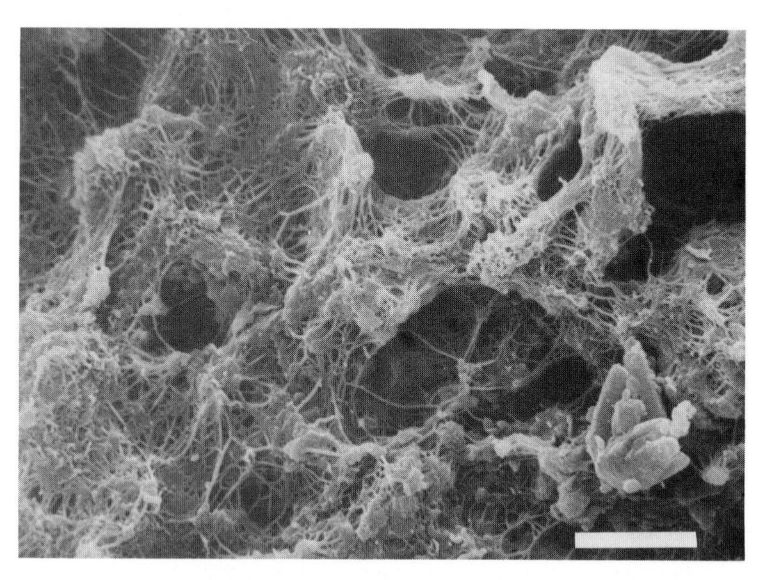

Fig. 2.22. The glycocalyx material of cyanobacteria remains, while cell material is completely degraded at 120 cm depth; bar = 6 μm.

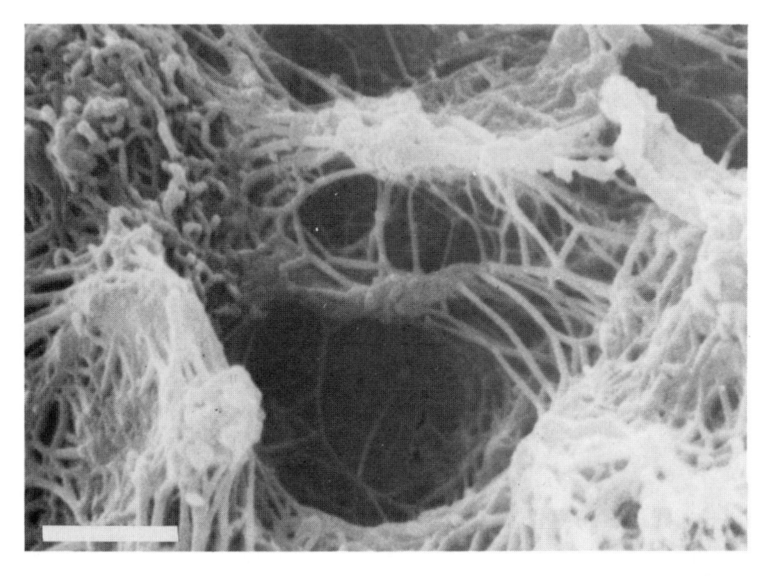

Fig. 2.23. Section of Fig. 2.22 enlarged, showing the spider's web type of the glycocalyx encasing a cavity delineating a colony of four cells; bar = 2 μm.

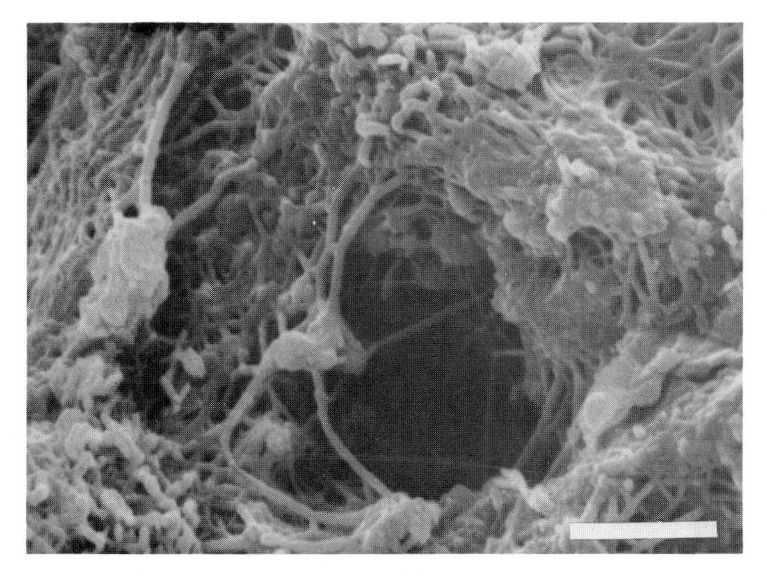

Fig. 2.24. Another section of Fig. 2.22 enlarged showing two former cells as cavity encased by spider's web remainders of the glycocalyx; bar = 2 μm.

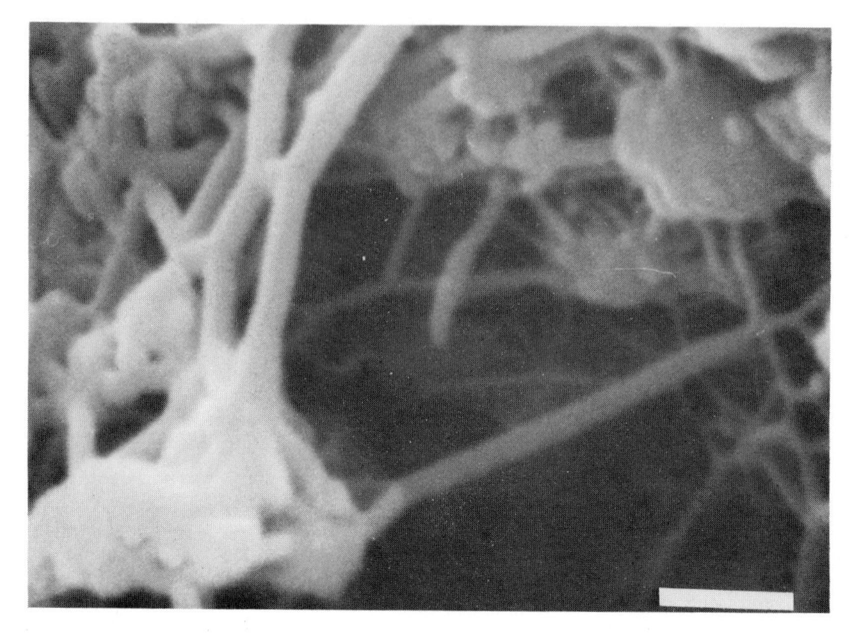

Fig. 2.25. Section of Fig. 2.24 enlarged. The strands of polysaccharide material are keeping the ancient colony form firmly intact. A thin veil of slime material still marks the former cell wall separation line of the two cells of the former microcolony; bar = 500 nm.

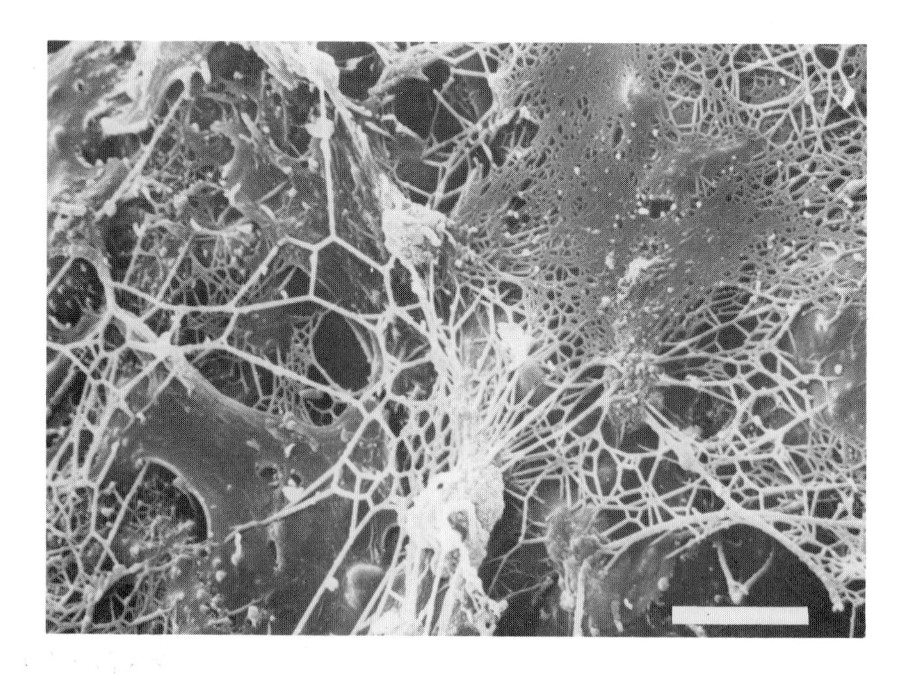

Fig. 2.27. Coccoid cyanobacteria of the Gavish Sabkha mixed with a LPP form (*Phormidium* sp.). Their glycocalyx consists almost exclusively of microfibrillar material; bar = 20 μm.

←

Fig. 2.26. Slime material originating from cyanobacteria in the Gavish Sabkha. The slime cannot be related to any defined species. It only indicates the sequence of events during the degradational processes within the mat. The sheath or capsule or glycocalyx material initially is a smooth film covering the cells. During the decay of the cellular material the sheath material fuses on to the frequent microfibrillar strands within the mucus thus creating the spider's web appearance. Some forms which produce less slime and more fibrillar material (usually of widths between 6 and 20 nm) even during their lifetime exhibit the spider's web-type capsules (see Fig. 2.27); bar = 6 μm.

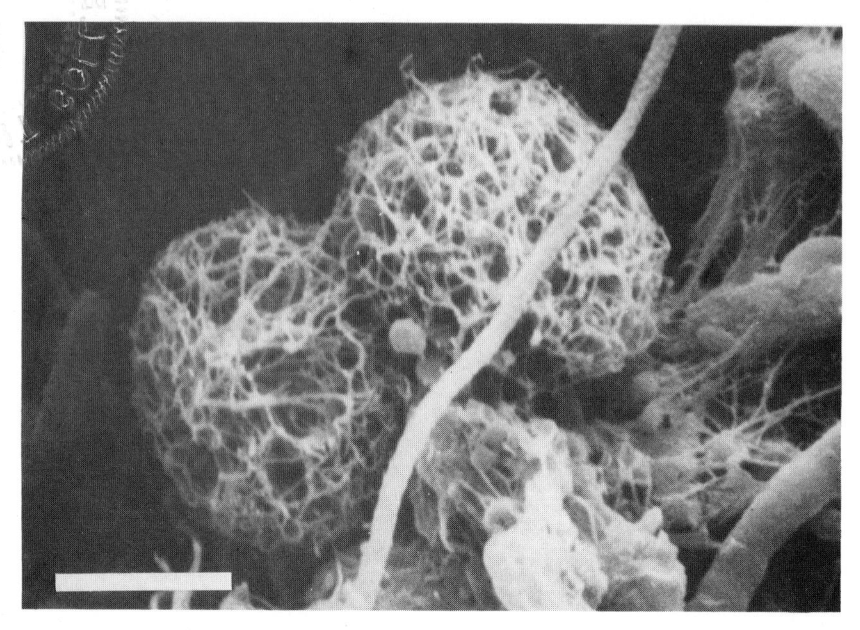

Fig. 2.28. A section of Fig. 2.27 enlarged. During cell division it can be clearly demonstrated that the capsule material in this case consists exclusively of fibrillar material. This phenomenon cannot be a fixation artefact inasmuch as the mucous sheath of other forms is completely smooth in the same preparations. The fibril material of the two cells gives the impression that one can almost look through the glycocalyx into the centre with the cell wall beneath. Figures 2.27–2.28 demonstrate that during the degradation of organic material within a microbial mat, cell wall and plasma as well as membranes are destroyed first and that the most resistant material of the primary production of a recent stromatolite is the fibrillar substance which is embedded in or makes up large portions of the glycocalyx. The rigidity and stability of the material is such that the form of the original colony of cyanobacteria is almost perfectly preserved as a cavity in the slowly transforming sediment. Such structures including some remains of organic material (kerogen) have been found by us not only in young stromatolites as in the Gavish Sabkha but also in palaeozoic sediments; bar = 20 μm.

→

Fig. 2.30. Fossil cyanobacterium from a stromatolitic carbonate rock of the Palaeozoic of Poland. (Samples were kindly given by J.Kazmierczak.) In thin sections and parallel SEM samples it was clearly demonstrated that the cavity is in fact the microfossil of a coccoid cyanobacterium; bar = 3 μm. The oval- and rod-shaped cavities in the calcified glycocalyx are without doubt (many reference micrographs do exist) the places of former bacteria cells and some organic matter of the bacteria, which were decomposing the cyanobacterium. These are preserved as kerogen within the small cavities (Kazierczak & Krumbein 1982).

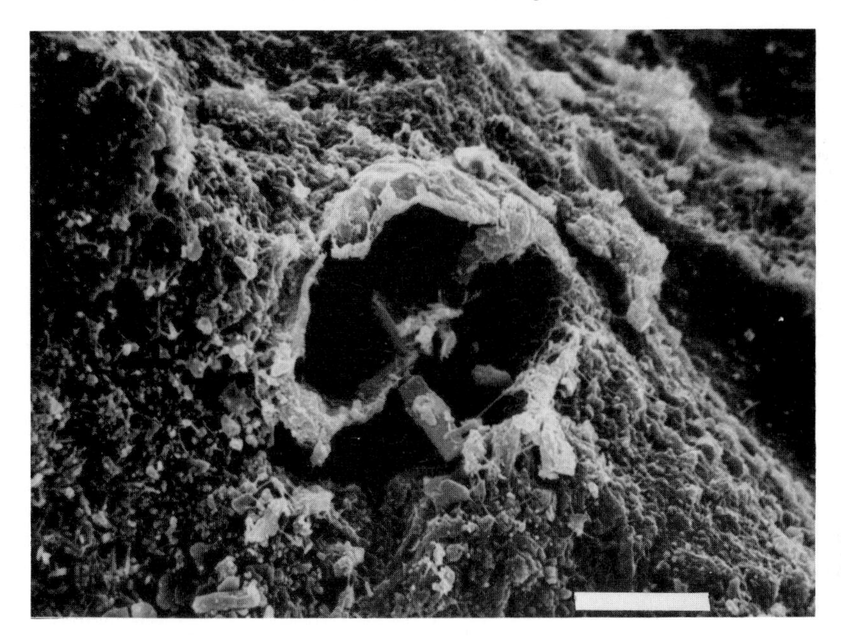

Fig. 2.29. Calcified colony of the Gavish Sabkha with some slime left and the cavity which in thin sections looks like a cell tetrad; bar = 20 μm.

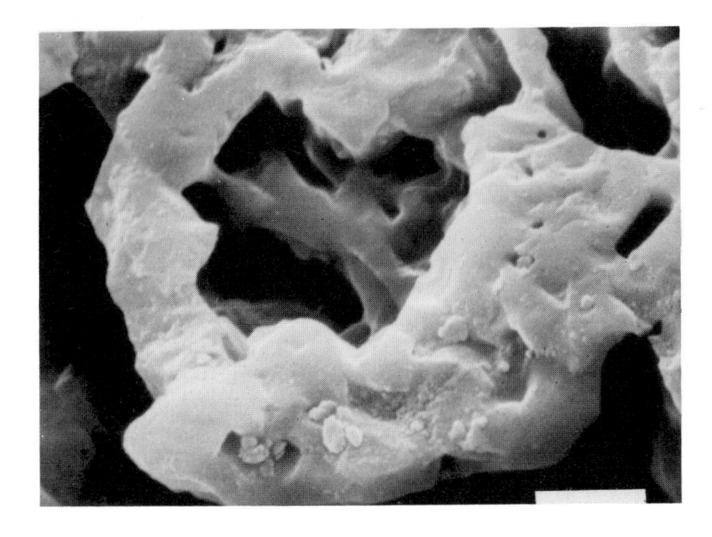

examples of the many such relationships which exist between unicellular organisms and plants and animals.

Recent studies on the members of the phylum Pogonophora as well as the larger related Vestimentifera, have shown that these organisms possess enzymes indicative of autotrophy, believed to be situated within endosymbiotic chemoautotrophic bacteria (Cavanaugh *et al.* 1981). The Pogonophora live typically in oligotrophic conditions and it has been a matter of speculation as to how they get enough nutrition, notwithstanding the fact that they lack a mouth, digestive tract and anus (Southward 1971). Therefore, the discovery of enzymes associated with the fixation of CO_2 by the Calvin cycle as well as extremely depleted $\delta^{13}C$ isotope values seems to indicate a complex metabolism including chemolithotrophic symbiosis.

It is a well-established principle when studying carbon isotope ratios that you are what you eat (deNiro & Epstein 1976). The extreme negative values of the tissues of the worms ($-46‰$), together with the presence of the enzymes implicates the RuBP-carboxylase-based restrictive pentose phosphate (Calvin)-cycle. There are at present two alternative hypotheses:
(1) Either that respirated CO_2 is being recycled through the Calvin cycle; respired CO_2 has a $\delta^{13}C$ similar to the respiring tissues and therefore subsequent fixation by RuBP-carboxylase and the accompanying fractionation result in extremely negative $\delta^{13}C$ values. Eventually, of course, a steady-state condition ensues with the fractionation of the respired CO_2 being balanced by the influx of inorganic, isotopically heavy CO_2.
(2) Alternatively, the unusual $\delta^{13}C$ might result from the utilization of C_1 compounds, such as methane. These compounds already possess very negative $\delta^{13}C$ values and it is likely that during their oxidation to CO_2 this isotopic composition will be retained either in the CO_2 incorporated by RuBP-carboxylase or in the assimilation of formaldehyde (see earlier discussion). In the case of the Pogonophora, methane can be postulated to be generated from anaerobic sediments. The related Vestimentifera, on the other hand, inhabit the hydrothermal vent systems such as those found in the Pacific oceans (Rau 1981). Although reduced carbon emanating from the vent systems has been implicated in the local carbon cycle, the $\delta^{13}C$ of the Vestimentifera is considerably heavier than that of the simultaneously occurring organisms such as the Pelycopods. Clearly, these worms, although containing endosymbiotic bacteria, have differing carbon cycling mechanisms.

One of the more significant symbiotic relationships in the marine environment is that between the endosymbiotic dinoflagellate *Gymnodinium microadriaticum* and hermatypic scleractinian corals. Scleractinians are the main framework and visual elements of modern coral reefs. Although the scleractinians emerged during the early Mesozoic, they did not become the dominant reef component until the Triassic (Stanley 1981). It is thought that the development of the symbiosis between the coral and the dinoflagellate (zooxanthellae) which inhabits its gastrodermis was responsible for the eventual dominance of the scleractinia. It is clear from the work of numerous researchers that the zooxanthellae positively affect the growth rate of the coral, although the precise mechanism remains uncertain. One of the simplest proposals is that CO_2 is removed during photosynthesis, thereby shifting the equilibria of the following equation (Goreau 1959):

$$Ca^{2+} + 2HCO_3^- = CaCO_3 + H_2O + CO_2.$$

Further theories are:
(1) photosynthate is provided by the zooxanthellae which is subsequently used in matrix formation (Muscatine & Cernichiari 1969); (2) energy is provided for the active transport of Ca^{2+} and HCO_3^- and for the synthesis of an organic matrix (Chalker & Taylor 1975); and (3) crystal poisons such as phosphates are removed by the zooxanthellae (Simkiss 1964).

Whatever the precise mechanism, both the carbon isotope composition of the skeleton and the tissues are affected by the activity of these microorganisms. Available evidence (Swart 1983) indicates that preferential uptake of $^{12}CO_2$ during photosynthesis causes the 'internal pool' of CO_2 to become $\delta^{13}C$ enriched. Ultimately, however, the overall composition of this 'pool' is determined by inputs of respired CO_2, inorganic CO_2 and, perhaps, other as yet unidentified sources. A scheme of the cycle with assigned $\delta^{13}C$ values is shown in Fig. 2.29. The net effect of such a coupling is to produce isotopically enriched skeletal material during periods of high zooxanthellar activity. Oxygen isotope values appear not to be affected by photosynthesis and are only influenced by the thermodynamic equilibrium in the H_2O-CO_2-CO_3 system. The result is that only under rarely achieved environmental conditions is a correlation between the carbon and oxygen isotopes seen in hermatypic corals. In ahermatypes or non-zooxanthellate corals, a positive correlation is always evident, thus enabling the two types to be distinguished by geochemical criteria.

2.6 GLOBAL CARBON CYCLE

Hydrogen, helium, oxygen and carbon are in this order, the most abundant elements in the cosmos. The earth, however, shows different abundances. Oxygen and silicon lead the list and carbon is only fourteenth in relative abundance. Despite the seemingly high relative abundances of carbon in extraterrestrial environments pressures are low, typically 10^{-5} atmospheres in dense interstellar clouds. Nevertheless, from astronomical observations over 30 simple 'organic' molecules have been recognized (Hayatsu & Anders 1981). More complex organic compounds such as alkanes, aromatics as well as amino acids have been isolated from chondritic meteorites and are truly recognized as abiogenic rather than vestiges of extraterrestrial life. A brief mention should perhaps be given to Lippman (1932), who claimed to have cultured bacteria from meteorites (these were probably contaminants), and more recent theories which propose that bacteria originate from outer space and are indeed responsible for the spread of common bacterial and viral-based diseases as well as life itself (Hoyle & Wickramasinghe 1978).

The origin of such complicated organic molecules as found in meteorites (see Hayes 1967, Hayatsu & Anders 1981) is believed to be through processes such as the Miller–Urey, or Fischer–Tropsch type synthesis. Of these two only the latter can adequately explain the large difference in $^{13}C/^{12}C$ ratios between carbonate and organic carbon found in C1 and C2 carbonaceous chondrites: 60 to 80‰. However, others point out that the high temperatures needed for the Fischer–Tropsch reaction would destroy many of the amino acids found in chondrites. A compromise may be that both reactions occur with isotope fractionation of carbon as well as other elements being enhanced by ion-molecule reactions and nuclear processes.

The Fischer–Tropsch and Miller–Urey reactions have also been suggested as forming the assumed terrestrial 'primordial soup'. Brief mention may, however, be given to Corliss *et al.* (1981) and their hydrothermal vent theory.

The analysis of meteorite data and calculations on the evolution of the atmosphere of the earth as compared to other planets have shown clearly that various models are still feasible and that further experimentation and computation of data is needed to understand the mechanisms of the early synthesis of organic materials. Many of the primordial carbon compounds may have reached the planet with meteorites, some of them may even be older than our galaxy (Swart *et al.* 1983). However, since the biological cycle of carbon was started on earth it has transformed our planet into a closed system which guarantees its continuation. Bolin *et al.* (1979) and Woodwell *et al.* (1978) have summarized a world carbon budget as influenced by the biota. Some of the most reliable calculations of the total carbon budget have been produced by taking account of the CO_2 problem caused by fossil burning, by increased destruction and burning of forests and by the increasing erosion and destruction of soils. Concern was initially aroused by the result of continuous measurements of the concentration of CO_2 in the atmosphere at the Mauna Loa observatory and at the South Pole by Keeling and co-workers (e.g. Keeling 1960, 1973, Ekdahl & Keeling 1973). Very roughly summarized it was deduced from these and other data, that the overall biologically controlled carbon budget of the earth may be completely disturbed (or even destroyed) by the human society. It can be questioned whether human influence such as fossil fuel burning, wood cutting and burning and soil erosion, as well as an increase in acid rains resulting from SO_2-emission during coal and petroleum burning may not alter the equilibrium between the major carbon pools on the earth which are cycled and controlled by the total biosphere. Golubic *et al.* (1979) have calculated that one carbon atom is cycled 10^7 times through the atmosphere via the biosphere while it cycles 10^5 times through the hydrosphere and only once through the lithosphere in the same timespan.

It has been demonstrated by carbon isotope analyses and organic geochemical analyses that carbon is fixed in the crust in a biologically controlled way either as biogenic organic carbon or as biogenic inorganic carbon with very small negligible amounts stored as evaporitic or magmatic abiogenic carbonates. This indicates the central importance of biological cycles in the turnover of carbon through lithosphere, hydrosphere and atmosphere.

Garrels *et al.* (1975) have calculated the residence time of individual carbon atoms as 11 years in the biosphere, 4 years in the atmosphere, 385 years in the hydrosphere (the upper part of the ocean) up to 100 000 years in the deep oceanic waters and 342 million years in the earth's crust. The theory of homeostatically regulated cycles and equilibria of organic and inorganic carbon compounds has been supported by the analysis of isotopic fractionation of carbon isotopes in different biological processes (Benedict 1978), by the relatively light isotopic fraction in organic material in rocks as compared to the relatively heavy isotopic fraction in carbonates (Schidlowski 1980), by the ratio of organic carbon to inorganic carbon in all sedimentary rocks throughout the biologically influenced history of the Earth for 3.5 billion years (Schidlowski *et al.* 1979, Golubic *et al.* 1979, Holland & Schidlowski 1982, Hayes *et al.*, 1983) and by the computation of present day primary production and degradation of organic materials which is constantly producing small amounts of organic carbon compounds which are buried for geological periods in sediments and sedimentary rocks.

Golubic *et al.* (1979) conclude that—as during the mid-Cretaceous event or in former periods of change—we may also witness today the capability of the global biosphere to buffer fluctuations by its own homeostatic equilibrium. The net increase of CO_2 in the atmosphere in the past 100 years has reached about 15%. If all CO_2 which is brought into the atmosphere by human influences remained there, i.e. if no other regulatory means existed, then the level should already have reached more than 30%. Thus there must be several other factors in the biosphere which exert control over the biological cycle of carbon to ensure equilibration of the atmospheric CO_2 pool. One suggestion is a net increase in rates of global photosynthesis, another is a net increase in organic and inorganic carbon burial, e.g. by differences in the degradation and remineralization of organic carbon compounds and in weathering rates. Unfortunately neither theory can be proven with our present day methods. The net increase in photosynthesis needed to balance fossil fuel burning, wood cutting and forest fires approximates only 1%. The determination of primary production on a global scale has an error larger than 5%. Furthermore, the interlocking of the carbon and silica cycle which was started some 2.2 billion years ago may help to buffer human influences on the natural biological cycle of carbon. Phosphorus availability via burial and weathering equilibria may be of primary importance as well. Bolin *et al.* (1979) give a clear picture of our present uncertainty about the problems involved in man-made perturbations of the global carbon budget. The same position is outlined in Stumm (1977) and in other summaries. We are able to calculate a theoretical atmospheric CO_2 increase. This does not correspond to the measured increase. The suggested pathways to explain the discrepancy cannot be demonstrated satisfactorily by existing ecological methods.

The biological cycling of carbon and its impact on the present day biogeochemistry of carbon is subject to speculation inasmuch as most of the necessary data to solve the problem satisfactorily are lacking or are not precise enough (Krumbein 1982).

(1) Approximately 140×10^{15} g C have been emitted so far by fossil fuel burning (Bolin *et al.* 1979).

(2) Anything between 40×10^{15} and 120×10^{15} g C may have been added by deforestation and expansion of intensive agriculture (Bolin *et al.* 1979, Woodwell *et al.* 1978).

(3) Models of hereby modified rates of erosion and consequences for atmospheric CO_2 have been calculated and may predict an increase of atmospheric CO_2 from 320 ppm presently to 800 ppm, with many extreme consequences (Garrels *et al.* 1976).

(4) The data on which the first three statements are based have a range of error of 150% (Woodwell *et al.* 1978).

(5) The deep waters of the ocean may play a relatively minor role in the short-term balancing of increased CO_2 release (Broecker *et al.* 1971, Broecker 1975). An increase in primary productivity and therefore organic carbon storage in recent years may be possible but essential proof cannot be gained, because the methodology is inadequate (Woodwell *et al.* 1978, Bolin *et al.* 1979, Golubic *et al.* 1979).

(6) The role of dissolved and particulate organic carbon in the ocean is not clear and analytical errors in determination of the two is approximately 50%. Fenchel and Jørgensen (1977) have overcome the problem of estimation of dissolved and particulate organic carbon by defining detritus as the sum of non-living dissolved and particulate organic carbon and incorporating both reservoirs in one box of their cyclic model.

(7) The roles of silica, sulphur and phosphorus in relation to productivity and degradation as well as in burial and weathering are not completely understood.

Hence it is possible to calculate many different models for carbonate, organic carbon and atmospheric carbon dioxide pools and fluxes and actually several different ones do exist. We know, that the global carbon cycle is driven by endogenic forces (internal heat flow and plate tectonics) and exogenic forces (solar energy, bond energy of reduced chemical compounds) and controlled by pool sizes, fluxes and availability of associated elements such as sulphur, nitrogen and phosphorus. These parameters enable several different models of exchange and cycling which may be verified in past and future geobiological events and by further analytical work.

References

Alekseev F.A. & Lebedev V.S. (1975) The carbon isotope composition of carbon dioxide and methane in the bottom sediments of the Black Sea. In *Dispersed gases and biochemical conditions of sediments and rocks.* pp. 49–53. Nedza, Moscow (in Russian).

Arnon. D.I., Tsujimoto H.Y. & Tang G.M.S. (1981) Proton transportin photooxidation of water: a new perspective on photosynthesis. *Proc. Natl. Acad. Sci. USA* **78,** 2942–6.

Awramik S.M. (1981) The pre-phanerozoic biosphere—three billion years of crises and opportunities. In *Biotic crises in ecological and evolutionary time* (Ed. M.H.Nitecki) pp. 83–102. Academic Press, New York.

Awramik S.M. (1982) Biogeochemical evolution of ocean atmospheric systems, group report. In *Biospheric evolution and precambrian metallogeny* (Eds H.Holland & M.Schidlowski). Springer, New York.

Awramik S.M., Schopf J.W. & Walter M.R. (1983) Filamentous fossil bacteria from the archean of western Australia. *Precambrian Research* (in press).

Baltscheffsky H. (1979) Evolutionary and functional aspects on prokaryotic energy coupling in photosynthesis, respiration and nitrogen fixation. In: Abstracts III. Internat. Symposium on photosynthetic prokaryotes, Oxford.

Belkin S. & Padan E. (1978) Hydrogen Metabolism in the Facultative Anoxygenic Cyanobacteria (Blue-Green Algae) *Oscillatoria limnetica* and *Aphanothece halophytica. Arch. Microbiol.* **116,** 109–111.

Benedict C.R. (1978) The fractionation of stable carbon isotopes in photosynthesis. *What's New in Plant Physiology* **9,** 13–16.

Bolin B., Degens E.T., Duvigneaud P. & Kempe S. (1979) The global biogeochemical carbon cycle. In *The global carbon cycle* (Eds B.Bolin, E.T.Degens, S.Kempe & P.Ketner), pp. 1–56. John Wiley & Sons, Chichester.

Boon J.J., de Leeuw J.W. & Krumbein W.E. (in press) Biogeochemistry of the Gavish Sabkha sediments II. Pyrolysis mass spectrometry of the laminated mats in the permanently water covered zone before and after the desert sheet floods of 1979 and 1980. In *The Gavish Sabkha—A model of a hypersaline ecosystem* (Eds G.M.Friedman, W.E.Krumbein & G.Gerdes). Ecological Studies. Springer Verlag, Berlin.

Brock T.D. (1979) Biology of microorganisms, 3e, pp. 802. Prentice-Hall Inc., Englewood Cliffs, N.J.

Broecker W.S. (1975) Climatic change: Are we on the brink of a pronounced global warming? *Science* **189,** 460–3.

Broecker W.S., Liu Y.H. & Peng T.H. (1971) Carbon dioxide—man's unseen artefact. In *Impingement of Man on the Oceans* (Ed. D.W.Hood), pp. 287–324. Wiley-Interscience, New York.

Bromfield S.M, (1954) Reduction of ferric compounds by soil bacteria. *J. gen. Microbiol.* **11,** 1–6.

Cavanaugh C.M., Gardiner S.L., Jones M.L., Jannasch H.W. & Waterbury J.B. (1981) Prokaryotic cells in the hydrothermal vent tube worm *Riftia pachyptila* Jones: Possible chemoautotrophic symbionts. *Science* **213,** 340–2.

Chalker B.E. & Taylor D.L. (1975) Light enhanced calcification, and the role of oxidative phosphorylation in calcification of the coral *Acroopora cervicornis. Proc. R. Soc. Lond., Ser. B.*, **190,** 323–31.

Claypool G.E. & Kaplan I.R. (1974) The origin and distribution of methane in marine sediments. In *Natural Gases in Marine Sediments* (Ed. I.R.Kaplan), pp. 99–139. Plenum Press, New York.

Claypool G.E., Presley B.J. & Kaplan I.R. (1973) Gas analyses in sediment samples from legs 10 to 19. *Initial Rep. of the Deep Sea Drilling Project* **19,** 879–84. US Gov. Printing Office, Washington D.C.

Clemmey H. & Badham N. (1982) Oxygen in the Precambrian atmosphere: an evaluation of the geological evidence. *Geology* **10,** 141–6.

Cloud P.E. & Gibor A. (1974) The oxygen cycle. In *Planet Earth* (Eds F.Press & R.Siever), pp. 67–78. W.H.Freeman, San Francisco.

Cohen Y., Krumbein W.E. & Shilo M. (1977) Solar Lake (Sinai). 2. Distribution of photosynthetic microorganisms and primary production. *Limnology and Oceanography* **22,** 609–20.

Cohen Y., Padan E. & Shilo M. (1975) Facultative anoxygenic photosynthesis in the cyanobacterium *Oscillatoria limnetica. J. Bacteriol.* **123,** 855–61.

Colby J., Dalton H. & Whittenbury R. (1979) Biological and biochemical aspects of microbial growth on C_1 compounds, *Ann. Rev. Microbiol.* **33,** 481–517.

Corliss J.B., Baross J.A. & Hoffman S.E. (1981) An hypothesis concerning the relationship between submarine hot springs and the origin of life on earth. *Oceanologica Acta N^O Speciale*, 59–69.

Craig H. (1953) The geochemistry of stable carbon isotopes. *Geochim. Cosmochim. Acta* **3,** 53–92.

Craig H. (1957) Isotopic standards for carbon and oxygen and correction factors for mass spectrometric analysis of carbon dioxide. *Geochim. Cosmochim. Acta* **12,** 133–49.

Deines P. (1980a) The carbon isotopic composition of diamonds: relationship to diamond shape, color, occurrence and vapor composition. *Geochim. Cosmochim. Acta* **44,** 943–61.

Deines P. (1980b) The isotopic composition of reduced organic carbon. In: *Handbook of Environmental Geochemistry*, vol. I (Ed. P.Fritz & J.Ch.Fontes). Elsevier, Amsterdam.

deNiro M.J. & Epstein S. (1976) You are what you eat (plus a few ‰): the carbon isotope cycle in food chains. *Geol. Soc. Am., Abstr. Prog.* **8,** 834–5.

Deuser W.G., Degens E.T., Harvey G.R. & Rubin M. (1973) Methane in Lake Kivu: new data on its origin. *Science* **181,** 51–9.

Edmond J.M. (1982) Ocean hot springs: A status report. *Oceanus* **25,** 22–7.

Eigen M. & Schuster P. (1979) The hypercycle. A principle of natural self-organization, pp. 92. Springer, Berlin.

Ekdahl C.A. & Keeling C.D. (1973) Atmospheric carbon dioxide and radiocarbon in the natural carbon cycle. I. Quantitative deductions from the records at Mauna Loa Observatory and the South Pole. In *Carbon and the biosphere* (Eds G.M.Woodwell & E.V.Pecan), pp. 51–85. Springfield, Virginia (Atomic Energy Commission CONF-720510 National Technical Information Services).

Fenchel T.M. & Jørgensen B.B. (1977) Detritus food chains of aquatic ecosystems: The role of bacteria. In *Advances in Microbial Ecology* (Ed. M.Alexander), pp. 1–58. Plenum Press, New York.

Fisher C.R. & Trench R.K. (1980) In vitro carbon fixation by *Prochloron* sp. isolated from *Diplosoma virens*. *Biol. Bull.* **159,** 636–48.

Fox G.E., Stackebrandt E., Hespell R.B., Gibson J., Maniloff J., Dyer T.A., Wolfe R.S., Balch W.E., Tanner R.S., Magrum L.J., Zablen L.B., Blakemore R., Gupta R., Bonen L., Lewis B.J., Stahl D.A., Luehrsen K.R., Chen K.N. & Woese C.R. (1980) The phylogeny of prokaryotes. *Science* **209,** 457–63.

Friedman G.M. & Krumbein W.E. (in press) *The Gavish Sabkha—A model of a hypersaline ecosystem.* Ecological Studies. Springer Verlag, Berlin.

Friedman I. & O'Neil J.R. (1977) In *Data of Geochemistry,* 6e. (Ed. M.Fleischer). Chapter KK Compilation of Stable isotope Fractionation Factors of Geochemical Interest. U.S. Geol. Surv. Prof. Pap. 440-KK.

Froelich P.N., Klinkhammer G.P., Bender M.L., Luedtke N.A., Heath G.R., Callen D. & Dauphin P. (1979) Early oxidation of organic matter in pelagic sediments of the eastern equatorial Atlantic: suboxic diagnosis. *Geochim. Cosmochim. Acta* **43,** 1075–90.

Fuchs G. & Stupperich E. (1981) Wege der autotrophen CO_2-Fixierung in Bakterien. *Forum Mikrobiologie* **4,** 198–201.

Fuchs G. & Stupperich E. (1982) Autotrophic CO_2-fixation pathway in Methanobacterium thermoautotrophicum. *Zbl. Bakt. Hyg. I. Orig. C.* **3,** 277–88.

Fuchs G., Stupperich E. & Eden G. (1980) Autotrophic CO_2-fixation in *Chlorobium limicola.* Evidence for the operation of a reductive tricarboxylic acid cycle in growing cells. *Arch. Microbiol.* **128,** 64–71.

Games L.M. & Hayes J.M. (1974) Carbon in ground water at the Columbus, Indiana landfill. In *Solid waste disposal by land burial in Southern Indiana* (Eds D.B.Waldrip & R.V.Ruhe), pp. 81–110. Indiana University Water Resources Research Center, Bloomington, Indiana.

Games L.M. & Hayes J.M. (1976) On the mechanisms of CO_2 and CH_4 production in natural anaerobic environments. In *Proc. 2nd Int. Symp.* (Ed. J.O.Nriagū), pp. 51–73. Ann Arbor Science, Ann Arbor, Michigan.

Garrels R.M., Lerman A. & Mackenzie F.T. (1976) Controls of atmospheric O_2 and CO_2: Past, present, and future. *Am. Sci.* **64,** 306–15.

Garrels R.M., Mackenzie F.T. & Hunt C. (1975) *Chemical Cycles and the Global Environment.* Kaufmann, Los Altos, Ca.

Ghiorse W.C. & Ehrlich L. (1976) Electron transport components of the MNO_2 Reductase system and the location of the terminal reductase in a marine Bacillus. *Appl. Environm. Microbiol.* **31,** 977–85.

Golubic S., Krumbein W.E. & Schneider J. (1979) The carbon cycle. In *Biogeochemical cycling of mineral-forming elements* (Eds P.A.Trudinger & D.J.Swaine), pp. 29–45. Elsevier, New York.

Goreau T.F. (1959) The physiology of skeleton formations in corals. I. A method for measuring the rate of calcium deposition by corals under different conditions. *Biol. Bull.*, **116**, 59–75.

Goreau T.J. (1977) Carbon metabolism in calcifying and photosynthetic organisms theoretical models based on stable isotope data. *Proc. 3rd Int. Coral Reef Symp., Miami* **2**, 395–401.

Grady M.M., Swart P.K. & Pillinger C.T. (1982) The variable carbon isotopic composition of type 3 ordinary chondrites. *J. Geophys. Res. Suppl.* **87**, A289–96.

Grosovsky B., Kretzschmar D.D.M., Krumbein W.E. & Lorenz M.G. (submitted) Possible Microbial Pathways playing a role in the formation of Precambrian ore Deposits. (Submitted to *Geology Journal.*)

Gross G.G. (1977) Biosynthesis of lignin and related monomers. In *Recent Advances in Phytochemistry vol II. The Structure, Biosynthesis, and Degradation of Wood* (Eds F.A.Loewus & V.C.Runeckles). Plenum Press, New York.

Hammann R. & Ottow J.C.G. (1974) Reductive dissolution of Fe_2O_3 by Saccharolytic Clostridia and *Bacillus polymyxa* under anaerobic conditions. *Z. Pflanzenern. Bodenk.* **137**, 108–15.

Hartmann M., Müller P., Suess E. & van der Wejden C.H. (1973) Oxidation of organic matter in recent marine sediments. *Meteor-Forschungsergebnisse* c(**12**), 74–86.

Hayatsu R. & Anders E. (1981) Organic compounds in meteorites and their origins. In *Topics in Current Chemistry* (Ed. F.L.Boschke). Springer Verlag, Berlin.

Hayes J.M. (1967) Organic constituents of meteorites—a review. *Geochim. Cosmochim. Acta* **31**, 1395–440.

Hayes J.M. (1983) Geochemical evidence bearing on the origin of aerobiosis, a speculative interpretation. In *Origin and evolution of the Earth's Earliest Biosphere: An Interdisciplinary Study* (Ed. J.W.Schopf). Princeton University Press, Princeton.

Hayes J.M., Kaplan I.R. & Wedeking K.W. (1983). Precambrian organic geochemistry. In: *Origin and evolution of Earth's Earliest Biosphere: An Interdisciplinary Study* (Ed. J.W.Schopf). Princeton Press, New Jersey.

Holland H.D. & Schidlowski M. (Eds) (1982) *Mineral deposits and the evolution of the biosphere.* Dahlem Konferenzen, Springer, Berlin.

Hoyle F. & Wickramasinghe N.C. (1978) *Life Cloud: The origin of life in the universe.* J.M.Dent, London.

Iversen N. & Blackburn T.H. (1981) Rates of methane oxidation in anoxic marine sediments. *Appl. environm. Microbiol.* **41**, 1295–1300.

Jørgensen B.B. (1978a) A comparison of methods for the quantification of bacterial sulfate reduction in coastal marine sediments. I. Measurements with radiotracer techniques. *Geomicrobiol. J.* **1**, 11–28.

Jørgensen B.B. (1978b) A comparison of methods for the quantification of bacterial sulfate reduction in coastal marine sediments. II. Calculations from mathematical models. *Geomicrobiol. J.* **1**, 29–51.

Jørgensen B.B. (1978c) A comparison of methods for the quantification of bacterial sulfate reduction in coastal marine sediments. III. Estimation from chemical and bacteriological field data. *Geomicrobiol. J.* **1**, 53–68.

Jørgensen B.B. & Cohen Y. (1977) Solar Lake (Sinai). 5. The sulfur cycle of the benthic cyanobacterial mats. *Limnol. Oceanogr.* **22**, 657–66.

Jørgensen B.B., Hansen M. & Ingvorsen K. (1978) Sulfate reduction in coastal sediments and the release of H_2S to the atmosphere. In *Environmental*

Biogeochemistry and Geomicrobiology 1: The Aquatic Environment (Ed. W.E. Krumbein), pp. 245–53. Ann Arbor Science, Ann Arbor, Michigan.

Karl D.M., Wirsen C.O. & Jannasch H.W. (1980) Deep Sea primary production at the Galapagos Hydrothermal vents. *Science* **207,** 1345–7.

Keeling C.D. (1960) The concentration and isotopic abundances of CO_2 in the atmosphere. *Tellus* **12,** 200–3.

Keeling C.D. (1973) Industrial production of carbon dioxide from fossil fuels and limestone. *Tellus* **25,** 174–98.

Keltjens J.T. & Vogel G.D. (1981) Novel coenzymes of methanogens. In *Microbial Growth on C1 compounds* (Ed. G.Dalton), pp. 152–8. Heyden Publishers, London.

Knoll A.H. (1979) Archaean photoautotrophy: Some alternatives and limits. *Origin of Life* **9,** 313–27.

Knoll A.H. & Golubic S. (1979) Anatomy and taphonomy of a precambrian algal stromatolite. *Precambrian Research* **10,** 115–51.

Krumbein W.E. (1971) Sediment microbiology and grainsize distribution, as related to tidal movement, during the first mission of the West German Underwater Laboratory 'Helgoland'. *Marine Biology* **10,** 101–12.

Krumbein W.E. (1979) Öber die Zuordnung der Cyanophyten. In *Cyanobakterien—Bakterien oder Algen?* (Ed. W.E.Krumbein), pp. 33–48. I. Oldenburger Symposium über Cyanobakterien 1977—Taxonomische Stellung und Ökologie. Universität Oldenburg, Oldenburg.

Krumbein W.E. (1980) Microbial Geochemistry: A historical perspective. In *Interaction of the biota with the Atmosphere and sediments* (Eds M.N.Dastoor, L.Margulis & K.Nealson). NASA workshop on Global Ecology, final report, Washington, D.C.

Krumbein W.E. (1981) *Schwermetalltransfer und -anreicherung durch Mikroorganismen,* pp. 85–107. BMFT-Projektleitung (Hrsg.): Geomikrobiologie und Rohstoffsicherung (Expertengespräch 1980).

Krumbein W.E. (1982) *Biogeochemistry and geomicrobiology of lagoons and lagoonary environments.* In *Coastal Lagoon/Research Present and Future,* pp. 97–109. UNESCO Technical Papers Mar. Sci. 33, Paris.

Krumbein W.E. (1983). Stromatolites—the challenge of a term in space and time. *Precambrian Res.* **20,** 493–531.

Krumbein W.E., Buchholz H., Franke P., Giani D., Giele C. & Wonneberger C. (1979) O_2 and H_2S coexistence in stromatolites. A model for the origin of mineralogical lamination in stromatolites and banded iron formations. *Naturwissenschaften* **66,** 381–9.

Krumbein W.E. & Cohen Y. (1974) Biogene, klastische und evaporitische Sedimentation in einem mesothermen monomiktischen ufernahen See (Golf von Aqaba). *Geol. Rundschau* **63,** 1035–65.

Krumbein W.E. & Cohen Y. (1977) Primary production, mat formation and lithification: Contribution of oxygenic and facultative anoxygenic cyanobacteria. In: Fossil Algae (Ed. E.Flügel), pp. 37–56. Springer Verlag, Berlin.

Küster K. (1978) Microbiology of peat. In *Environmental Biogeochemistry and Geomicrobiology* (Ed. W.E.Krumbein), pp. 439–50. Ann Arbor Science, Ann Arbor, Michigan.

Lein A.Yu, Namsaraev B.B., Trotsyuk V.Ya. & Ivanov M.V. (1981) Bacterial methanagenesis in Holocene sediments of the Baltic Sea. *Geomicrobiol. J.* **2,** 299–315.

Lewin R.A. (1981) The Prochlorophytes. In *The Prokaryotes* (Eds M.P.Starr, H.Stolp, H.G.Trüper, A.Balows & H-G.Schlegel), pp. 257–66. Springer Verlag, Berlin.

Lippman C.B. (1932) Are there living bacteria in stony meteorites? *Am. Museum Novitates* **589.**

Lorenz M.G., Aardema B.W. & Krumbein W.E. (1981) Interaction of marine sediment with DNA and DNA availability to nucleases. *Marine Biology* **64,** 225–30.

Lovelock J.E. (1979) *Gaia. A new look at life on earth*, pp. 157. Oxford University Press, Oxford.

Mah R.A., Hungate R.E. & Ohwaki K. (1977) Acetate, a key intermediate in methanogenesis. In *Microbial Energy Conversion* (Eds H.G.Schlegel & J.Barnea), pp. 97–106. Pergamon Press, Oxford.

Margulis L. (1981) *Symbiosis in Cell Evolution.* W.H.Freeman & Co, San Francisco.

Margulis L. & Nealson K.H. (1983) *Planetary Biology and Microbial Ecology; Research Report.* NASA and BU, Boston.

Martens C.S. & Klump J.V. (1980) Biogeochemical cycling in an organic-rich coastal marine basin 1. Methane sediment-water exchange processes. *Geochim. Cosmochim. Acta* **44,** 471–90.

Monson K.D. & Hayes J.M. (1980) Biosynthetic Control of the natural abundance of carbon 13 at specific positions within fatty acids in *Escherichia coli. J. Biol. Chem.* **255,** 11435–41.

Monster J., Appel, P.W.U., Thode H.G., Schidlowski M., Carmichael C.M. & Bridgewater D. (1979) Sulfur isotope studies in early Archean sediments from Isua, West Greenland: Implications for the antiquity of bacterial sulfate reduction. *Geochim. Cosmochim. Acta* **43,** 405–14.

Muscatine L. & Cernichiari E. (1969) Assimilation of photosynthetic product of zooxanthellae by a reef coral. *Biol. Bull.* **137,** 506–23.

Neish A.C. (1968) Monomeric intermediates in the biosynthesis of lignin. In *Constitution and Biosynthesis of Lignin.* Springer, New York.

Nissenbaum A., Presley B.J. & Kaplan I.R. (1972). Early diagenesis in a reducing liquid. Saaurch—Inlet British Columbia—I. Chemical and isotopic changes in major components of interstitial water. *Geochim. Cosmochim. Acta* **36,** 1007–27.

Oana S. & Deevey E.S. (1960) Carbon 13 in lake waters and its possible bearing on paleolimnology. *Am. J. Sci.* **258-A,** 253–72.

Oehler J.H. (1982) Reduced carbon compounds in sediments, group report. In *Biospheric evolution and precambrian metallogeny* (Eds H.Holland & M.Schidlowski). Springer, New York.

O'Leary M.H. (1981) Carbon isotope fractionation in plants. *Phytochemistry* **20,** 553–67.

Orpen J.L. & Wilson J.F. (1981) Stromatolites at 3500 Myr and a green stone-granite unconformity in the Zimbabwean Archaean. *Nature* **291,** 218–20.

Ottow J.C.G. (1968) Evaluation of iron reducing bacteria in soil and the physiological mechanism of iron reduction in aerobacter aerogenes. *Z. allg. Mikrobiol* **8,** 441–3.

Ottow J.C.G. (1969) The distribution and differentiation of iron-reducing bacteria in gley soils. *Zbl. Bakt.* **123,** 600–15.

Ottow J.C.G. (1970) Selection, characterization and iron reductaseless (nit^-) mutants of iron-reducing bacteria. *Z. allg. Mikrobiol.* **10,** 55–62.

Ottow J.C.G. & Ottow H. (1970) Gibt es eine Korrelation zwischen der eisenredu-

zierenden und nitratreduzierenden Flora des Bodens? *Zbl. Bakt. Abt. II* **124,** 314–18.

Padan E. (1979) Impact of Facultatively Anaerobic Photoautotrophic Metabolism on Ecology of Cyanobacteria (Blue-Green Algae). *Advances in Microbial Ecology* **3,** 1–39.

Pichinoty F. (1973) La reduction bactérienne des composés oxygènes minéraux de l'azote. *Bull. Inst. Pasteur Paris* **71,** 317–95.

Por F.D. (1980) An ecological theory of animal progress: A revival of the philosophical role of zoology. *Perspectives in Biology and Medicine,* **23,** 389–99.

Rau G.H. (1981) Hydrothermal vent clam and tube work $^{13}C/^{12}C$: Further evidence of nonphotosynthetic food sources. *Science* **213,** 338–9.

Redfield A.C. (1958) The biological control of chemical factors in the environment. *Am. Sci.* **46,** 205–22.

Reeburgh W.S. (1982) A major sink and flux control for methane in marine sediments: anaerobic consumption. In *The dynamic environment of the ocean floor* (Eds K.Fanning & F.T. Manheim) 203–17. Heath, Lexington.

Reeburgh W.S. (in press) Rates of biogeochemical processes in anoxic sediments. *Ann. Rev. Earth and Planetary Sci.* **11,** 269–98.

Revsbach N.P., Sorensen J., Blackburn T.H. & Lomholt J.P. (1980) Distribution of oxygen in marine sediments measured with microelectrodes. *Limnol. Oceanogr.* **25,** 403–11.

Roberts J.L. (1947) Reduction of ferric hydroxide by strains of *Bacillus polymyxa. Soil. Sci.* **63,** 135–40.

Rosenfeld W.D. & Silverman S.R. (1959) Carbon isotope fractionation in bacterial production of methane. *Science* **130,** 1658–9.

Ruby E.G., Wirsen C.O. & Jannasch H.W. (1981) Chemolithotrophic sulfur-oxidizing bacteria from the Galapagos Rift hydrothermal vents. *Appl. Environm. Microbiol.* **42,** 317–24.

Rust F. (1981) Ruminant methane ($^{13}C/^{12}C$) values: relation to atmospheric methane. *Science* **211,** 1044–6.

Schidlowski M. (1980) Antiquity of photosynthesis: possible constraints from archaean carbon isotope record. In *Biogeochemistry of ancient and modern environments* (Eds P.A.Trudinger & M.R.Walter). Springer Verlag, Berlin.

Schidlowski M., Appel P.W.U., Eichmann R. & Junge C.E. (1979) Carbon isotope geochemistry of the 3.7×10^9-yr-old isua sediments, West Greenland: Implications for the archaean carbon and oxygen cycles. *Geochim. Cosmochim. Acta* **43,** 189–99.

Schidlowski M. & Matzigkeit U. (in press) Isotope geochemistry of organic carbon and carbonate from the Gavish Sabkha. In *The Gavish Sabkha—Model of a Hypersaline Ecosystem* (Eds G.M.Friedmann, W.E.Krumbein & G.Gerdes). Springer Verlag, Berlin.

Schlegel H.G. & Barnea J. (1977) *Microbial energy conversion.* Pergamon Press, Oxford.

Simkiss K. (1964) Phosphates as crystal poisons. *Biol. Rev.* **39,** 487–505.

Sorokin Y.I. & Kadota H. (1972) *Techniques for the assessment of microbial production and decomposition in fresh waters.* IBP handbook No 23. Blackwell Scientific Publications, Oxford.

Southward E.C.A. (1971) Recent researches on the Pogonophora. *Rev. Oceangr. mar. Biol.* **9,** 193–220.

Southward A.J. & Southward E.C. (1980) The significance of dissolved organic

compounds in the nutrition of *Siboglinum ekmani* and other small species of Pogonophora. *J. Mar. Biol. Ass. U.K.* **60,** 1005–34.
Stanley G.D. (1981) Early history of scleractinian corals and its geological consequences. *Geology*, **9,** 507–11.
Stumm W. (1977) Global chemical cycles and their alterations by man. *Physical and Chemical Sciences Research Report 2*, Dahlem Conference.
Swart P.K. (1982) Carbon and oxygen isotope fractionation in scleractinian corals: A review. *Earth Sci. Rev.* **19,** 51–80.
Swart P.K., Grady M.M., Pillinger C.T., Lewis R.S. & Anders E. (1983) Interstellar carbon in meteorites. *Science* **220,** 406–410.
Thauer R.K., Jungermann K. & Decker K. (1977) Energy conservation in chemotrophic anaerobic bacteria. *Bact. Rev.* **41,** 100–80.
Towe K.M. (1978) Early precambrian oxygen: A case against photosynthesis? *Nature* **274,** 657–61.
Trimble R.B. & Ehrlich H.L. (1970) Bacteriology of manganese nodules. IV. Induction of an MnO_2-reductase system in a marine bacillus. *Appl. Microbiol.* **19,** 966–72.
Troughton J.H. (1972) Carbon isotope fractionation by plants. In: *Proceedings of the 8th International Conference on Radiocarbon Dating*, pp. E20–E57. Wellington, New Zealand.
Trudinger P.A., Walter M.R. & Ralph B.J. (1980) Biogeochemistry of ancient and modern environments. Proceedings of the IV. Int. Symp. Environm. Biogeochemistry. Springer Verlag, New York.
Trüper H.G. (1982) Microbial processes in the sulfuretum through time. In: *Biospheric evolution and precambrian metallogeny* (Eds H.Holland & M.Schidlowski). Springer Verlag, New York.
Vollenweider R.A. (1971) *A manual on methods for measuring primary production in aquatic environments.* IBP handbook No 12. Blackwell Scientific Publications, Oxford.
Walter M.R., Buick R. & Dunlop J.S.R. (1980) Stromatolites 3400 to 3500 Myr. old from the North Pole area Western Australia. *Nature* **284,** 443–5.
Wasserburg G.S., Mazor E. & Zartman R.E. (1963) Isotopic and chemical composition of some terrestrial natural gases. In *Earth Science and Meteorites* (Eds J.Geis & E.D.Goldberg) pp. 219–240. North Holland, Amsterdam.
Waterbury J.B. & Stanier R.Y. (1978) Patterns of growth and development in pleurocapsalean cyanobacteria. *Baceteriological Rev.* **42,** 2–44.
Whitman W.B. & Wolfe R.S. (1980) Presence of nickel in factor F_{430} from *Methanobacterium bryanti. Biochim. Biophys. Res.* **92,** 1196–1201.
Williams N. (1982) Stratified sulfide deposits, group report. In: *Biospheric evolution and precambrian metallogeny* (Ed. by H.Holland & M.Schidlowski). Springer Verlag, New York.
Woodwell G.M., Whittaker R.H., Reiners W.A., Likens G.E., Delwiche C.C. & Botkin D.B. (1978) The biota and the world carbon budget. *Science* **199,** 141–6.
Zehnder A. & Brock T.D. (1979) Methane formation and methane oxidation by methanogenic bacteria. *J. Bacteriol.* **137,** 420–32.
Zeikus, J.G. (1977) The biology of methanogenic bacteria. *Bact. Rev.* **41,** 514–37.
Zeikus J.G. (1981) Lignin metabolism and the carbon cycle: Polymer biosynthesis, biodegradation and environmental recalcitrance. *Adv. Microb. Ecol.* **5,** 211–243.

CHAPTER 3
THE MICROBIAL NITROGEN CYCLE

T. HENRY BLACKBURN

3.1 INTRODUCTION

The main components of the microbial nitrogen cycle are shown in Fig. 3.1. The processes are discussed individually under separate headings, but it is perhaps useful to briefly review the cycle as a whole. Dead organic material is broken down by bacteria to ammonium (see section 3.2). In anoxic environments, much of the ammonium will be reassimilated by bacteria and plants. In environments where oxygen is present the ammonium is oxidized by nitrifying bacteria to nitrate (see section 3.3). This process involves a transition from an oxidation state of -3 to $+5$ (NH_4^+ to NO_3^-).

Nitrate may be denitrified to dinitrogen by bacteria which can use it as an electron acceptor. This, and similar processes, are discussed in section 3.4. Alternatively nitrate may be incorporated into bacteria and plants (see section 3.5). The loss of dinitrogen to the atmosphere, due to denitrification, is balanced by the fixation of dinitrogen by certain bacteria (see section 3.6).

The approximate dimensions of the reservoirs of nitrogen on earth are seen in Fig. 3.2. The largest reservoir of nitrogen (14×10^{21} g) is in igneous rocks, mostly as ammonium within potassium-rich primary minerals (Stevenson 1962). Bacteria are involved, together with physical factors, in the weathering and disintegration of igneous rocks. This degradation results in 'fixed' ammonium becoming potentially available. The large reservoir of nitrogen in sediments and sedimentary rocks (4×10^{21} g) is mostly in the form of fixed ammonium (ammonium in secondary silicate minerals, not exchangeable by potassium ions). This quantity of nitrogen is the same size as atmospheric dinitrogen. In contrast, the inorganic nitrogen reserve in the oceans and land are approximately one-tenth of the organic reserves which are themselves approximately one-thousandth of the atmospheric dinitrogen. The total biomasses-N in the oceans (5×10^{14} g) and land (1.5×10^{16} g) are approximately one-thousandth and one-hundredth of their respective organic reserves (Svensson & Söderlund 1976). The size of a reserve is not necessarily related to its importance, there may even be an inverse relationship, as in the case of the igneous-N and sedimentary-N pools, which are almost disconnected from the main biological cycle. Similarly, the rates of the processes, which involve the large N_2-pool (denitrification and N_2-fixation), are relatively small compared to the other biological processes (Figs 3.1 and 3.2).

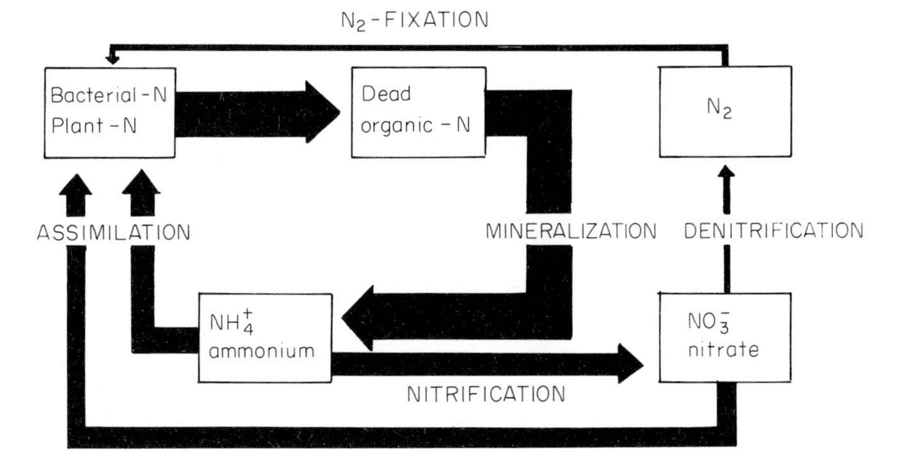

Fig. 3.1. The bacterial processes involved in nitrogen cycling. The width of each arrow is an approximation of process rate.

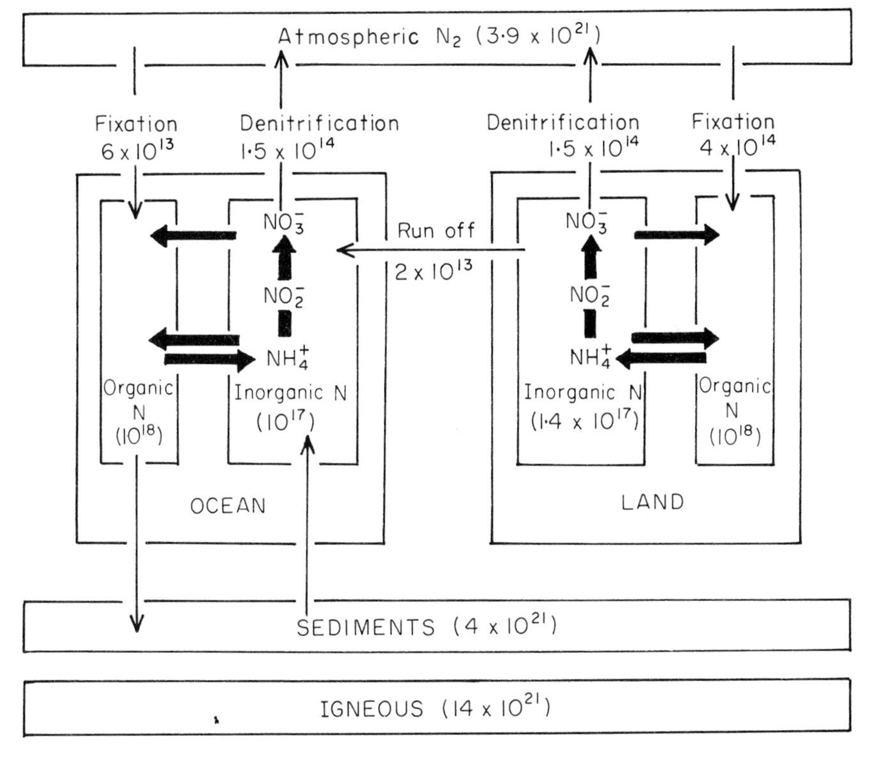

Fig. 3.2. The global nitrogen cycle. Pool sizes, in parentheses, are given as g N. The quantitatively most important transfer rates (g N per year) are indicated by thick arrows (based on data from Fenchel & Blackburn 1979).

Most reviews of N-cycling (Campbell & Lees 1967, Delwiche 1970, Svensson & Söderlund 1976) have emphasized nitrification, denitrification and N_2-fixation, at the expense of the mineralization, which is quantitatively greater. To redress the balance, I have emphasized mineralization in this chapter, rather than follow too closely these excellent reviews. The previous emphasis on nitrification, denitrification and N_2-fixation has partially been because of the considerable body of information that has been accumulated on the physiology of the bacterial species which mediate the processes. In contrast, relatively little is known as to which bacteria decompose complex organic detritus, although the process of detrital decomposition is well understood (Fenchel & Jørgensen 1977, Fenchel & Blackburn 1979). The general intention of this chapter is to emphasize the rates of processes mediated by unknown bacteria, rather than to discuss the potential of well-defined bacteria in ill-defined processes. Ideally, it would be better to combine both approaches so that the contribution of a known bacterium to a known process rate could be defined.

There is also another reason, in addition to its quantitative significance, for selecting mineralization for special emphasis. This is because of the possibility of relating the N-cycle to the S-cycle, through the mineralization of organic detritus in marine sediments (see Chapters 2 and 4). There are many similarities and potential interactions between the C-, S- and N-cycles (Fig. 3.3). The most reduced species of each cycle is separated from the most oxidized species by 8 electrons. There is, however, a very large difference in the $\Delta G_0'$ between the extremes in each cycle and between the electron donating and accepting capacities of the species. Nitrate can oxidize H_2S; SO_4^{2-} cannot oxidize NH_4^+, but can oxidize CH_4 and other reduced carbon compounds. As a result of these reactions, the carbon of detritus in anoxic marine sediments is oxidized to carbon dioxide and the nitrogen of detritus accumulates as ammonium. Therefore, the rates of ammonium production and accumulation can be related to the rates of sulphate reduction and carbon oxidation.

3.2 ORGANIC-N MINERALIZATION

Proteins and nucleic acids are the main nitrogen-containing components of plants and animals. These are relatively easily hydrolysed and degraded to a variety of carbon compounds (ultimately to CO_2) and to ammonium. The global yearly flux of ammonium is unknown but is equal to at least the yearly incorporation of nitrogen into primary producers. This must be true, since there is almost no net accumulation of organic matter on land, and only 4×10^{-13} g N per year accumulation in the oceanic sediments (Svensson & Söderlund 1976). Different components of organic plant litter have different

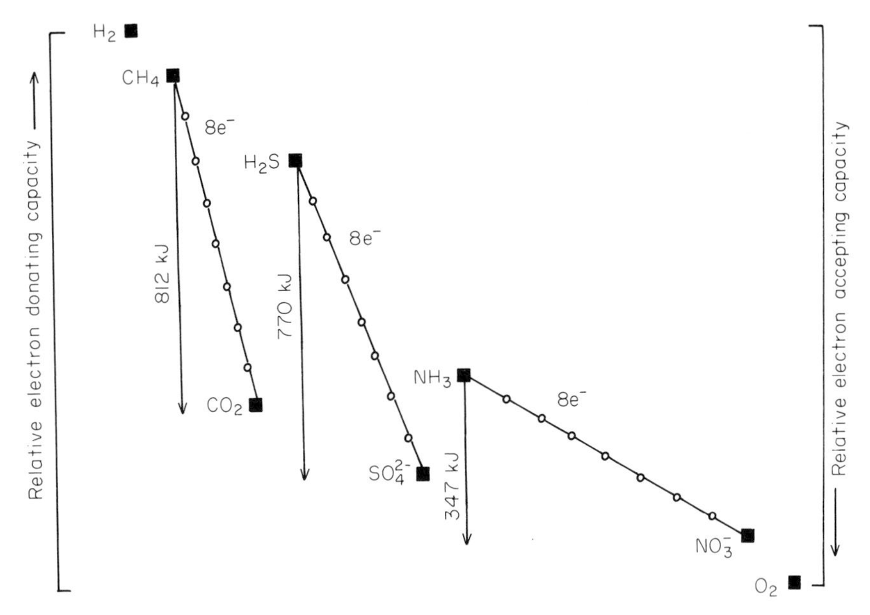

Fig. 3.3. Oxidation/reduction reactions in C, S and N-cycles. The most oxidized form of each species is separated from its most reduced form by a slope representing an 8 electron difference. The vertical distance represents the $\Delta G_0'$ between the two forms. The location of the cycles relative to each other has no fundamental basis, and is designed to illustrate the increasing capacity of CO_2, SO_4^{2-}, NO_3^- and O_2 to accept electrons, and of NH_3, H_2S, CH_4 and H_2 to donate electrons (redrawn from Fenchel & Blackburn 1979).

degrees of resistance to bacterial decomposition (Stout *et al.* 1976). The more resistant components accumulate in the largest pools, which are turned over at the slowest rates. The land and oceans are almost closed systems, separated from each other and connected only by river run-off (2×10^{13} g N per year) and by a mutual connection through atmospheric dinitrogen. This mutual connection to the atmosphere through denitrification and N_2-fixation, is quantitatively insignificant, since the N_2 reservoir is so large and the processes are relatively small. Organic-N is intimately associated with organic-C, and the mineralization rates of the two elements are closely related (Parnas 1975, Fenchel & Blackburn 1979, Blackburn 1980). There is a relationship between the N/C ratio in the organic substrate, the N/C ratio in the bacterial cells, which are synthesized, and the efficiency of incorporation of C into new cells. These relationships may be expressed as a model (Fig. 3.4). It is seen that $Cd = Co + Ci$; $Ci/Cd = E$, the efficiency of C incorporation; $d = Cd \times Ns$; and $i = Ci \times Nc \times E$. It is assumed that anaerobic bacteria in sediments assimilate nitrogen in the form of ammonium, thus allowing the following equations to be reached:

$$Ns = Nc \times E \times d/i \quad (3.1)$$
$$Co = d(1-E)/Ns \quad (3.2)$$
$$Co = i(1-E)/(Nc \times E) \quad (3.3)$$
$$Co = (d-i)(1-E)/(Ns - Nc \times E) \quad (3.4)$$
$$Co = 2Sr \quad (3.5)$$

where Sr is the rate of sulphate reduction by carbon at the oxidation/reduction level of CH_2O.

The net rate of ammonium production $(d-i)$ and the total rate of ammonium production in anoxic marine sediments may be measured by a $^{15}N\text{-}NH_4^+$ dilution technique (Blackburn 1979b). There is usually a positive net production of ammonium, due to the bacterial breakdown of organic

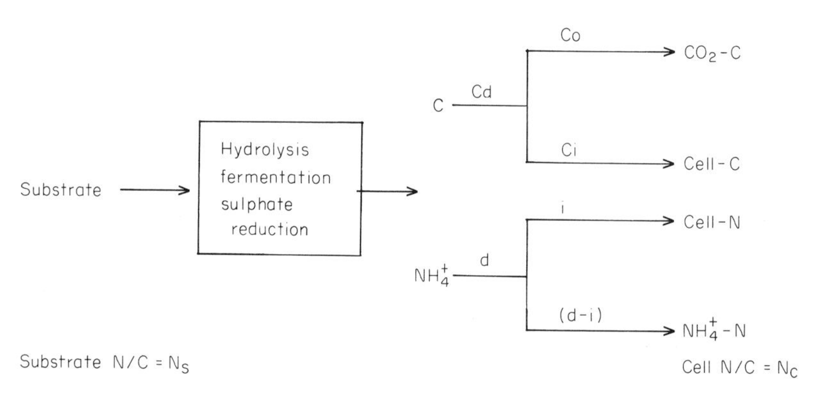

Fig. 3.4. Model of C and N mineralization rates in an anoxic marine sediment. Carbon is degraded at a rate Cd, oxidized at a rate Co and incorporated into bacterial cells at a rate Ci. Similarly organic nitrogen is degraded at a rate d, incorporated into cells at a rate i and has a net excess production rate of $(d-i)$ (redrawn from Blackburn 1979a).

matter in sediments. This is because the organic matter has a relatively high organic-N content ($N/C \sim 0.1$) and the efficiency of C-incorporation is low, $E \sim 0.3$, $Nc \sim 0.16$ (Blackburn 1979a) which results in a low efficiency of N-incorporation. The N/C ratio of sediment organic matter would, therefore, have to be less than 0.05 (equation 3.1) in order that there should be a net uptake of ammonium ($i > d$), rather than the normal net production ($d > i$). There are, however, situations where there is a net uptake of ammonium (Fig. 3.5). The net uptake of ammonium in May 1977, and the low rate of net production in June, resulted in a weakening of the gradient and an indentation in the interstitial ammonium gradient at 6 to 8 cm depth, paticularly obvious in May.

The relationship between bacterial ammonium production rates and the

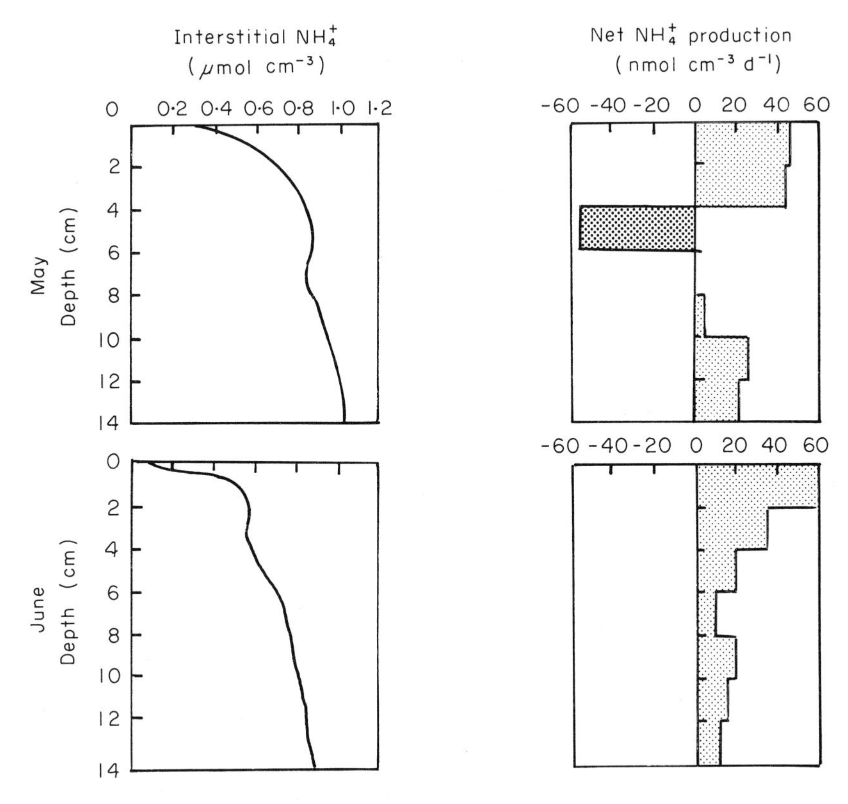

Fig. 3.5. Ammonium profiles and net rates of production in an anoxic marine sediment (redrawn from Blackburn 1979a).

resulting ammonium gradients is more easily seen in isopleth plots (Fig. 3.6) than in individual concentration and activity plots (Fig. 3.5). When ammonium concentrations increase with time or temperature (spring to summer), gradients become steeper and ammonium isoconcentration curves slope upwards towards the sediment surface. These isopleths slope down again during autumn to winter. These effects are seen in the upper layers of the sediment (0 to 4 cm, Fig. 3.6(a)). As might be expected, the total rate of ammonium production increased during spring to summer. This resulted in isopleths sloping downwards during this time, and then sloping up again from autumn to winter (Fig. 3.6(c)). The net rate of ammonium production (Fig. 3.6(b)) had similar isopleths in the 0 to 4 cm layer, but there were times of marked ammonium uptake in $\approx$4 to 10 cm layer for most of the year, except August. This resulted in a weakening of the ammonium gradient in these layers, indicated by a wide separation of the isoconcentration curves (4 to 10 cm, Fig. 3.6(a)). There was thus an obvious

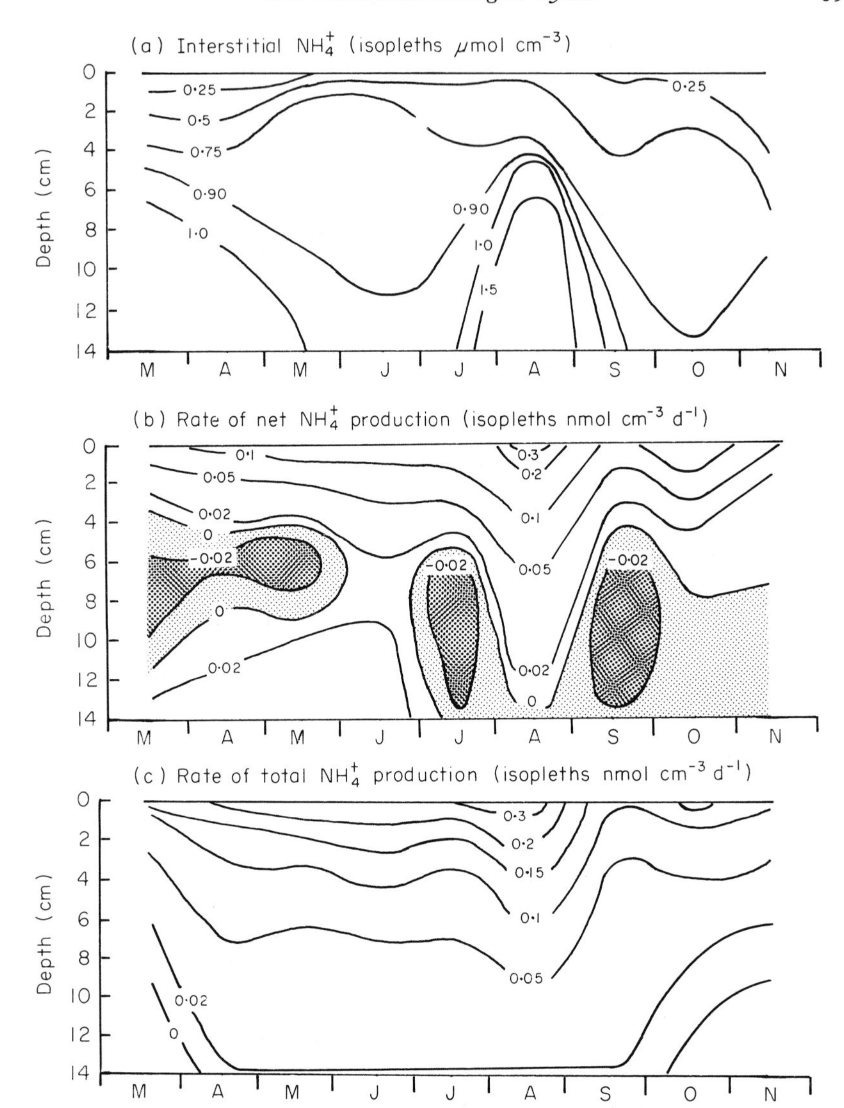

Fig. 3.6. Isopleths of (a) interstitial ammonium concentrations, (b) net rates of ammonium production, and (c) total rates of ammonium production for an anoxic marine sediment, 1977 (redrawn from Blackburn 1980).

relationship between cause (net rate of ammonium production by bacteria, and effect (interstitial ammonium concentration). This type of relationship could also be demonstrated for sediments from four other stations (Blackburn 1980).

The excess ammonium cannot be oxidized in the anoxic sediments. It

reacts with ion exchange sites and can enter into fixed ammonium pools (Blackburn 1979b, 1980, Rosenfeld 1979). Normally the exchangeable ammonium pool is approximately equal to the free interstitial pool and is rapidly exchanged with it. This is of some importance in measuring changes in the available pool of ammonium. The exchangeable pool can, however, be

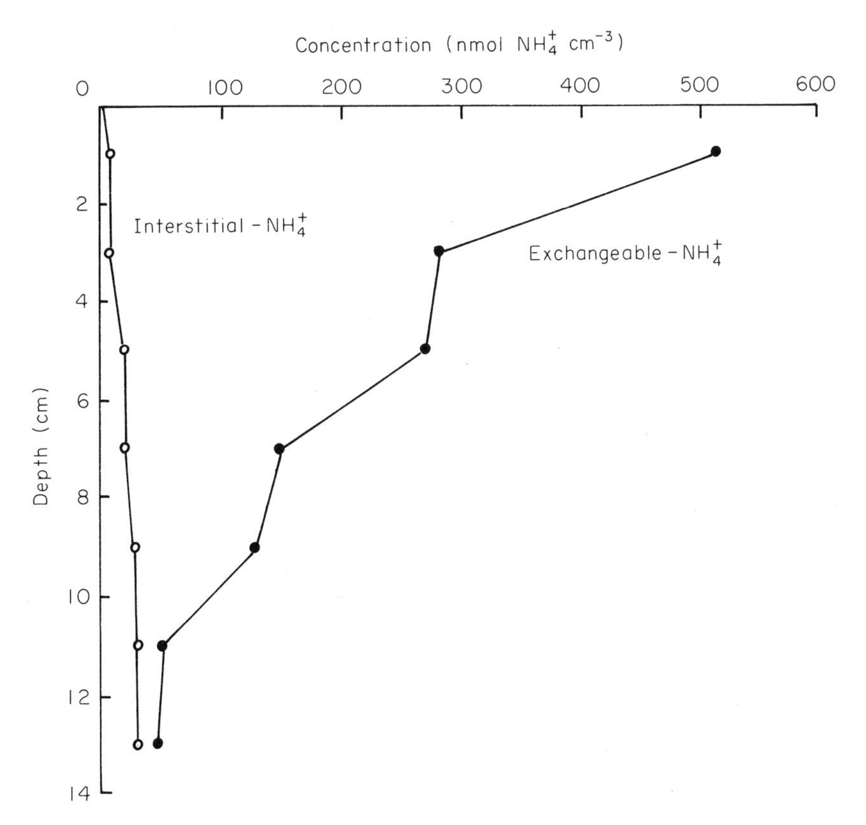

Fig. 3.7. Ammonium profiles in Aalborg Bugt. The exchangeable ammonium was measured as KCl-extractable minus interstitial ammonium, July 1979 (Blackburn, unpublished observations).

very much larger than the interstitial pool. The type of ammonium distribution (Fig. 3.7) must have a very marked effect on the diffusion of ammonium to the surface. It could act as a passive mechanism, preventing the loss of ammonium from the sediment to the overlying water, without greatly affecting its availability to sediment microorganisms.

A cumulative plot of net ammonium production showed that 80% of all the anoxic N-mineralization had occurred by 4 cm depth (Blackburn 1980), in contrast to only 65% anoxic organic C-mineralization in the same

sediments by 10 cm depth (Jørgensen 1977). These mineralization data, together with the N and C content of the sediment, indicated that considerable quantities of organic material had been introduced into the sediment after initial sedimentation. This is not surprising, since there is considerable bioturbation and physical mixing of these and other inshore sediments (Goldhaber *et al.* 1977). The sediment bacteria degrade this fresh organic material more easily than the old, resistant sediment detritus. Evidence for this selective degradation of fresh organic material in the anoxic sediment is seen in Fig. 3.8. The total rate of ammonium production

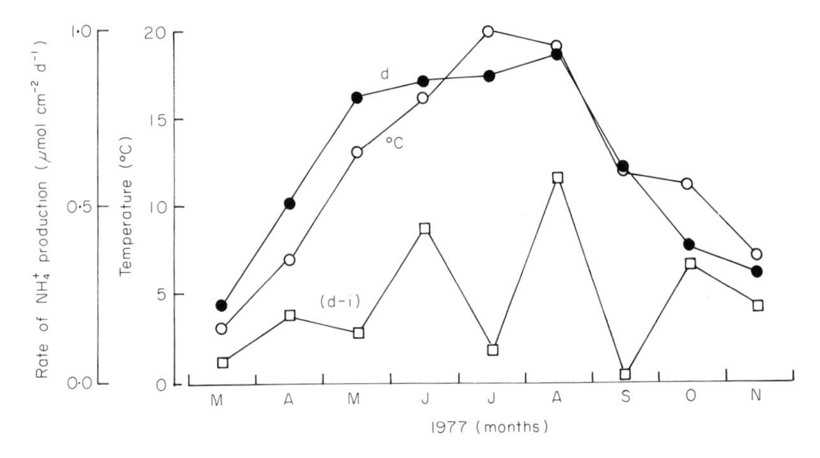

Fig. 3.8. The integrated mean rates (Σ0 to 14 cm depth) for 5 stations of net ammonium production (d – i), total ammonium production (d), and sediment temperature were plotted against time of sampling, March to November, 1977 (redrawn from Blackburn 1980).

followed sediment temperature very closely, with maxima for both in July to August. The mean integrated values for the net rate of ammonium production were, however, very variable, with minima in May, July and September. The mixing downwards and subsequent bacterial degradation of N-depleted, but readily degradable organic material could have produced these effects. The rate of bacterial ammonium uptake (i) probably increased relative to the total rate of production (d), resulting in reduced rates of net ammonium production (d – i). This type of measurement is thus very useful, as it informs on both bacterial activity and substrate quality.

Further information may be derived from the weighted average values for the year (Fig. 3.9). There was a regular decrease in all rates from the top of the sediment downwards. This is consistent with the presence of more

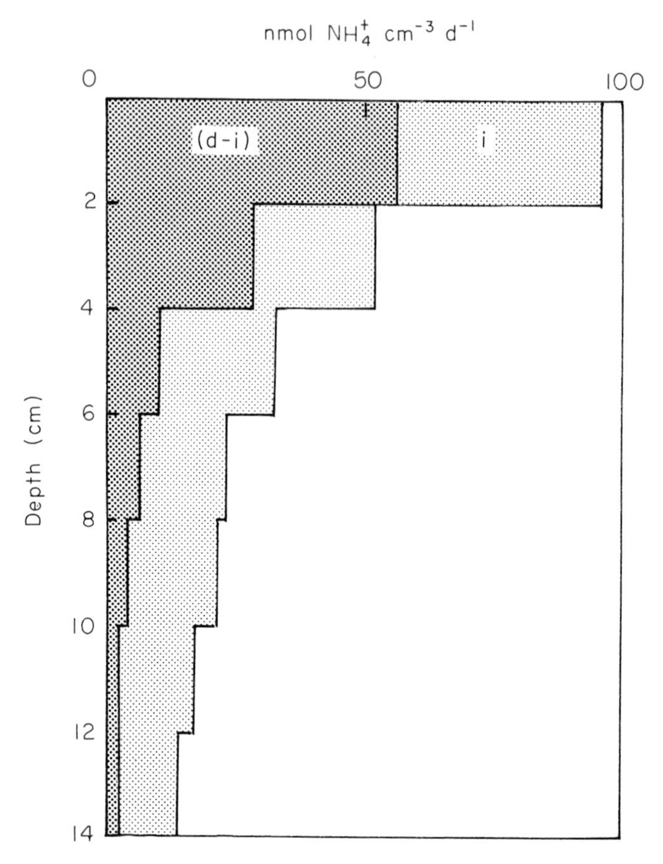

Fig. 3.9. Mean weighted values of 5 stations for 1977. Net rate of ammonium production (d − i) is dark-shaded; rate of ammonium incorporation (i) is light-shaded, and total rate of ammonium production is the total shaded area (redrawn from Blackburn 1980).

degradable organic material in the upper sediment layers, which maintains a larger and more active bacterial population. Equation 3.1 may be used to calculate the N/C ratio (Ns) of the organic material undergoing decomposition, taking Nc = 0.16, E = 0.3 and the measured d and i values. The calculated Ns values were higher than the sediment N/C ratio from 0 to 4 cm depth, but below 4 cm the Ns was lower than the actual N/C ratio. An alternative possibility is that bacteria were degrading material of the sediment N/C composition but were synthesizing material with a variable N/C ratio. The calculated Nc increased from 0.13 (0 to 2 cm) to 0.21 (12 to 14 cm), instead of the Nc = 0.16 previously used. The implication is that bacteria in the lower sediment synthesize cell material of a higher than average N-content (protein, nucleic acid). This is compatible with the hypothesis that bacteria in the lower sediment layers are energy-starved and have a maintenance metabolism based on protein turnover. The rate of carbon oxidation (Co), which was calculated according to equation 3.2 using Ns = N/C ratio of the sediment, gave an activity profile similar to that of carbon oxidation by sulphate reduction (Jørgensen 1977). The integrated rate calculated for Co was 39 mmol C m^{-2} d^{-1}; the weighted average for carbon oxidation, by sulphate reduction, was 19 nmol C m^{-2} d^{-1} (Jørgensen

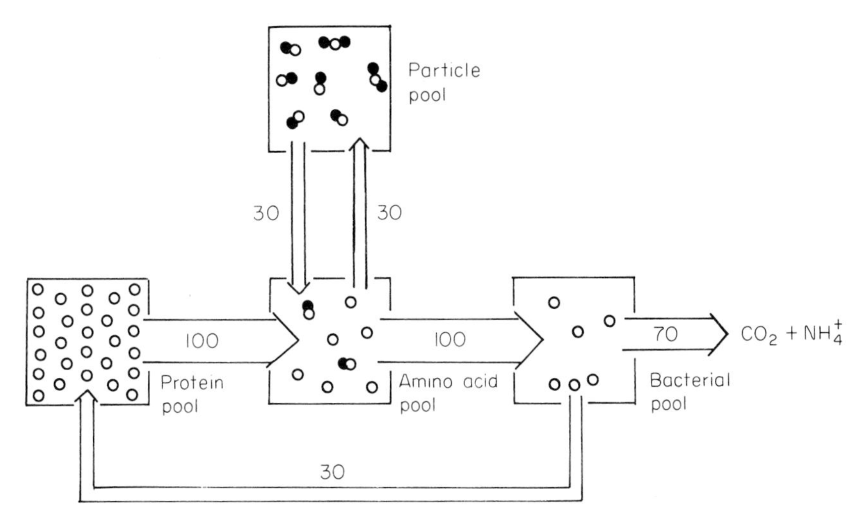

Fig. 3.10. Model of protein mineralization and amino acid turnover in sediment. ○ = an amino acid; ● = a binding site. The relative velocities of the reactions are indicated (redrawn from Christensen & Blackburn 1979).

1977). The anoxic N-mineralization rates possibly include the mineralization of some organic matter by electron acceptors other than sulphate (e.g. ferric Fe). This could partially explain why the rate of carbon oxidation, calculated from ammonium turnover, is higher than that calculated from sulphate reduction. There are good opportunities for exploring the true values for the N/C ratio in cells (Nc), in substrate (Ns) and for the efficiency of bacterial carbon incorporation (E), by simultaneous measurements of rates of ammonium turnover and sulphate reduction.

The mineralization of organic-N may be examined more directly by measuring the rate of breakdown of specific organic molecules. Since the composition of the macromolecules in most detritus is unknown, these experiments are often limited to small molecules, such as amino acids. We have used ^{14}C-alanine and ^{3}H-alanine turnover in anoxic marine sediments as a model for the probable mineralization of protein (Fig. 3.10). Protein

amino acids entered an interstitial pool and left this free pool by two processes:
(1) adsorption ($k = 0.05 \text{ min}^{-1}$);
(2) biological uptake ($k = 0.22 \text{ min}^{-1}$).
The amino acids were desorbed at a rate equal to the adsorption rate, but because of very different turnover rate constants, it was deduced that the adsorbed pool was 1000-fold greater than the free pool; some complexing of the soluble amino acid also occurred. Biological uptake resulted in a 30% incorporation and a 70% mineralization of amino acids. The bacterial protein becomes part of the general protein pool. The actual concentration of alanine in the free pool was 8000 nmol l^{-1}, but the very high turnover rate constant indicated that the concentration of available alanine must be ≈ 10 nmol l^{-1} pore water. Relatively high *measured* concentrations of amino acids were found in salt marsh sediments (Hanson & Gardner 1978), in coastal waters (Clark *et al.* 1972, Jørgensen *et al.* 1980) and sea-grass sediments (Jørgensen *et al.* 1981), but it is not always certain that all the measured amino acid was biologically available to bacteria. Amino acid turnover has also been measured in freshwater sediments (Hall *et al.* 1972) and in water (Hobbie & Crawford 1969). The production of interstitial ammonium was also used to measure organic nitrogen mineralization in a coastal sediment (Billen 1978) and lake sediments (Graetz *et al.* 1973). Rates of organic nitrogen mineralization by sediment bacteria have been calculated from ammonium gradients (Martens *et al.* 1978).

There are many more difficulties in measuring the rate of organic mineralization in aerobic environments, e.g. oxidized water bodies and soils. This is because ammonium pools are generally low, and difficult to trace-label with ^{15}N-ammonium. More complex transformations of ammonium occur in the presence of oxygen, which causes additional problems. Mineralization rates have, however, been measured with considerable success in Japanese soils (Hiura *et al.* 1976, Hiura *et al.* 1977, Marumoto *et al.* 1977, Yoneyama & Yoshida 1977a, b, c) and soils in the United States (Focht *et al.* 1980).

3.3 AMMONIUM OXIDATION

The ammonium, which is released by the mineralization of organic matter, is oxidized in aerobic environments to nitrate. The ammonium in anoxic environments diffuses to the aerobic interface, where again it is oxidized. The bacteria which perform this oxidation are usually autotrophic lithotrophs, but heterotrophic nitrification can also occur, resulting in the transient accumulation of, for example, hydroxylamine in natural waters and sewage (Verstraete & Alexander 1972). The oxidation of ammonium occurs in two steps. The first is an oxidation to nitrite by *Nitrosomonas* or related species:

$$NH_4^+ + 1.5\ O_2 \rightleftharpoons NO_2^- + H_2O + 2H^+ \quad \Delta G_0' = -276\ kJ\ mol^{-1}.$$

Hydroxylamine is probably an intermediate product but does not accumulate. The nitrite is oxidized in the second step to nitrate, by *Nitrobacter* or related species:

$$NO_2^- + 0.5\ O_2 \rightleftharpoons NO_3^- \quad \Delta G_0' = -73\ kJ\ mol^{-1}.$$

The energy yield from these two oxidations is small and much energy is required for the reduction of carbon dioxide. As a result there is only a small cell yield from the oxidation of a large quantity of substrate (Fenchel & Blackburn 1979). Their cell biomass, therefore does not make a significant contribution to the organic nitrogen or carbon pools. There is, however, some connection between nitrification and the carbon cycle, since there is competition between the nitrifying bacteria and the heterotrophs for oxygen, which is often a limiting nutrient. There is a similar connection with the sulphur cycle, as there is competition for oxygen between the sulphide-oxidizing bacteria and the nitrifying bacteria.

Nitrifying bacteria are found in organic-rich environments, where ammonium is produced by organic-N mineralization. They are not inhibited by organic matter and probably adhere to organic particles in marine sediment (Vanderborght & Billen 1975), in soil (Gray & Parkinson 1968) and in river bottoms (Tuffey *et al.* 1974). Nitrifying bacteria grow best at neutral to alkaline pH and are inhibited by the acid which they produce (Winogradsky & Winogradsky 1933). A temperature of 30 to 35 °C produces most growth with poor growth below 5 °C (Buswell *et al.* 1954, Deppe & Engel 1960). They are reported to survive for long periods in anaerobic conditions and to be capable of growth when given oxygen (Painter 1970). The following observations, relevant to these growth and activity characteristics, were made by our group in Aarhus (Hansen 1980, Henriksen 1980, Henriksen *et al.* 1981, Hansen *et al.* 1981). Nitrifying bacteria can survive under anoxic conditions, but 90% are rapidly killed by anaerobic incubation when hydrogen sulphide is produced. The numbers of nitrifying bacteria were determined by incubating marine sediment with saturating concentrations of ammonium and oxygen at 23 °C. The rate of nitrate production was proportional to the number of nitrifying bacteria, since for many ammonium-oxidizing bacteria, the maximum activity per cell is relatively constant (Belser 1979, Belser & Schmidt 1980). This potential maximum rate of nitrate production (PN) was greater in winter (January) and fell to a minimum in July and August (Fig. 3.11). The number (PN) rose again in the autumn. It is likely that the decrease in numbers was due to sulphide

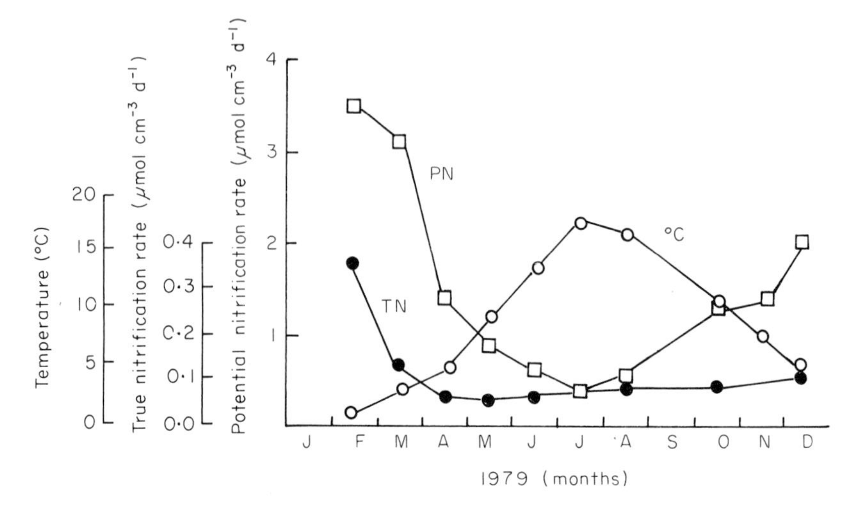

Fig. 3.11. Potential nitrification rate (PN) for the upper 1 cm of sandy sediment was measured throughout 1979. The true nitrification rate (TN) was calculated by correcting for temperature and depth of oxygen penetration into the sediment (redrawn from Hansen 1980, Hansen *et al.* 1981).

poisoning, but oxygen and ammonium deficiency might also have restricted growth and population size. The possible effect of ammonium depletion on the numbers of nitrifying bacteria is seen in Fig. 3.12. There was a lower population in the upper 1 cm layer of sediment (ammonium depleted) than in the lower layers (stations II, III and V). This might even have been due to inhibition by oxygen, an effect that we have observed in these potential nitrification experiments. The presence of nitrifying bacteria in the deep sediment layers, in all the stations (Fig. 3.12) cannot be explained. Oxygen penetration by diffusion did not reach below 5 mm but oxygen may have been introduced to the lower strata by sediment fauna (Grundmanis & Murray 1977). Alternatively, the nitrifying bacteria might have migrated downwards into the anoxic layers, or have been mixed downwards by sediment disturbance. An interesting possibility is that some oxygenated molecule (perhaps an iron or manganese oxide) might be able to oxidize ammonium. These sediments (Fig. 3.12) were probably oxidized to 8 cm depth, since they did not contain hydrogen sulphide, but neither did they contain free oxygen. An estimate of the true nitrification rate may be made from the potential rate by correcting for temperature ($Q_{10} \sim 3$) and depth of oxygen penetration (Fig. 3.11). Oxygen did not penetrate below 5 mm (Sørensen *et al.* 1979, Revsbech *et al.* 1980a, b). No correction was made for ammonium concentration, although the kinetic parameters V_{max} and K_m for ammonium were measured. The true nitrification rates calculated in this way (0.1 to 0.4 μmol cm^{-3} d^{-1}) corresponded very well with the nitrification

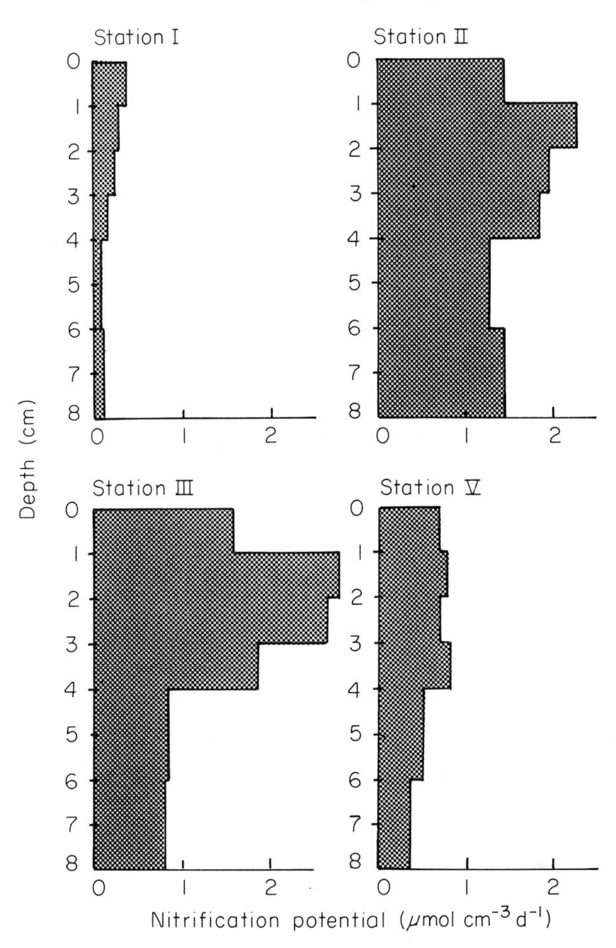

Fig. 3.12. The potential nitrification rate (NP) at different depths in four marine sediments from Danish inshore waters, November 1978 (Hansen 1980).

rates which were measured by a direct method (Fig. 3.13). N-Serve did not affect organic-N mineralization or NH_4^+ assimilation, as confirmed by $^{15}NH_4^+$ turnover. N-Serve, however, completely inhibited nitrification. The accumulation of ammonium in sediment cores, which had been inhibited by N-Serve, could be used as a measure of the rate of nitrification. The main peak of nitrifying activity was located at 2 cm depth, and the integrated rate was 0.3 mmol m^{-2} d^{-1}. The dependence of nitrification on the availability of ammonium and oxygen is illustrated in Fig. 3.14. The peak of nitrifying activity was located where the gradients of ammonium and oxygen intersected. It is likely that the availability of oxygen and ammonium, more than temperature, regulates the rate of nitrification in sediments. Maximum

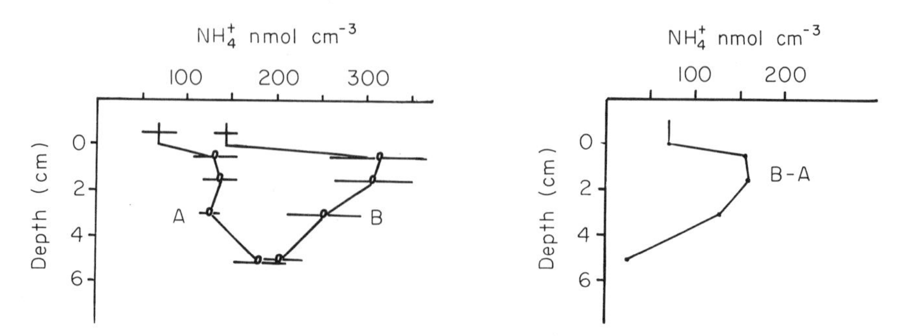

Fig. 3.13. Ammonium concentrations in undisturbed sandy sediment cores after incubation for three days at in-situ temperature (Kysing Fjord, 12 °C. A Minus N-Serve; B plus N-Serve. SD of five cores from A and B are indicated. The difference between curves A and B is shown on the right (redrawn from Henriksen 1980).

oxygen is available in the sediment during the winter months, when competition from heterotrophic processes and hydrogen sulphide is least. This coincides with the period of least competition for ammonium at the sediment surface by aerobic autotrophs and heterotrophic bacteria. The maximum rate of nitrification was found at 1°C (Fig. 3.11), a temperature which is normally considered to be inhibitory. This corresponded well with the calculation of high winter nitrification rates (Table XVII in Fenchel & Blackburn 1979).

There have been surprisingly few measurements of nitrification rates. Billen (1976) measured nitrifying activity in sediments by a dark ^{14}C-bicarbonate incorporation, with and without N-Serve, an inhibitor of nitrification. His rates may have been overestimated due to the presence of excess oxygen and mixed ammonium profiles. More recently Vincent and Downes (1981), using a similar technique, have measured nitrification rates in a lake and concluded that much of the nitrate in the water column originates from the sediment.

Viable nitrifying bacteria can be enumerated by the most-probable-number technique (MPN). There are problems in separating bacteria from sediment particles and from each other, and of finding a universal medium, suitable for all ammonium- or nitrite-oxidizing bacteria (Belser & Schmidt 1978a, b, c). A more specific and elegant method for enumerating nitrifying bacteria is by using fluorescent antibody, that has been prepared against isolated strains (Schmidt 1973, 1978). The great potential of this method has been reviewed by Belser (1979). Inhibition of nitrite oxidation by chlorate is an excellent method for assessing nitrification rates in soils and sediments (Belser & Mays 1980).

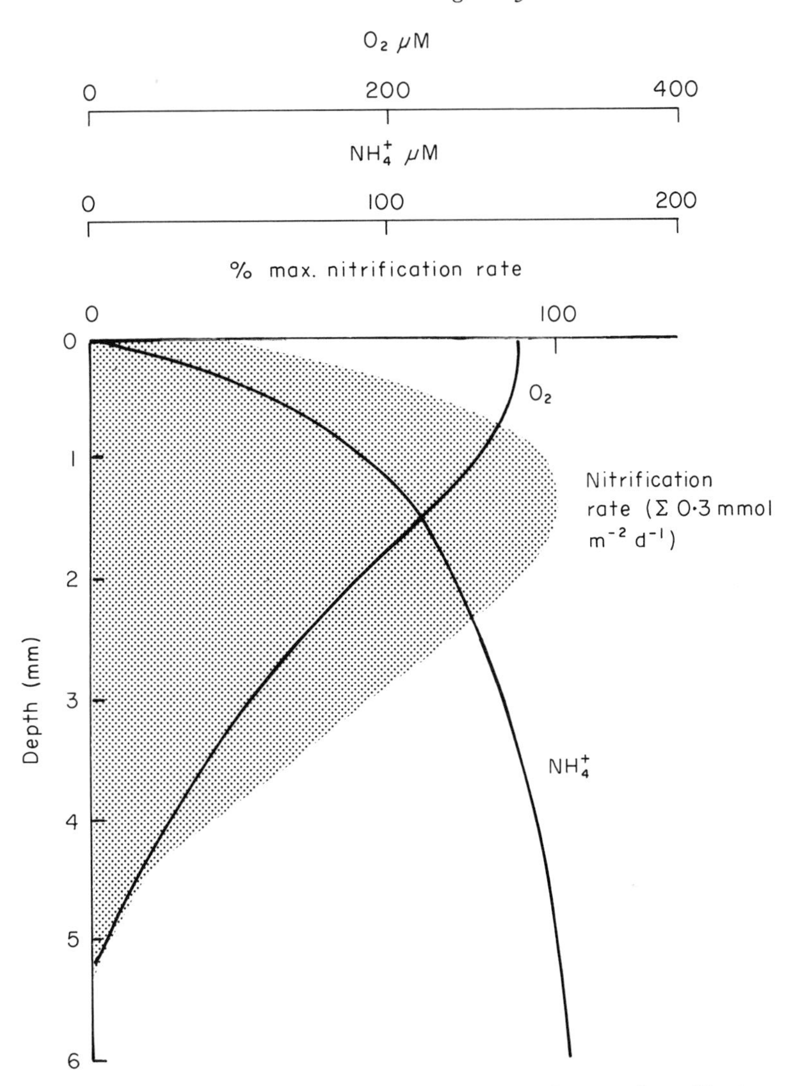

Fig. 3.14. The dependence of nitrification on oxygen and ammonium. Maximum rate of nitrification occurs where oxygen and ammonium interstitial gradients intersect in a sandy sediment, Kysing Fjord, 12 °C (based on data from Henriksen 1980 and Revsbech *et al.* 1980b).

3.4 DISSIMILATIVE NITRATE REDUCTION

The biochemistry and physiology of nitrate reduction has been reviewed by Payne (1973), Focht and Verstraete (1977), Fenchel and Blackburn (1979) and Knowles (1981). In this chapter emphasis will be placed on rates of denitrification, and more paticularly these rates in sediments.

Nitrate reduction is an anaerobic process in which a reduced substrate (e.g. CH_2O, H_2S or H_2) is oxidized at the expense of nitrate. When the product of such a reaction is a nitrogenous gas (N_2, N_2O) the process is defined as 'denitrification'. When the product is ammonium, the process is defined as a 'nitrate fermentation' (Broda 1975). These are the two processes which appear to be the most important in marine sediments (Koike & Hattori 1978, Sørensen 1978a, b, c, Sørensen *et al.* 1979, Delwiche 1981). The capacity of a marine sediment to reduce nitrate to ammonium was equal to, or greater than the denitrifying activity (Fig. 3.15(a)).

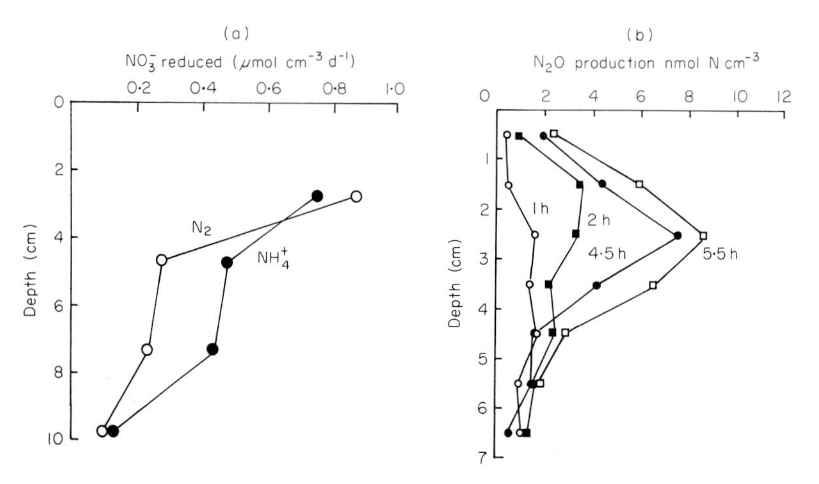

Fig. 3.15. (a) The capacity of the sediment to reduce nitrate to dinitrogen and ammonium (redrawn from Sørensen 1978a). (b) The production of nitrous oxide, after varying time intervals, in undisturbed sandy cores which had been treated with the inhibitor acetylene (redrawn from Sørensen 1978b).

There was less tendency for nitrate to be reduced to ammonium in soils (Tiedje *et al.* 1981) and in lake sediment (Madsen 1979), and this activity was associated with Clostridia in soil (Caskey & Tiedje 1979). Reduction of nitrate to ammonium was stimulated by N-limitation in a strain of *Klebsiella* isolated from the Tay Estuary (Herbert *et al.* 1980), and ferrous iron appeared to affect the process in soils (Sahrawat 1980). *Vibrio succinogenes*, which was grown on formate and nitrate, was also capable of reducing nitrate to ammonium, and of coupling each reductive step to yield energy (Yoshinari 1980).

The ecological significance of denitrification in various habitats has recently been reviewed (Focht and Verstraete 1977, Knowles 1981). I will supplement these excellent reviews with a single example of denitrification, that of denitrification in *Cladophera* mats (periphyton), reported by Triska and Oremland (1981). This is exactly the type of habitat where denitrification might be expected: much available decomposing carbohydrate, little available oxygen beneath the *Cladophera*, and occasionally nitrate in high concentrations.

The use of acetylene as a blocking agent for nitrous oxide reductase has been used in the measurement of denitrification rates, from the rate of nitrous oxide accumulation (Balderston *et al.* 1976, Yoshinari *et al.* 1977, Sørensen *et al.* 1979, and reviewed by Knowles 1981). The great advantage of the method for use with marine sediments is that it can be used with intact sediment cores (Fig. 3.15(b)). The gradients of oxygen, ammonium, nitrifying activity, denitrifying activity, and nitrate gradients are maintained undisturbed. The disadvantage of the method is that acetylene also blocks ammonium oxidation (Hynes & Knowles 1982). When there are high concentrations of nitrate in the sediment, the nitrate pool is not depleted during the incubation. The method thus works well for winter denitrification measurements, when nitrate pools are high, but it may underestimate summer measurements, when nitrate concentrations are low.

There are great uncertainties regarding measurements of denitrification rates, and other methods using ^{15}N-nitrate (Koike & Hattori 1978, Oren & Blackburn 1979), ^{13}N-nitrate (Gersberg *et al.* 1976, Tiedje *et al.* 1979), natural abundance of ^{15}N (Wada *et al.* 1975), N_2 evolution (Seitzinger *et al.* 1980) or model systems (Vanderborght & Billen 1975) for measuring sediment denitrification, are also open to criticism. Global rates of denitrification are usually estimated as being equivalent to rates of nitrogen fixation, which are themselves suspect.

Nitrous oxide (N_2O), one of the gaseous products of denitrification in sediments (Sørensen 1978c), is the subject of much discussion as it can cause the destruction of the stratospheric ozone layer. There is considerable speculation as to N_2O fluxes, sources and sinks (Svensson & Söderlund 1976). It has recently been shown that in the eastern tropical North Pacific, nitrification is a probable source of N_2O, and that denitrification is a sink (Cohen & Gordon 1978). A similar situation exists in lakes (Lemon & Lemon 1981), but denitrification may also be a source of nitrous oxide in fresh water (Knowles *et al.* 1981). Much N_2O is lost from land immediately after ammonia fertilization, when nitrification is most active, possibly indicating that N_2O is a product of nitrification in soil (Mosier & Hutchinson 1981). A number of factors can affect the ratio of N_2O/N_2 in denitrification; sulphide increases the ratio by inhibiting N_2O reduction (Sørensen *et al.* 1980) and oxygen and high nitrate concentrations also increase the ratio N_2O/N_2 (Firestone *et al.* 1979, Letey *et al.* 1980a, b). The presence of nitrate may inhibit the synthesis of nitrous oxide reductase.

3.5 ASSIMILATIVE NITRATE REDUCTION

Plants and aerobic bacteria can assimilatively reduce nitrate for incorporation into cells. It is unlikely that anaerobic bacteria are ever in a situation where nitrate concentrations exceed ammonium concentrations. There can be little doubt, however, that many soil and ocean bacteria must use nitrate as a major source of nitrogen. Their conversion of nitrate to cell-N can have a beneficial effect, as it prevents the leaching of nitrate from the soil.

3.6 ASSIMILATIVE N_2-REDUCTION (FIXATION)

Nitrogen fixation is very important since it is the only biological process in which atmospheric N_2 is made available to the biosphere. It thus compensates for the loss, by denitrification, of N_2 and N_2O to the atmosphere.

Nitrogen fixation can only be carried out by bacteria. Some representative species of the various physiological groups are shown in Table 3.1. Although some of these bacterial species are aerobic, the process of nitrogen reduction is essentially anaerobic. Aerobic cells adopt various strategies to combat the inhibitory effects of oxygen (Postgate 1979). Nitrogen fixation requires the expenditure of large quantities of ATP and it is common to find that this ATP is generated by a photosystem. The free-living bacterial

Table 3.1. Representative types of N_2-fixing bacteria (based on data from Fenchel & Blackburn 1979).

	Phototrophic	*Chemotrophic*
Free-living, aerobic	Cyanobacteria	*Azotobacter* *Mycobacterium* Methane oxidizers *Thiobacillus*
Free-living, anaerobic	Cyanobacteria *Chromatium* *Chlorobium* *Rhodospirillum*	*Clostridium* *Klebsiella* *Bacillus* *Desulfovibrio* *Desulfotomaculum* Methanogenic bacteria
Symbiotic, aerobic	Cyanobacteria (+fungi, ferns)	*Rhizobium* (+legumes, grass) *Spirillum* (+grass) Unknown sp. (+alder, hawthorn, etc.)
Symbiotic, anaerobic	None known	*Citrobacter* (+termites)

phototrophs (Table 3.1) are active nitrogen fixers, both aerobically and anaerobically. It is also usual to find that chemotrophic nitrogen fixers are in association with a phototrophic partner, which supplies the bacterium with reduced carbon, for use in nitrogen reduction.

Nitrogen fixation is usually determined by measuring acetylene reduction. This can be inaccurate unless the acetylene reduction rates are calibrated against $^{15}N_2$ reduction rates (Hauck & Bremner 1976). For this reason many measurements of nitrogen fixation may be in error, and the calculated global fluxes may also be inaccurate. The yearly nitrogen fixation rates (Fig. 3.2) of 6×10^{13} g for ocean and 1.5×10^{14} for land, indicate that $\approx 10\%$ oceanic biomass-N and $\approx 1\%$ land biomass-N could be supplied by nitrogen fixation.

Nitrogen fixation probably plays a minor role in anoxic marine sediments, where ammonium concentration is high and energy is limiting. It is, however, strange that many sulphate-reducing bacteria (*Desulfovibrio* and *Desulfotomaculum*, Table 3.1) possess nitrogenase. There is a situation where ammonium is depleted and energy becomes available in anoxic marine sediments. This situation results from the uptake of ammonium by plant roots, and the excretion by these roots of soluble organic compounds. Nitrogen fixation in the rhizosphere of *Thalassia testudinum*, is an example of such an event (Capone *et al.* 1979).

A more complete survey of nitrogen fixation is given by Postgate (1979).

3.7 A SEDIMENT N-CYCLE

A model annual N-cycle in a typical anoxic inshore sediment has been constructed, using our values for organic-N mineralization, ammonium oxidation, and nitrate reduction (Fig. 3.16). The relationship between the N, C, and S-cycles has been discussed but some further comparisons may be made. The anoxic oxidation of carbon by sulphate is 19 mmol m^{-2} d^{-1} (see Chapter 4 and Fig. 3.13) while the net rate of ammonium mineralization is 2.19 mmol m^{-2} d^{-1} (Fig. 3.16). This suggests that the N/C ratio of the sediment organic matter was 0.12. This is a little higher than the measured value, but is lower than the ratio of 0.16 which is often used in diagenetic calculations.

The rate of sulphide oxidation (8.8 mmol m^{-2} d^{-1}) was much higher than the rate of ammonium oxidation (1.0 mmol m^{-2} d^{-1}). Sulphide oxidation would thus create a greater oxygen demand and could be expected to generate a greater biomass of chemolithotrophic autotrophic cells.

The rate of sulphate reduction (9.5 mmol m^{-2} d^{-1}) was approximately 20 times greater than the rate of nitrate reduction (0.5 mmol m^{-2} d^{-1}). A

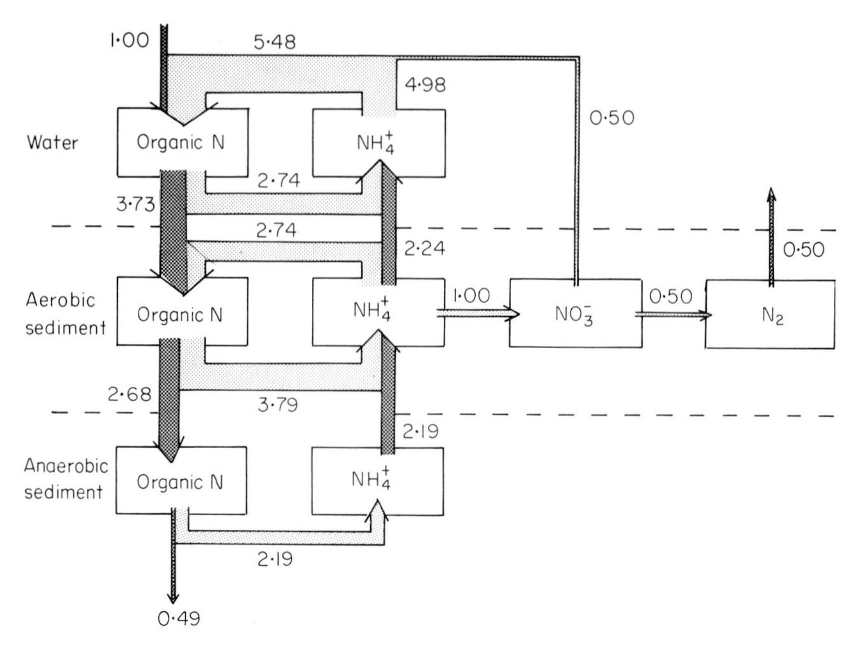

Fig. 3.16. Model of a N-cycle in an inshore marine sediment. The transformation rates are in mmol $m^{-2} d^{-1}$. The anaerobic net mineralization rate is based on data from Blackburn (1980). The pelagic primary productivity of 150 g C $m^{-2} y^{-1}$ (5.48 mmol N $m^{-2} d^{-1}$) is typical of such an inshore water body and the benthic primary productivity (2.74 mmol N $m^{-2} d^{-1}$) is approximately half the pelagic productivity under 10 m water (Jørgensen & Revsbech, unpublished information).

proportionately smaller quantity of carbon would be oxidized by the latter process.

It is anticipated that some of the rates in Fig. 3.16 will be changed, as methods improve, but it is a useful working model of N-cycling. More complex models of N-cycling in soil have been described (Frissal & van Veen 1981). These models can simulate all aspects of the temporal changes in rates and pools that result from any perturbation. This type of model has very great potential but should be used with discretion: 'Our mathematical skills exceed by far our knowledge of the biology of the N-cycle' (A.J. van Veen, personal communication).

References

Balderston W.L., Sherr B. & Payne W.J. (1976) Blockage by acetylene of nitrous oxide reduction in *Pseudomonas perfectomarinus*. *Appl. Environ. Microbiol.* **31,** 504–8.

Belser L.W. (1979) Population ecology of nitrifying bacteria. *Ann. Rev. Microbiol.* **33,** 309–33.

Belser L.W. & Mays E.L. (1980) Specific inhibition of nitrite oxidation by chlorate and its use in assessing nitrification in soils and sediments. *Appl. Environ. Microbiol.* **39,** 505–10.

Belser L.W. & Schmidt E.L. (1978a) Diversity in the ammonia oxidizing nitrifier population of a soil. *Appl. Environ. Microbiol.* **36,** 584–8.

Belser L.W. & Schmidt E.L. (1978b) Serological diversity within a terrestrial ammonia-oxidizing population. *Appl. Environ. Microbiol.* **36,** 589–93.

Belser L.W. & Schmidt E.L. (1978c) Nitrification in soils. In *Microbiology 1978* (Ed. D.Schlessinger) pp. 348–51. Am. Soc. Microbiol, Washington, D.C.

Belser L.W. & Schmidt E.L. (1980) Growth and oxidation of ammonia by three genera of ammonium oxidizers. *FEMS Microbiol. Letters* **7,** 213–16.

Billen G. (1976) Evaluation of nitrifying activity in sediments by dark C-14 bicarbonate incorporation. *Water Res.* **10,** 51–7.

Billen G. (1978) A budget of nitrogen recycling in North Sea sediments off the Belgian coast. *Estuarine Coastal Mar. Sci.* **1,** 127–46.

Blackburn T.H. (1979a) Nitrogen/carbon ratios and rates of ammonia turnover in anoxic sediments. In *Proceedings of the Workshop: Microbial Degradation of Pollutants in Marine Environments* (Eds A.W.Bourquin & D.H.Pritchard) pp. 148–53. National Technical Information Service, Springfield, VA.

Blackburn T.H. (1979b) A method for measuring rates of NH_4^+ turnover in anoxic marine sediments using a ^{15}N–NH_4^+ dilution technique. *Appl. Environ. Microbiol.* **37,** 760–5.

Blackburn T.H. (1980) Seasonal variations in the rate of organic-N mineralization in anoxic marine sediments. In *Biogeochimie de la Matière Organique à l'Interface Eau-Sediment Marin. Colloque International C.N.R.S.* Edition du centre national se la recherche scientifique, Paris.

Broda E. (1975) The history of inorganic nitrogen in the biosphere. *J. Mol. Evol.* **7,** 87–100.

Buswell A.M., Shiota T., Lawrence N. & Meter I.W. (1954) Laboratory studies on the kinetics of growth of *Nitrosomonas* with relation to the nitrification phase of the BOD test. *Appl. Microbiol.* **2,** 21–5.

Campbell N.E.R. & Lees H. (1967) The nitrogen cycle. In *Soil Biochemistry* (Eds A.D.McLaren & G.H.Peterson) pp. 194–215. Marcel Dekker Inc., New York.

Capone D.J., Penhale P.A., Oremland R.S. & Taylor B.F. (1979) Relationship between productivity and $N_2(C_2H_2)$ fixation in a *Thalassia testudinum* community. *Limnol. Oceanogr.* **24,** 117–25.

Caskey W.H. & Tiedje J.M. (1979) Evidence for Clostridia as agents of dissimilatory reduction of nitrate to ammonium in soils. *Soil Sci. Soc. Am. J.* **43,** 931–6.

Christensen D. & Blackburn T.H. (1980) Turnover of alanine anoxic marine sediments. *Mar. Biol.* **58,** 97–103.

Clark M.E., Jackson G.A. & North W.J. (1972) Dissolved free amino acids in Southern California coastal waters. *Limnol. Oceanogr.* **17,** 749–58.

Cohen Y. & Gordon L.I. (1978) Nitrous oxide in the oxygen minimum of the eastern tropical North Pacific: evidence for its consumption during denitrification and possible mechanisms for its production. *Deep-Sea Res.* **25,** 504–24.

Delwiche C.C. (1970) The nitrogen cycle. *Sci. Am.* **223,** 137–46.

Delwiche C.C. (Ed.) (1981) *Denitrification, Nitrification and Atmospheric Nitrous Oxide.* Wiley-Interscience, New York.

Deppe K. & Engel H. (1960) Untersuchurgen über die Temperabhängkeit der Nitratbildung durch *Nitrobacter winogradski* Buck. bei ungehemmten und gehemmten Wachstum. *Zentbl. Bakt. Parasitkde* **11,** 113, 561–8.

Fenchel T. & Blackburn T.H. (1979) *Bacteria and Mineral Cycling*. Academic Press, London.

Fenchel T. & Jørgensen B.B. (1977) Detritus food chains of aquatic ecosystems: the role of bacteria. *Adv. Microbiol. Ecol.* **1,** 1–57.

Firestone M.K., Smith M.S., Firestone R.B. & Tiedje J.M. (1979) The influence of nitrate, nitrite, and oxygen on the composition of the gaseous products of denitrification in soil. *Soil Sci. Soc. Am. J.* **43,** 1140–1144.

Focht D.D., Valoras N. & Letey J. (1980) Use of interfaced gas chromatography-mass spectrometry for detection of concurrent mineralization and denitrification in soil. *J. Environ. Qual.* **9,** 218–23.

Focht D.D. & Verstraete W. (1977) Biochemical ecology of nitrification and denitrification. *Adv. Microbial. Ecol.* **1,** 135–214.

Frissal M.J. & van Veen J.A. (Eds) (1981) *Simulation of nitrogen behaviour of soil-plant systems*. Centre for Agricultural Publishing and Documentation, Wageningen.

Gersberg R., Krohn K., Peek N. & Goldman C.R. (1976) Denitrification studies with ^{13}N-labelled nitrate. *Science* **192,** 1229–31.

Goldhaber M.B., Aller R.C., Cochran J.K., Rosenfeld J.K., Martens C.S. & Berner R.A. (1977) Sulfate reduction, diffusion, and bioturbation in Long Island Sound sediments: report of the FOAM group. *Am. J. Sci.* **277,** 193–237.

Graetz D.A., Keeney D.R. & Aspiras R.B. (1973) Eh status of lake sediment–water systems in relation to nitrogen transformations. *Limnol. Oceanogr.* **18,** 908–17.

Gray T.R.G. & Parkinson D. (Eds.) (1968) *The Ecology of Soil Bacteria*. University of Toronto Press, Toronto.

Grundmanis V. & Murray J.W. (1977) Nitrification and denitrification in marine sediments from Puget Sound. *Limnol. Oceanogr.* 804–13.

Hall K.J., Kleiber P.M. & Yesaki I. (1972) Heterotrophic uptake of organic solutes by microorganisms in the sediment. *Mem. Ist. Ital. Idrobiol.* **29,** Suppl. 441–71.

Hansen J.I. (1980) Potential nitrifikation i marine sedimenter. M.Sc. thesis. Aarhus University, Aarhus, Denmark.

Hansen J.I., Henriksen K. & Blackburn T.H. (1981) Seasonal distribution of nitrifying bacteria and rates of nitrification in coastal marine sediments. *Microb. Ecol.* **7,** 297–304..

Hanson R.B. & Gardner W.S. (1978) Uptake and metabolism of two amino acids by anaerobic microorganisms in four diverse salt marsh soils. *Mar. Biol.* **46,** 101–7.

Hauck R.D. & Bremner J.M. (1976) Use of tracers for soil and fertilizer research. *Adv. Agron.* **28,** 219–66.

Henriksen K. (1980) Measurement of *in situ* rates of nitrification in sediment. *Microb. Ecol.* **6,** 329–37.

Henriksen K., Hansen J.I. & Blackburn T.H. (1981) Rates of nitrification, distribution of nitrifying bacteria and nitrate fluxes in different types of sediments from Danish Waters. *Marine Biol.* **61,** 299–304.

Herbert R.A., Dunn G.M. & Brown C.M. (1980) The physiology of nitrate dissimilatory bacteria from the Tay Estuary. *Proc. Roy. Soc. Edin.* **788,** 79–87.

Hiura K., Hattori T. & Furusaka C. (1976) Bacteriological studies on the mineralization of organic nitrogen in paddy soils. I. Effects of mechanical disruption of soils on ammonification and bacterial number. *Soil Sci. Plant Nutr.* **22,** 459–65.

Hiura K., Sato K., Hattori T. & Furusaka C. (1977) Bacteriological studies on the mineralization of soil organic nitrogen in paddy soils. II. The role of anaerobic isolates on nitrogen mineralization. *Soil Sci. Plant Nutr.* **23,** 201–5.

Hobbie J.E. & Crawford C.C. (1969) Respiration corrections for bacterial uptake of dissolved organic compounds in natural waters. *Limnol. Oceanogr.* **14,** 528–32.

Hynes R.K. & Knowles R. (1982) Effect of acetylene on autotrophic and heterotrophic nitrification. *Can. J. Microbiol.* **28,** 334–340.

Jørgensen B.B. (1977) The sulfur cycle of a coastal marine sediment (Limfjorden, Denmark). *Limnol. Oceanogr.* **22,** 814–32.

Jørgensen N.O.G., Blackburn T.H., Henriksen K. & Day D. (1981) The importance of *Posidonia oceanica* and *Cymodocea nodusa* as contributors of free amino acids in water and sediments of seagrass beds. *Mar. Ecol.* **2,** 97–112.

Jørgensen N.O.G., Mopper K. & Lindroth P. (1980) Occurrence, origin, and assimilation of free amino acids in an estuarine environment. *Ophelia Suppl.* **1,** 179–92.

Knowles R. (1981) Denitrification. In *Soil Biochemistry,* V.5 (Eds E.A.Paul & J.Ladd) pp. 323–69. Marcel Dekker Inc., New York.

Knowles R., Lean D.R.S. & Chan Y.K. (1981) Nitrous oxide concentration in lakes: variations with depth and time. *Limnol. Oceanogr.* **26,** 855–66.

Koike I. & Hattori A. (1978) Denitrification and ammonia formation in anaerobic coastal sediments. *Appl. Environ. Microbiol.* **35,** 278–82.

Lemon E. & Lemon D. (1981) Nitrous oxide in freshwaters of the Great Lakes Basin. *Limnol. Oceanogr.* **26,** 867–79.

Letey J., Hadas A., Vaoras N. & Focht D.D. (1980a) Effect of preincubation treatments on the ratio of N_2O/N_2 evolution. *J. Environ. Qual.* **9,** 232–5.

Letey J., Valoras N., Hasas A. & Focht D.D. (1980b) Effect of air-filled porosity, nitrate concentration, and time on the ratio of N_2O/N_2 evolution during denitrification. *J. Environ. Qual.* **9,** 227–31.

Madsen P.P. (1979) Seasonal variation of denitrification rate in sediment determined by use of ^{15}N. *Water Res.* **13,** 461–5.

Martens C.S., Berner R.A. & Rosenfeld J.K. (1978) Interstitial water chemistry of anoxic Long Island Sound sediments: 2. Nutrient regeneration and phosphate removal. *Limnol. Oceanogr.* **23,** 607–17.

Marumoto T., Kai H., Yoshida T. & Harada T. (1977) Drying effect on mineralization of microbial cells and their cell walls in soil and contribution of microbial cell walls as a source of decomposable soil organic matter due to drying. *Soil Sci. Plant Nutr.* **23,** 9–19.

Mosier A.R. & Hutchinson G.L. (1981) Nitrous oxide emissions from cropped fields. *J. Environ. Qual.* **10,** 169–73.

Oren A. & Blackburn T.H. (1979) Estimation of sediment denitrification rates at in situ nitrate concentrations. *Appl. Environ. Microbiol.* **37,** 174–6.

Painter H.W. (1970) A review of literature on inorganic nitrogen metabolism in microorganisms. *Water Res.* **4,** 393–450.

Parnas H. (1975) Model for decomposition of organic material by microorganisms. *Soil Biol. Biochem.* **7,** 161–9.

Postgate J.R. (1979) Nitrogen fixation. In *Microbial Ecology: A Conceptual Approach* (Eds J.M.Lynch & N.J.Poole) pp. 191–213. Blackwell Scientific Publications, Oxford.

Revsbech N.P., Sørensen J., Blackburn T.H. & Lomholt J.P. (1980a) Oxygen distribution in sediments measured with microelectrodes. *Limnol. Oceanogr.* **25,** 403–416.

Revsbech N.P., Jørgensen B.B. & Blackburn T.H. (1980b) Oxygen in the sea bottom measured with a microelectrode. *Science* **207,** 1355–6.

Rosenfeld J.K. (1979) Ammonium adsorption in nearshore anoxic sediments. *Limnol. Oceanogr.* **24,** 356–64.

Sahrawat K.L. (1980) Is nitrate reduced to ammonium in waterlogged acid sulfate soil? *Plant and Soil* **57,** 147–9.

Schmidt E.L. (1973) Fluorescent antibody techniques for the study of microbial ecology. *Bull. Ecol. Res. Comm. (Stockholm)* **17,** 67–76.

Schmidt E.L. (1978) Nitrifying microorganisms and their methodology. In *Microbiology 1978* (Ed. D.Schlessinger) pp. 288–91. Am. Soc. Microbiol., Washington, D.C.

Seitzinger S., Nixon S.A., Pilson M.E.Q. & Burke S. (1980) Denitrification and N_2O production in near-shore marine sediments. *Geochem. Cosmochim. Acta* **44,** 1858–61.

Sørensen J. (1978a) Capacity for denitrification and reduction of nitrate to ammonia in a coastal marine sediment. *Appl. Environ. Microbiol.* **35,** 301–5.

Sørensen J. (1978b) Denitrification rates in a marine sediment as measured by the acetylene inhibition technique. *Appl. Environ. Microbiol.* **36,** 139–43.

Sørensen J. (1978c) Occurrence of nitric and nitrous oxides in a coastal marine sediment. *Appl. Environ. Microbiol.* **36,** 809–13.

Sørensen J., Jørgensen B.B. & Revsbech N.P. (1979) A comparison of oxygen, nitrate, and sulfate respiration in coastal marine sediments. *Microbial Ecol.* **5,** 105–15.

Sørensen J., Tiedje J.M. & Firestone R.B. (1980) Inhibition by sulfide of nitric and nitrous oxide reduction by denitrifying *Pseudomonas fluorescens*. *Appl. Environ. Microbiol.* **39,** 105–8.

Stevenson F.J. (1962) Chemical state of the nitrogen in rocks. *Geochim. Cosmochim. Acta.* **26,** 797–809.

Stout J.D., Tate K.R. & Molloy L.F. (1976) Decomposition processes in New Zealand soils with particular respect to rates and pathways of plant degradation. In *The Role of Terrestrial and Aquatic Organisms in Decomposition Processes* (Eds J.M.Anderson & A.Macfadyen) pp. 97–144. Blackwell Scientific Publications, Oxford.

Svensson B.H. & Söderlund R. (Eds) (1976) *Nitrogen, Phosphorus and Sulfur-global cycles. Ecol. Bull.* 22. Swedish Natural Science Council, Stockholm.

Tiedje J.M., Firestone R.B., Firestone M.K., Betlach M.R., Smith M.S. & Caskey W.H. (1979) Methods for the production and use of ^{13}N in studies of denitrification. *Soil Sci. Soc. Am. J.* **43,** 709–16.

Tiedje J.M., Sørensen J. & Chang Y. (1981) Assimilatory and dissimilatory nitrate reduction: perspectives and methodology for simultaneous measurement of several nitrogen cycle processes. In *Terrestrial Nitrogen Cycles. Processes,*

Ecosystem Strategies and Management Impacts (Eds F.E.Clark & T.Rosswall) *Ecol. Bull. (Stockholm)* **33,** 331–42.

Triska F.J. & Oremland R.S. (1981) Denitrification associated with periphyton communities. *Appl. Environ. Microbiol.* **42,** 745–8.

Tuffey T.J., Hunter J.V. & Matulewich W.A. (1974) Zones of nitrification. *Bull. Am. Water Res. Assoc.* **10,** 1–10.

Vanderborght J-P. & Billen G. (1975) Vertical distribution of nitrate concentration in interstitial water of marine sediments with nitrification and denitrification. *Limnol. Oceanogr.* **20,** 953–61.

Verstraete W. & Alexander M. (1972) Heterotrophic nitrification in samples from natural environments. *Naturwissenschaften* **59,** 79–80.

Vincent W.F. & Downes M.T. (1981) Nitrate accumulation in aerobic hypolimnia: relative importance of benthic and planktonic nitrifiers in an oligotrophic lake. *Appl. Environ. Microbiol.* **42,** 565–73.

Wada E., Kadonga T. & Matsuo S. (1975) ^{15}N abundance in nitrogen of naturally occurring substances and global assessment of denitrification from an isotopic viewpoint. *Geochem. J.* **9,** 139–48.

Winogradsky S. & Winogradsky H. (1933) Etudes sur la microbiologie du sol. Nouvelles recherches sur les organismes de la nitrification. *Ann. Inst. Pasteur, Paris* **50,** 350–432.

Yoneyama T. & Yoshida T. (1977a). Decomposition of rice residue in tropical soils. I. Nitrogen uptake by rice plants from straw incorporated, fertilizer (ammonium sulfate) and soil. *Soil Sci. Plant Nutr.* **23,** 33–40.

Yoneyama T. & Yoshida T. (1977b) Decomposition of rice residue in tropical soils. II. Immobilization of soil and fertilizer nitrogen by intact rice residue in soil. *Soil Sci. Plant Nutr.* **23,** 41–8.

Yoneyama T. & Yoshida T. (1977c) Decomposition of rice residue in tropical soils. III. Nitrogen mineralization and immobilization of rice residue during its decomposition in soil. *Soil Sci. Plant Nutr.* **23,** 175–83.

Yoshinari T. (1980) N_2O reduction by *Vibrio succinogenes*. *Appl. Environ. Microbiol.* **39,** 81–4.

Yoshinari T., Hynes R. & Knowles R. (1977) Acetylene inhibition of nitrous oxide reduction and measurement of denitrification and nitrogen fixation in soil. *Soil Biol. Biochem.* **9,** 177–83.

CHAPTER 4
THE MICROBIAL SULPHUR CYCLE

BO B. JØRGENSEN

4.1 INTRODUCTION

Sulphur is among the ten most abundant elements on earth. It occurs in a large number of chemical compounds of which sulphate and iron sulphides are the quantitatively dominating forms. The sulphur atom ranges in oxidation steps from +6 as in sulphate (SO_4^{2-}) to −2 as in hydrogen sulphide (H_2S). Among the many inorganic intermediates, sulphite (SO_3^{2-}), thiosulphate ($S_2O_3^{2-}$), polythionates ($S_nO_6^{2-}$), and elemental sulphur (S^0) are mainly of relevance to the sulphur cycle in nature. The ability of sulphur to form a great variety of chemical compounds and to undergo changes in oxidation state over an eight electron shift bears resemblance to the properties of carbon and nitrogen. It is, therefore, not surprising that sulphur also has many important functions in the living organisms. It occurs mainly in cysteine and methionine bound in proteins and constitutes 0.1–2% dry weight of the biomass. Sulphur also plays a functional role in several coenzymes such as biotin, coenzyme A and lipoic acid.

Sulphur undergoes cyclic transformations which can be viewed from different levels of organization and complexity. Different aspects of the sulphur cycle become important depending on whether one considers the individual organisms, the ecosystems, or the whole globe:

Sulphur is assimilated by most bacteria as well as by algae and other plants in the form of sulphate. Within the cells, sulphate is activated by ATP to form adenylyl sulphate (APS and its phosphorylated derivative, PAPS). The sulphate then undergoes an assimilatory reduction via sulphite to sulphide which is ultimately transferred to amino acids as sulphydryl groups (Roy & Trudinger 1970, Siegel 1975). Animals require sulphur in organic form via their food. Many anaerobic bacteria, which also lack a sulphate-reducing pathway, assimilate sulphide instead, for example the green sulphur bacteria. The reduced organic sulphur is again released to the environment after the death and decomposition of the organisms. Among the degradation products are the sulphur gases H_2S, CH_3SH, and $(CH_3)_2S$, which are soon oxidized back to sulphate in the oxygen-containing environment.

The assimilatory sulphur transformations in the living organisms create a cycle between inorganic and organic states of sulphur. In addition to this, some bacteria can also perform a dissimilatory metabolism of sulphur

compounds. These organisms incorporate only a small fraction of the metabolized sulphur into the cells. The majority of the sulphur is used in energy metabolism as an electron acceptor or donor in a similar manner to that in which oxygen is used in aerobic organisms. Some specialized bacteria perform sulphate respiration and release sulphide. Others reoxidize the sulphide, either phototrophically with CO_2 or chemotrophically with O_2 or NO_3^-. In combination, all these different organisms drive the sulphur cycle of the ecosystems.

On a global scale, there is a mass transformation and transport of sulphur within and between the continents, the oceans and the atmosphere. Sulphur is mobilized from the continents by uplifting, weathering and leaching of rocks. It is transported to the sea by rivers, mainly in the form of dissolved sulphate. A small amount of sulphur is ejected from the earth's crust by volcanic activity. In the oceans there is a vast pool of dissolved sulphate which is slowly assimilated and recycled by the marine organisms. Some of the sulphate diffuses into the sea-bottom and becomes entrapped in the porewater of the sediment or is reduced and precipitated as iron sulphides. Sulphur exchange takes place between the atmosphere and the land or sea. Sea-spray and sulphur gases reach the atmosphere and the sulphur returns via rain or dry precipitation (Kellogg *et al.* 1972, Friend 1973, Granat *et al.* 1976).

In this chapter we will mainly discuss pathways of the sulphur cycle within aquatic sediments, and the groups of bacteria which carry out the processes will be presented.

4.2 SULPHATE-REDUCING BACTERIA

Bacteria which are able to perform respiratory sulphate reduction have been isolated from a wide variety of anoxic environments. They have been grouped in two genera, *Desulfovibrio* and *Desulfotomaculum* (LeGall & Postgate 1973, Buchanan & Gibbons 1974). Both comprise strictly anaerobic, motile rods. They contain cytochromes and utilize sulphate and several other oxidized sulphur compounds (SO_3^{2-}, $S_2O_3^{2-}$, some also S^0) as electron acceptors which are reduced to H_2S. The H_2S is excreted as the respiratory end-product with only a few percent being assimilated by the cells. Only *Desulfotomaculum* species form endospores. They are generally thermophiles which may grow at temperatures up to 70 °C. *Desulfovibrio* have optimum growth at around 35 °C.

The sulphate-reducing bacteria were, until recently, believed to be quite restricted in the organic compounds which they can use as energy sources for growth. The well-known strains of *Desulfovibrio* can utilize mainly lactate, pyruvate, and C-4 dicarboxylic acids such as malate and succinate,

but rarely carbohydrates (Postgate 1965, LeGall & Postgate 1973). The bacteria have an incomplete enzyme system of the tricarboxylic acid cycle as they cannot convert α-ketoglutarate to succinate (Lewis & Miller 1977). The carbon of the organic substrates is, therefore, not oxidized completely to CO_2 by *Desulfovibrio* but is excreted partly as acetate. As an example, the respiratory metabolism of lactate is:

$$2\ CH_3CHOHCOO^- + SO_4^{2-} \rightarrow 2\ CH_3COO^- + 2\ HCO_3^- + H_2S.$$

Sulphate here undergoes an eight electron shift. The organic substrate is oxidized to form two mol of HCO_3^- per mol of SO_4^{2-}. This stoichiometry is important to remember in the later considerations of the mineralization activity of the sulphate-reducing bacteria.

Most *Desulfovibrio* strains have hydrogenase and can also utilize H_2 as an energy substrate:

$$4\ H_2 + SO_4^{2-} + 2\ H^+ \rightarrow 4\ H_2O + H_2S.$$

Hydrogen is an important product in anaerobic fermentations and H_2 formation may be stimulated by its uptake by *Desulfovibrio*. When growing mixotrophically on H_2 and acetate, *Desulfovibrio* may obtain 30% of its carbon reqirements from assimilation of CO_2 (Sorokin 1966, Badziong *et al.* 1978). In the absence of sulphate, or if iron is limiting for the formation of cytochrome c_3, some strains may ferment pyruvate to acetate, hydrogen, and CO_2 (Vosjan 1975):

$$CH_3COCOO^- + H_2O + OH^- \rightarrow CH_3COO^- + HCO_3^- + H_2.$$

Also some species of *Desulfotomaculum* oxidize lactate and pyruvate to acetate and CO_2. Others, however, may oxidize short-chain fatty acids, such as acetate, completely to CO_2 (Widdel & Pfennig 1977):

$$CH_3COO^- + SO_4^{2-} \rightarrow 2\ HCO_3^- + HS^-.$$

Many new strains of sulphate-reducing bacteria with a wide spectrum of metabolism and morphology were recently isolated by Widdel (1980). For the classification of these new isolates five new genera were suggested (*Desulfobacter*, *Desulfobulbus*, *Desulfococcus*, *Desulfonema*, *Desulfosarcina*) as well as a revision of the genus *Desulfovibrio*. The work of Widdel has revealed the potential versatility in the substrate utilization of sulphate-reducing bacteria in nature. The organic substrates may be oxidized completely to CO_2, either by individual strains or by incompletely oxidizing strains which produce acetate and provide substrate for acetate oxidizing sulphate reducers. Thus, all the saturated N-fatty acids ranging from formate, acetate, propionate, butyrate, etc., up to stearate, as well as benzoate, hydroxybenzoates and urate can now be added to the list of substrates. None of the isolates utilize carbohydrates or hydrocarbons such as methane.

A bacterium which reduces elemental sulphur to H_2S, but cannot reduce sulphate, was recently isolated from mixed cultures in which it was growing together with phototrophic green sulphur bacteria (Pfennig & Biebl 1976). The organism, *Desulfuromonas acetoxidans*, is found in anaerobic, marine and freshwater environments. It uses acetate, ethanol or propanol as energy and carbon sources and oxidizes these to CO_2 with S^0 as the electron acceptor:

$$CH_3COO^- + 4\ S^0 + 2\ H_2O + H^+ \rightarrow 2\ CO_2 + 4\ H_2S.$$

The H_2S is then reoxidized to S^0 in the mixed culture by the green sulphur bacteria.

The biology of the sulphate-reducing bacteria has been reviewed by Postgate (1979).

4.3 SULPHATE-REDUCING BACTERIA IN SEDIMENTS

The sulphate-reducing bacteria are widely distributed in marine, freshwater, and terrestrial environments. Some of their main requirements are the presence of:

(1) reducing conditions;
(2) sulphate ions;
(3) suitable energy sources.

In combination, the three requirements largely determine the growth and metabolism of sulphate-reducing bacteria in nature.

(1) Reducing conditions are generally found in the sea-bottom, in lake sediments, in waterlogged soils, and in some stagnant water bodies of lakes and marine basins and fjords. In sediments, the reducing environment is underlying an oxidized surface layer of varying thickness. The oxidized zone is brownish due to the oxidized state of the iron compounds. Below this layer, the sediment colour is greyish or black due to black iron sulphides. The transition between the oxidized and the reduced environment can also be determined from the redox potential (Eh). At the transition layer, the redox potential changes from positive to negative values (Fenchel 1969). In pure culture, the sulphate-reducing bacteria require potentials below -100 mV (Postgate 1959).

The thickness of the oxidized zone in sediments generally increases with the distance from land. It is determined by the balance between the metabolic rate of the benthic organisms consuming oxygen, and the rate of oxygen influx. In lakes and in protected, coastal marine sediments it is only a few millimetres or centimetres thick. In the continental shelf and slopes it increases to several tens of centimetres and reaches several metres in the deep-sea sediments of which the central parts are completely oxidized (e.g. Fenchel & Riedl 1970, Volkov *et al.* 1972, 1976, Murray & Grundmanis 1980).

(2) Sulphate is the second-most abundant anion in sea water, which contains 2.7 g SO_4^{2-} l^{-1}. This corresponds roughly to a hundred times higher molar concentration than that of oxygen at air saturation, and sulphate consequently penetrates over a hundred times deeper into sediments than does oxygen. Coastal marine sediments may become sulphate depleted in a depth of one to a few metres while pelagic sediments have sulphate down to several hundred metres (Goldhaber & Kaplan 1974). In lake sediments, sulphate generally reaches only one to a few cm below the oxidized zone due to its low concentration in the water (Cappenberg 1974, Winfrey & Zeikus 1977). Bacterial sulphate reduction is, therefore, of less significance in the anaerobic sediment metabolism of lakes. This is a fundamental difference between the microbial geochemistry of freshwater and marine environments. In marine sediments, sulphate reduction is the dominating, terminal step in the anaerobic mineralization of organic matter. In freshwater sediments, methane formation ('CO_2-reduction') plays an equal or more important role.

(3) It is only very recently that information has become available on the energy sources for natural populations of sulphate-reducing bacteria. The bacteria are involved in the last stages of anaerobic mineralization of organic detritus. They seem to mediate a nearly complete oxidation of organic matter within their reducing environment. Bacteria of the genus *Desulfovibrio* have been considered to be the organisms mainly responsible for respiratory sulphate reduction in marine and freshwater sediments. *Desulfovibrio* spp. are, however, quite restricted in their metabolism both with respect to the range of utilizable energy substrates and to the fact that acetate is an end-product of the organic substrate oxidation.

The new strains of sulphate-reducing bacteria which were recently isolated (Widdel 1980) may, however, fill in the metabolic niche required to perform the complete mineralization. The first quantitative study of their distribution in marine sediments has shown that, for example, *Desulfobacter* and *Desulfobulbus* may be equally abundant as *Desulfovibrio* spp. (Laanbroek & Pfennig 1981).

The role of thiosulphate and sulphur-reducing bacteria also needs to be studied. Many strains have been isolated from sea water. They can oxidize a large spectrum of organic compounds to CO_2 (Tuttle & Jannasch 1973a).

Direct rate measurements of the organic substrate metabolism of the sulphate reducers are scarce. Radiotracer studies of lactate and acetate turnover in lake sediments (Cappenberg 1975) led to the conclusion that lactate is a very important substrate for the sulphate reducers. It was found to be converted into acetate which was in turn utilized by the methanogenic bacteria to produce CH_4 and CO_2. This syntrophism may be possible in lake sediments due to the close zonation of sulphate reduction and methane production (Fig. 4.1(b)). The significant role of lactate has also been indicated by the observation that addition of lactate to sediments generally

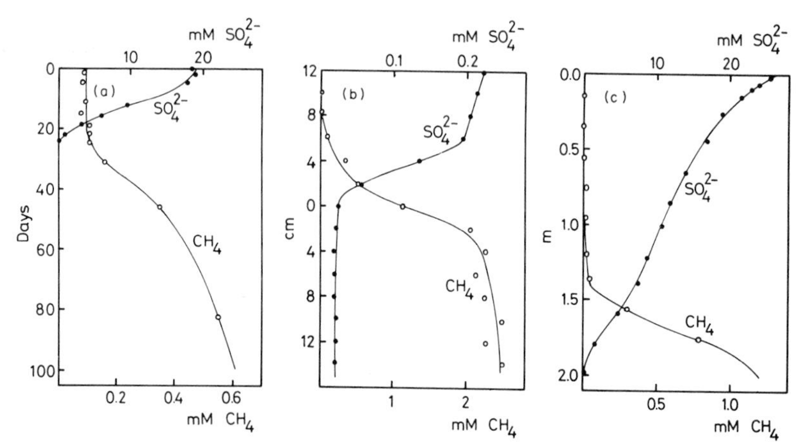

Fig. 4.1. (a) Time-course experiment of sulphate depletion and methane accumulation in enclosed marine sediment samples. (b) and (c) Vertical distribution of sulphate and methane in a lake sediment (b) and a marine sediment (c). Zero depth indicates the sediment surface. ((a) and (b) redrawn from Martens & Berner 1974 and Winfrey & Zeikus 1977; (c) T.H.Blackburn and B.B.Jørgensen, unpublished information.)

causes immediate stimulation of the sulphate reduction rate (Sorokin 1962, Oremland & Silverman 1979).

Recent laboratory studies on marine sediment slurries to which substrates or inhibitors of sulphate reduction or methanogenesis have been added have, however, demonstrated a competitive rather than a syntrophic relation between the two metabolic processes (Winfrey and Zeikus 1977, Abram & Nedwell 1978a, b, Oremland & Taylor 1978). The two types of bacteria seem to compete for both acetate and hydrogen as electron donors. In sulphate-rich sediments such as the sea-bottom, methane is generally produced only at a low or insignificant rate. If sediment samples are incubated anaerobically in the laboratory, methane does not begin to accumulate in the porewater until the sulphate has been depleted (Fig. 4.1(a)). If sulphate reduction is experimentally inhibited by addition of molybdate or β-fluorolactate, methane starts to accumulate. Methane production can also be stimulated by addition of excess acetate or hydrogen thereby releasing the competition for the energy substrates. The competitive advantage of sulphate reducers with respect to acetate and hydrogen utilization is in accordance with their higher affinity for the two substrates (lower K_s) as well as the higher energy yield from dissimilatory sulphate reduction (Fenchel & Blackburn 1979).

One further important interrelation seems to exist between the sulphate and the methane metabolisms in the sea-bottom. Below the depth of sulphate penetration, methane often builds up to high concentrations. The methane constantly diffuses upwards due to the concentration gradient. It does not reach the sediment surface, however, but only the lowermost part of the sulphate zone (Fig. 4.1 (c)) (Barnes & Goldberg 1976, Martens & Berner 1977, Reeburgh & Heggie 1977). Consequently, the sulphate zone must be a sink for methane. The explanation for this seems to be that methane is oxidized by bacteria in the presence of sulphate. Such an anaerobic methane oxidation has been demonstrated in samples of lake water and marine sediment (Kosiur & Warford 1979, Panganiban *et al.* 1979). Pure cultures of sulphate-reducing bacteria have generally been found not to oxidize methane, however, but a few exceptions to this are reported (Davis & Yarbrough 1966, Panganiban & Hanson 1976).

The microbial transformations of organic matter in sediments in relation

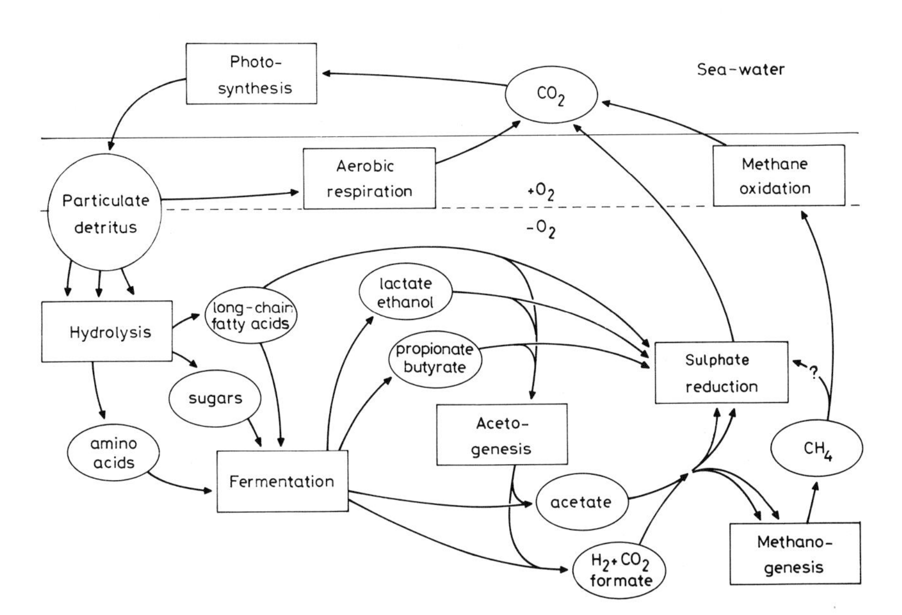

Fig. 4.2. Degradation and cycling of organic carbon in sediments in relation to bacterial sulphate reduction and methanogenesis.

to the bacterial sulphate reduction and methanogenesis are illustrated in Fig. 4.2. A fraction of the photosynthetically fixed carbon in lakes and seas ends up as detritus in the sediment. Much of this is decomposed and respired at the surface by aerobic bacteria and animals but some is buried deeper in the anoxic sediment. There it is attacked by bacteria which hydrolyse the structural components into monomers and oligomers of sugars, amino acids and long-chain fatty acids. These are, in turn, taken up by fermenting microorganisms and degraded into smaller molecules.

It is the end-products of the various fermentation processes which are energy sources for the sulphate-reducing bacteria as well as for the methanogens. Our knowledge of the fermentation pathways and their regulation in sediments is still rather inadequate and the scheme in Fig. 4.2 should, therefore, be considered only as preliminary.

Among the important fermentation products are the volatile fatty acids, formate, acetate, propionate and butyrate, as well as lactate, ethanol, hydrogen and CO_2. The reduced fermentation products such as fatty acids (propionate, butyrate and longer-chain fatty acids), lactate, etc., may be oxidized to acetate, H_2 and CO_2 by a still rather unknown group of acetogenic bacteria (Bryant 1979). This pathway seems to be regulated by the H_2 level in the sediment. By increasing H_2 pressures fermentation leads to more reduced products such as propionate. The very low H_2 pressure found in most anaerobic environments is maintained by a close coupling via interspecies hydrogen transfer between the acetogens and the sulphate reducers or methanogens. The acetogenic pathway may, therefore, be important both in anaerobic marine environments and in fresh water. It is required for the formation of the methane precursors, acetate and hydrogen, but it does not seem necessary for the sulphate reducers.

Fermentation studies in sulphate-free systems such as anaerobic sewage sludge digesters and the rumen have indeed shown acetate and hydrogen to be the main intermediate products preceding methanogenesis (Hungate 1966, Chynoweth & Mah 1971, Bryant 1979). A similar study in a marine sediment showed that acetate accounted for 30–50% of the substrate of the sulphate-reducing bacteria, while propionate, butyrate, isobutyrate and hydrogen each accounted for 5–10% (Sørensen *et al.* 1981). Lactate seems to be of less importance than acetate. Formate, which is also utilized, is split into H_2 and CO_2 by *Desulfovibrio* and other bacteria.

Acetate and hydrogen thus play a special role in being substrates for both the sulphate-reducing and the methane-producing bacteria. The observation that the sulphate reducers compete favourably with the methanogens for the two substrates gives an indication that they do indeed utilize them efficiently. In fresh water, sulphate availability is low and energy flows to a large extent to the methanogens. In sea water, sulphate reducers successfully compete with methanogenesis and dominate as the terminal step in the mineralization process.

Neither of the two bacterial groups can exploit all the chemical energy of their substrates. Much of the reducing power is transferred into H_2S and CH_4 which can later be utilized by specialized groups of aerobic bacteria living near the sediment surface (Fenchel & Jørgensen 1977).

4.4 SULPHATE REDUCTION IN SEDIMENTS

The activity of sulphate-reducing bacteria in sediments is demonstrated by a decrease in the sulphate concentration of the porewater with depth and by the accumulation of H_2S and iron sulphides (Goldhaber & Kaplan 1974). Both of these changes have been used to estimate the metabolic rate of the bacteria in the sea-bottom (Kaplan *et al.* 1963, Berner 1964). Calculations of sulphide accumulation rates will, however, tend to underestimate the reduction rate as some or even most of the produced sulphide may have been lost from the sediment by diffusion and oxidation (Jørgensen 1978c). Mathematical models based on observed sulphate profiles, sulphate diffusion and sediment accumulation provide an indirect tool to calculate sulphate reduction rates (Berner 1964, 1974).

A direct and quite sensitive method to measure the actual metabolic rates is the use of radioactively labelled sulphate (S-35) as a tracer (Sorokin 1962, Ivanov 1968, Jørgensen 1978a). The label is introduced into samples of sediments and the production of labelled sulphide is detected after a short incubation period. An example of results obtained with this method is given in Fig. 4.3. During winter, the sediment was oxidized down to a depth of 3 cm and maximum rates of sulphate reduction were found at a depth of 6–8 cm. During summer, the whole level of metabolism in the sediment increased. The reducing environment expanded upwards and sulphate reduction reached its maximum quite close to the surface. In more offshore sediments with a lower microbial activity, sulphate reduction is restricted to much deeper layers, for example from one metre below the surface of the sediment in the eastern Pacific Ocean at 5–6000 m water depth (Volkov *et al.* 1972).

Even in seemingly oxidized sediments, sulphate reduction may be detected in spite of the strictly anaerobic requirements of the bacteria. The organisms are living here in reduced microenvironments protected from the surroundings, for example inside the tests of dead foraminifera, within the walls of a dead plant cell, or just in a local accumulation of easily degradable organic matter (Jørgensen 1977b).

The population sizes of the sulphate-reducing bacteria living near the sediment surface have been studied in a variety of environments by means of viable counting techniques. The applied media have been suitable and

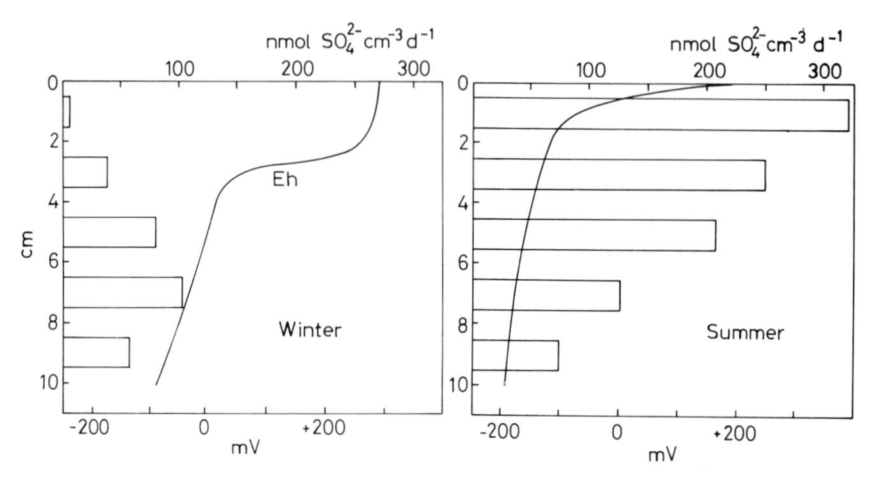

Fig. 4.3. Rates of bacterial sulphate reduction in a marine sediment during summer and winter (redrawn from Jørgensen 1977b).

selective for the strains of *Desulfovibrio* most commonly isolated from nature. The numbers of bacteria (or rather of colony-forming units) generally lie within 10^3–10^5 cm^{-3} of sediment (Oppenheimer 1960, Sorokin 1962, Sandkvist 1968). It has been calculated that this is less than 0.1‰ of the total bacterial population and it is probably more than 1000-fold lower than the true number of sulphate-reducing bacteria (Jørgensen 1978c). This gross underestimation is mainly due to the inherent errors of all such viable counting techniques in sediments but also due to the presence of sulphate reducers other than *Desulfovibrio* (Laanbroek & Pfennig 1981).

Even as a relative number, the colony counts of sulphate-reducing bacteria give an indication of their wide distribution and potential activity in nature. Sulphate reducers have been reported from environments such as deep-sea sediments at 10 km depth, from oil wells and from deep-ground waters (ZoBell 1963, Ivanov 1968). Their numbers in sediments decrease rapidly with depth along with a decrease in available carbon source. An example is shown in Fig. 4.4. The bacterial population at a depth of 1 m is less than 1% of that at a depth of 1 cm. The metabolic rate of the sulphate reducers in Fig. 4.4 decreases in proportion to their numbers down to 70 cm. Below that depth, the bacteria have depleted the porewater of dissolved sulphate. This is the zone of methane accumulation.

An important aspect of the bacterial sulphate reduction is its role in the mineralization of organic matter. A comparison of sulphate reduction and oxygen uptake rates in sediments shows the relative importance of the two compounds as oxidizing agents. In Fig. 4.5 the two processes are compared

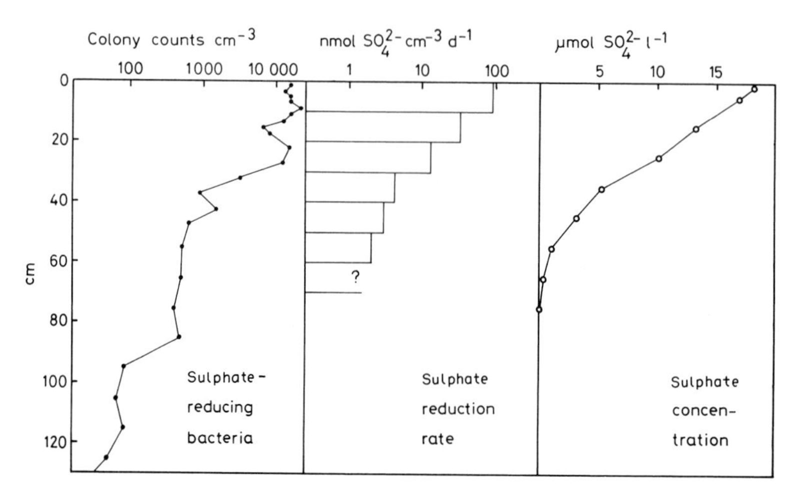

Fig. 4.4. Vertical distribution of sulphate-reducing bacteria, their metabolic rate and the resulting sulphate gradient in a sediment from Limfjorden, Denmark (redrawn from Jørgensen 1978c).

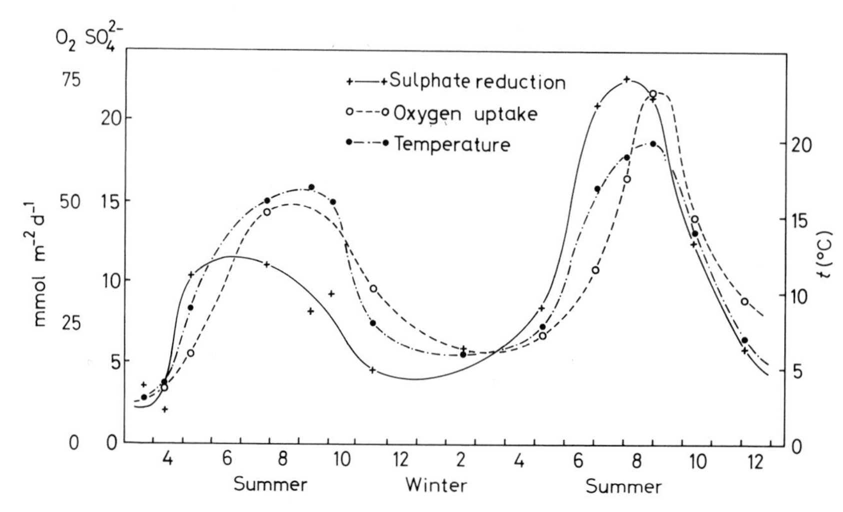

Fig. 4.5. Seasonal variation of sediment oxygen uptake rate, sulphate reduction rate in the whole sediment column, and sediment temperature in Limfjorden, Denmark (redrawn from Jørgensen 1977c).

in a coastal marine sediment throughout the seasons. The oxygen uptake is generally four times higher than the sulphate reduction when calculated per unit area (mmol m^{-2} d^{-1}). However, 1 mol of sulphate is equivalent to 2 mol of oxygen with respect to oxidizing capacity. It was, furthermore, calculated that half of the oxygen was used for H_2S reoxidation. So, in conclusion, the

sulphate reducers were responsible for about half of the organic carbon oxidation in the sediment. This result seems to be typical also of many other coastal areas. In more offshore sediments, sulphate respiration is of less relative significance. On the other hand, a complete dominance of sulphate reduction may be found in sulphureta as well as in salt marshes where a high root production by the marsh grasses provides a direct input of organic matter into the reducing zone (Howarth & Teal 1979, Skyring *et al.* 1979).

A survey of sulphate reduction rates in sediments is given in Fig. 4.6.

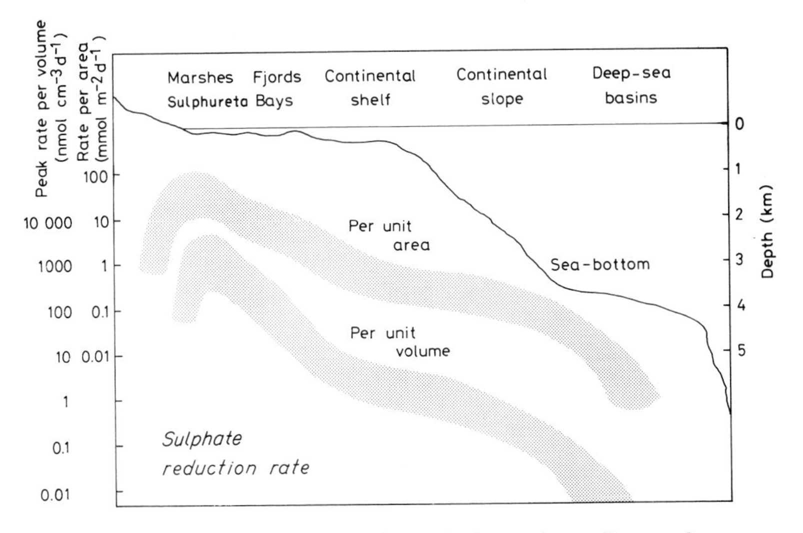

Fig. 4.6. Estimated sulphate reduction intensity in marine sediments from radiotracer measurements. Rates are calculated both per unit sediment volume at the depth of peak activity and per unit area for the whole sediment column.

The sulphate reduction is shown both as the rate per unit volume at the depth of highest activity (peak rate) and as the rate per unit area integrated over the whole sediment column. The graph is based on published data on sulphate reduction rates measured by radiotracer techniques. As the numbers of such measurements in the continental shelf and slopes are few, and are subject to methodological uncertainties, the quoted rates should be considered only as crude estimates. Fig. 4.6 does, however, demonstrate the great span of activities which have been measured. The peak rates range from a maximum of over 5000 nmol SO_4^{2-} cm^{-3} d^{-1} in some marshes and sulphureta to less than 0.1 in sediments of the continental slopes, i.e. a 100 000-fold range. The corresponding range in sulphate reduction rates calculated per unit area is only 1000-fold. This relative difference in the two estimates is due to the fact that, for example, in sulphureta most sulphate

reduction takes place within a thin surface layer of a few cm whereas in the shelf and continental slopes sediment strata of many metres contribute significantly to the total sulphate reduction.

4.5 FACTORS REGULATING SULPHATE REDUCTION

The rate of sulphate reduction in reduced marine sediments seems to be limited mainly by the rate at which suitable energy substrates become available. This rate is a function of both the concentration of organic matter and its degradability. Sediments often show little or no decrease in the organic content in different layers, but the organic compounds are modified and become increasingly refractory with depth. Since fresh organic matter is introduced by sedimentation, the rate of sulphate reduction often shows a positive correlation with the sedimentation rate (Berner 1978).

It is not clear to what extent the sulphate concentration in marine sediments regulates its reduction rate. In lakes, the sulphate level is only in the order of 0.1 mM and its depletion within the upper few cm of the lake bottom clearly restricts the overall sulphate reduction. In pure culture studies, 10 mM SO_4^{2-} was found to be the level below which the reduction rate became proportional to the sulphate concentration (Postgate 1951). Laboratory studies with artificially sulphate-depleted sediments have shown that the reduction kinetics depend on the pre-existing sulphate level. Sulphate saturation could be obtained over a 1000-fold range of concentrations from a few mM and down to a few μM. The latter values were obtained in marine sediments only after a week of adaptation to limiting sulphate concentrations (Jørgensen & Ingvorsen, unpublished results).

The H_2S, which is produced by the sulphate reducers, is toxic to aerobic organisms and in higher concentrations also to the anaerobic ones. *Desulfovibrio* has a very high tolerance to its metabolic product. Thus, pure cultures have been claimed to grow at up to about 50 mM H_2S (Miller 1950). Experimental addition of H_2S to a coastal marine sediment with an initial H_2S concentration of 0.1 mM showed that 10 mM H_2S inhibited the sulphate reduction by 50%, while 20 mM stopped it completely. In decomposing masses of sea-grasses in a small lagoon, sulphate reduction was, however, still taking place at a H_2S concentration of 25 mM (Jørgensen, unpublished results).

Apart from a direct toxic effect, H_2S may also inhibit the growth of the sulphate reducers in the sediment by binding iron which the bacteria need for their synthesis of functional cytochromes. In pure culture it is possible to grow cytochrome deficient cells on pyruvate which they ferment to acetate and H_2 (Vosjan 1975). Pyruvate is, however, not likely to be released in

significant quantities in sediments. The production of H_2S makes the sulphate-reducing bacteria very resistant to high levels of heavy metals (Hg, Cd, Pb, etc.) in their surroundings as these are precipitated as metal sulphides with extremely low solubility products.

Temperature has a regulating influence on the rate of sulphate reduction in any given environment. The reduction rate in sediments responds to temperature with a Q_{10} of about 3. In a marine sediment the temperature range at which the bacteria are active was found to be 0–40°C with an optimum around 30–35°C (Jørgensen, unpublished results). In a hot salt lake, where thermophilic *Desulfotomaculum* spp. may live, the upper temperature limit was found to be 65–70°C (Jørgensen *et al.* 1979a).

4.6 COLOURLESS SULPHUR BACTERIA

Two physiologically different groups of bacteria have specialized on the oxidation of H_2S. One of these groups contains the colourless sulphur bacteria which have in common the ability to oxidize sulphide or other reduced sulphur compounds with oxygen. Energy and reducing power may be gained from the process and enable some of the species to fix CO_2 and grow autotrophically (Sokolova & Karavaiko 1968, Kuenen 1975).

The colourless sulphur bacteria may be grouped as shown in Table 4.1. They comprise morphologically and physiologically very diverse forms. The best known are the thiobacilli which can oxidize a range of reduced sulphur compounds. They are small rods which occur widely in sediments and soils. Most of the other sulphur bacteria are large cells which are specialized in the oxidation of H_2S with oxygen. As O_2 and H_2S also undergo a rapid chemical reaction, these bacteria are typically gradient types. They are restricted in their distribution to the narrow zone where O_2 and H_2S

Table 4.1. Examples of colourless sulphur bacteria.

Family	Characteristic features	Examples of genera
Genera of uncertain affiliation	Rods or spheres, polar or peritrichous flagellate when motile, some deposit S^0 droplets in the cells	*Thiobacillus* *Thiomicrospira* *Thiovulum* *Macromonas*
Beggiatoaceae Leucotrichaceae	Filamentous, segmented, gliding when motile, S^0 droplets in the presence of H_2S	*Beggiatoa* *Thiothrix*
Achromatiaceae	Large ovoid cells, S^0 droplets and $CaCO_3$ inclusions	*Achromatium*

gradients meet and where the two compounds coexist in dynamic equilibrium (Lackey *et al.* 1965).

The oxidation of H_2S often proceeds via elemental sulphur. In the gradient forms such as *Beggiatoa*, the sulphur can be seen in the light microscope as small, refractile droplets in the cells which make a large part of the sulphur bacteria easy to recognize. Closer examination of the sulphur droplets in *Beggiatoa* under the electron microscope has revealed that the sulphur is actually not intracellular. It lies in invaginations of the cell membrane and is thus situated between that and the cell wall (Strohl & Larkin 1978). The sulphur may be oxidized to sulphate at a low rate and the overall stoichiometry is thus:

$$2\,H_2S + O_2 \rightarrow 2\,S^0 + 2\,H_2O$$
$$2\,S^0 + 3\,O_2 + 2\,H_2O \rightarrow 2\,H_2SO_4.$$

The biochemical pathways of sulphur oxidation are quite complex. They have been studied mostly in the thiobacilli which do not form sulphur

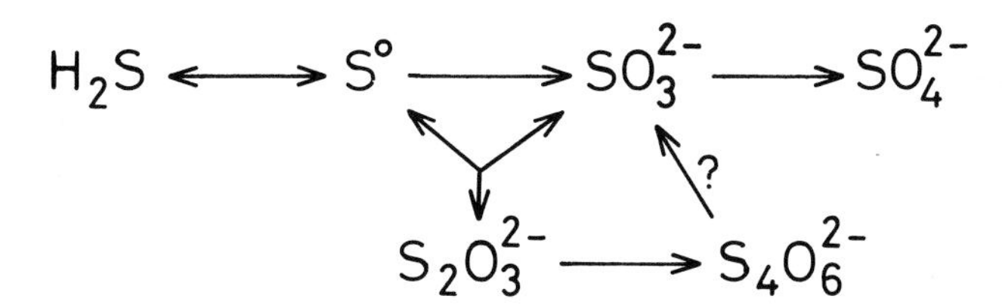

Fig. 4.7. Metabolic pathways of sulphur during sulphide oxidation in thiobacilli.

droplets in the cells. As they can utilize reduced sulphur compounds other than H_2S, it is possible to study them in media which are chemically stable. A simplified scheme of the sulphur pathway is shown in Fig. 4.7. Thiosulphate, polythionates, and sulphite appear as intermediates in the oxidation of H_2S and S^0 to sulphate (Roy & Trudinger 1970, Siegel 1975).

Some of the thiobacilli, for example *Thiobacillus thioparus* and *T. thiooxidans* are obligate autotrophs which fix CO_2 via the Calvin cycle. From the oxidation of sulphur compounds via the electron transport chain they can generate ATP. Reduced NAD is formed by an ATP-driven reverse electron flow in the electron transport chain. Most species are strict aerobes which use oxygen as electron acceptor. An interesting exception is *T. denitrificans* which may grow anaerobically using nitrate instead of oxygen (Baalsrud & Baalsrud 1954, Peeters & Aleem 1970):

$$5\,S_2O_3^{2-} + 8\,NO_3^- + H_2O \rightarrow 10\,SO_4^{2-} + 4\,N_2 + 2\,H^+.$$

The sulphide-oxidizing bacteria are a very diverse group ranging from

Table 4.2. Physiological diversity of H_2S-oxidizing bacteria (adapted from Kuenen 1975).

Physiological group	Organisms
Obligate chemolithotrophs	*T. thioparus* *T. thiooxidans* *T. denitrificans*
Mixotrophs	*T. intermedius*
Chemolithotrophic heterotrophs	*T. perometabolis*
?	*Beggiatoa*
Heterotrophs (which may or may not benefit from oxidation of reduced sulphur compounds)	*Pseudomonas*, etc.

obligate autotrophs to pure heterotrophs (Table 4.2). Among the thiobacilli, the chemolithotrophic autotrophs are already mentioned. *Thiobacillus intermedius* is a mixotroph which has a very flexible metabolism. Depending on the environmental conditions it can grow as an autotroph or as a heterotroph or as an intermediate. *T. perometabolis* is a chemolithotrophic heterotroph which assimilates organic carbon compounds but which can derive energy from the oxidation of reduced sulphur. Some *Beggiatoa* species and others of the gradient forms may also belong to this group. Such intermediate types may be the most important sulphur oxidizers in the chemocline of stratified waters (Tuttle & Jannasch 1973b). Finally, many heterotrophic bacteria can oxidize sulphide without gaining energy from it but may benefit in other ways. It has been suggested that sulphide oxidation in *Beggiatoa* serves mainly to detoxify intracellular hydrogen peroxide which otherwise accumulates due to the lack of catalase in the organism (Burton & Morita 1964).

4.7 PHOTOTROPHIC SULPHUR BACTERIA

The other physiological group which is specialized to oxidize H_2S are the phototrophic sulphur bacteria (Kondratjeva 1965, Pfennig 1967, 1975, 1978). These are anaerobic organisms which can use reduced sulphur compounds such as H_2S or S^0 as electron donors to assimilate CO_2 in the light. This type of photosynthesis is called anoxygenic, i.e. oxygen is not evolved. Only one photosystem is involved, in contrast to the oxygenic photosynthesis of plants which requires two photosystems. The H_2S is first oxidized to elemental sulphur which is then concomitantly oxidized to sulphate:

$$CO_2 + 2\ H_2S \xrightarrow{h\nu} [CH_2O] + 2\ S^0 + H_2O$$
$$3\ CO_2 + 2\ S^0 + 5\ H_2O \xrightarrow{h\nu} [CH_2O] + 2\ H_2SO_4.$$

The first process is faster than the second one which results in the accumulation of elemental sulphur (Gemerden 1967). The phototrophic sulphur bacteria are divided into the green and the purple bacteria (Pfennig 1977). Most of the purple forms store elemental sulphur as small droplets within the cells while sulphur is excreted as extracellular granules by the green sulphur bacteria. Some of the phototrophic bacteria which oxidize sulphide are presented in Table 4.3. They all have intracytoplasmic membrane structures which carry the photopigments. In the purple bacteria these are invaginations of the cytoplasmic membrane, while in the green they are special chlorobium vesicles attached to the membrane. The two

Table 4.3. Examples of phototrophic sulphide-oxidizing bacteria.

Family	Characteristic features	Examples of genera
Chromatiaceae	Purple sulphur bacteria, spheres, rods, spirals, etc., bacteriochlorophyll *a*, *b*, S^0 mostly inside cells, anaerobic, some are motile	*Chromatium* *Thiopedia* *Lamprocystis* *Thiospirillum*
Chlorobiaceae	Green and brown sulphur bacteria, varying morphology, bacterochlorophyll *a*, *c*, *d*, *e*, S^0 outside cells, anaerobic, non-motile	*Chlorobium* *Pelodictyon*
Chloroflexaceae	Filamentous, gliding green bacteria, bacteriochlorophyll *a*, *c*, S^0 outside cells, facultatively anaerobic	*Chloroflexus*
Cyanobacteria	Blue-green bacteria, chlorophyll *a* and phycobiliproteins, some facultatively anaerobic, S^0 outside cells	*Oscillatoria*

groups also differ in pigment types. Only the purples have bacteriochlorophyll *b* in addition to *a* while only the green have bacteriochlorophyll *a* plus *c*, *d*, or *e* (Pfennig 1978). Many of the phototrophic bacteria assimilate CO_2 via the Calvin cycle (reductive pentose cycle) like plants. According to present knowledge, in the green bacteria a reductive carboxylation in the ('reversed') tricarboxylic acid cycle is the major pathway for CO_2 fixation.

Filamentous green bacteria were only recently isolated from hot springs in different parts of the world (Castenholz 1973). Although they are of minor importance to the sulphur cycle they are physiologically very interesting. Under anoxic conditions they may perform anoxygenic photosynthesis with H_2 as electron donor, or they may grow as photoheterotrophs. Under oxic conditions they grow as aerobic heterotrophs.

The cyanobacteria (formerly called 'the blue-green algae') were also recently shown to have a very versatile metabolism (Cohen *et al.* 1975, Garlick *et al.* 1977). Like the algae, they have oxygenic photosynthesis with two photosystems and H_2O as the electron donor, but in H_2S environments, the water-cleaving photosystem II is uncoupled and many of the species switch to anoxygenic photosynthesis. The H_2S is oxidized to S^0 which precipitates as granules outside the cells. In the dark, the cyanobacteria may utilize the S^0 for anaerobic respiration and reduce it back to H_2S (Oren & Shilo 1979).

4.8 SULPHIDE OXIDATION IN SEDIMENTS

There are many chemical and bacterial processes by which sulphide may become oxidized in sediments. The oxidation is already initiated within the reducing environment of the sulphate-reducing bacteria. During the precipitation of iron sulphide and pyrite, sulphide may be the electron donor for reduction of ferric to ferrous iron (Berner 1970). This leads to the formation of elemental sulphur which has been found to be present in many sediments at concentrations of up to 1 μmol S^0 cm^{-3} or more, i.e. in similar

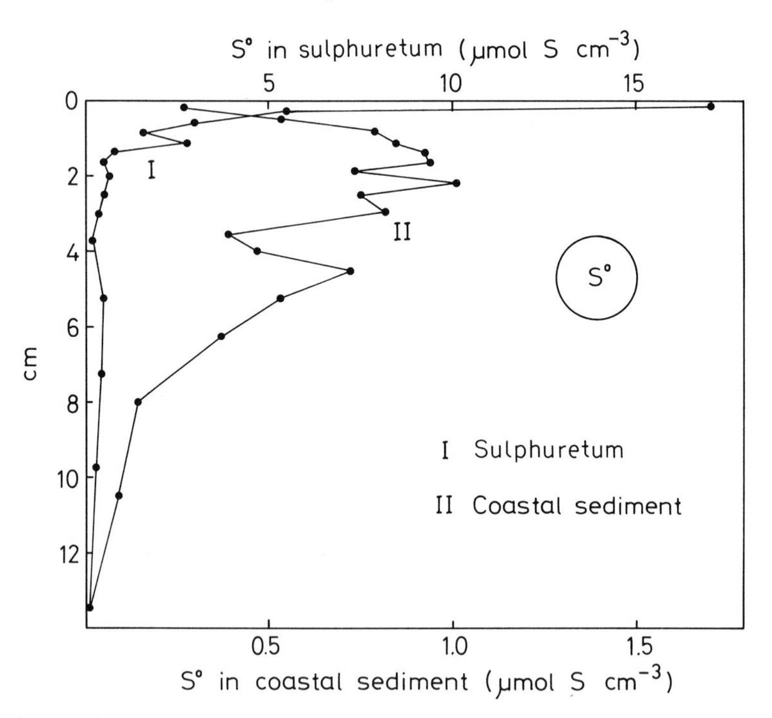

Fig. 4.8. Vertical distribution of elemental sulphur in two contrasting sediments of Danish fjords. Note the difference of scales. (Redrawn from Troelsen & Jørgensen 1982.)

concentrations to H_2S (Kaplan *et al.* 1963, Volkov *et al.* 1976). Fig. 4.8 shows the S^0 distribution in a coastal marine sediment and in an organically very rich sediment coated with sulphur bacteria (a 'sulphuretum'). The elemental sulphur present in the reduced sediment can potentially be further oxidized by bacteria using ferric iron (Brock & Gustafson 1976) or it can be reduced again to H_2S by sulphur-reducing bacteria, for example *Desulfuromonas* (Pfennig & Biebl 1976). It is not known to what extent this takes place.

In the upper part of the anoxic sediment, nitrate is often present to allow growth of denitrifying bacteria such as *Thiobacillus denitrificans* which oxidize the H_2S or S^0 to sulphate. Although the activity of *T. denitrificans* has not been directly demonstrated in sediments, the bacteria can be isolated from there. Furthermore, active denitrification has been demonstrated in layers where sulphate-reducing bacteria were producing H_2S (Sørensen *et al.* 1979).

Much of the H_2S in the porewater is transported upwards within the sediment towards the oxidized surface. When the H_2S comes into contact with the oxygen of the sea water above, it is rapidly oxidized. At the low concentrations at which the two compounds normally coexist, they react spontaneously with a half-life of a few minutes to several hours (Chen & Morris 1972, Almgren & Hagström 1974, Jørgensen *et al.* 1979a, b). The products of chemical oxidation in sea water are mainly thiosulphate and sulphate with smaller amounts of sulphite and polythionates (Cline & Richards 1969). If the H_2S reacts with oxygen at high concentrations, a large part will be converted into elemental sulphur.

Mass developments of elemental sulphur may be found in sulphur springs where sulphide-rich groundwater enters the oxic environment. Thick crusts of elemental sulphur deriving from such non-biogenic sulphide can be found in a variety of hydrothermal and volcanic environments. The hot springs of Yellowstone National Park, USA, are one example. The initial oxidation of H_2S to S^0 is here purely chemical. The following step from S^0 to SO_4^{2-} is carried out by thermophilic bacteria such as *Sulfolobus* which can grow at temperatures up to the boiling point (Brock 1978).

An example of the distribution of O_2 and H_2S in sediments is shown in Fig. 4.9. In this case the sediment is rather special: a compact cyanobacterial mat from a hypersaline lake. The sulphate reduction is very intensive and produces a high level of H_2S. The oxygen distribution is governed by its diffusion from above and its consumption or photosynthetic production by cyanobacteria within the uppermost few mm. During the day, oxygen builds up to a maximum in the photic zone and O_2 and H_2S coexist over a 0.2 mm depth interval where they react rapidly with a half-life of about 20 s. At night, H_2S production predominates and the O_2–H_2S interface ascends to the surface of the sediment. The pH gradient follows these changes. It reaches a maximum of 9.2 at the peak of photosynthesis and falls to 7.3 in the deeper, heterotrophic layers.

Generally, the interface between oxygen and sulphide in sediments is much more heterogeneous than in Fig. 4.9 due to the activity of burrowing

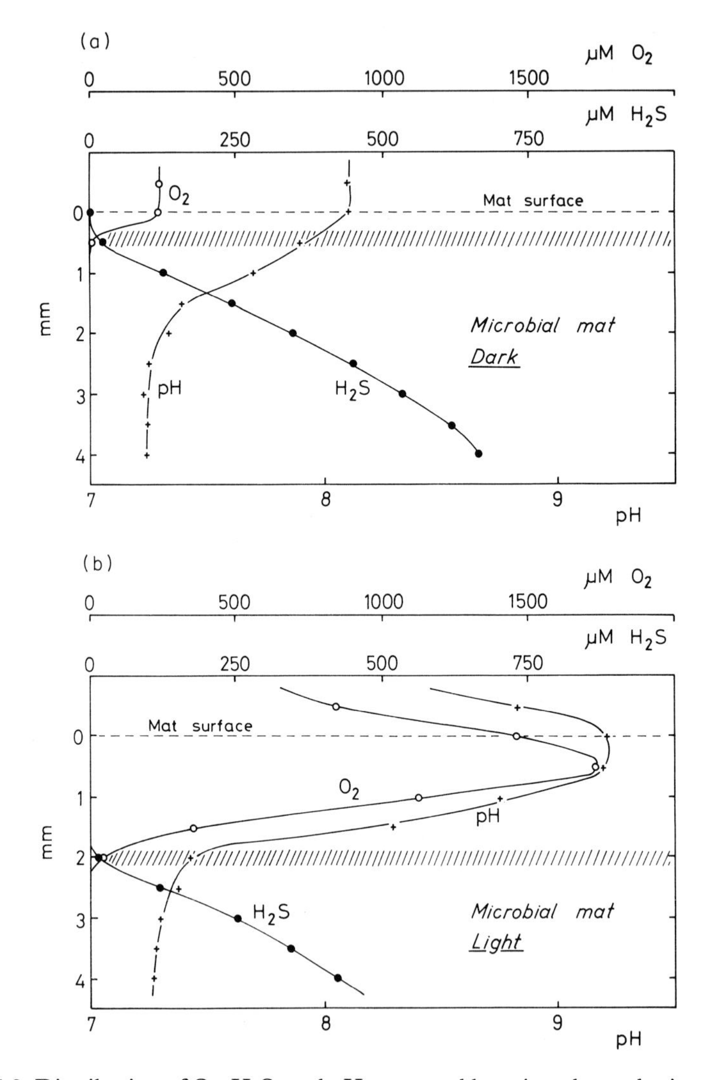

Fig. 4.9. Distribution of O_2, H_2S, and pH measured by microelectrodes in a cyanobacterial mat sediment at night (a) and during the day (b). The hatched areas indicate the zones of dynamic coexistence of O_2 and H_2S. (Based on data from N.P.Revsbech, B.B.Jørgensen, T.H.Blackburn and Y.Cohen, in preparation.)

animals. In the bioturbated zone the two compounds may not coexist in detectable concentrations and it is not clear how the sulphide oxidation then takes place. An example from a coastal marine sediment is shown in Fig. 4.10. Only the upper few mm are completely oxygenated while free H_2S becomes detectable only below a depth of several cm.

The gradient zone between O_2 and H_2S is the restricted environment of many of the colourless sulphur bacteria (Lackey *et al.* 1965). One of the most widespread types is *Beggiatoa* which may reach high population densities in both marine and freshwater sediments as well as in waterlogged soils (cf.

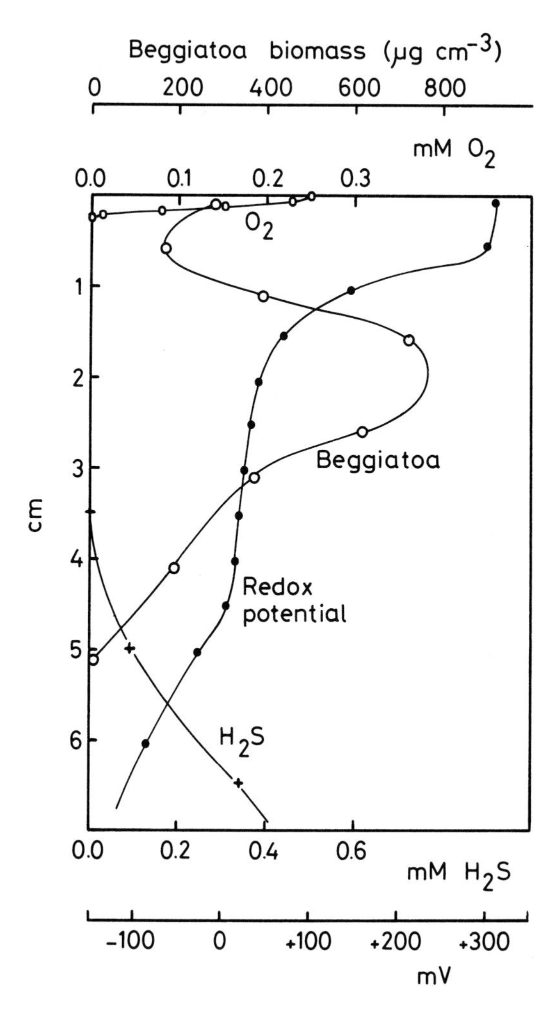

Fig. 4.10. Distribution of *Beggiatoa* filaments in a marine sediment in relation to the chemical zonation (redrawn from Jørgensen 1977c; O_2 data from N.P.Revsbech, unpublished information).

Fig. 4.10). Sometimes these bacteria form dense coatings on the sediment surface. Such white *Beggiatoa* mats have been observed to form a coherent cover on the mud over an area of 50–100 km^2 in a Danish fjord. In rice fields, *Beggiatoa* grows in the rhizosphere where oxygen is introduced into the sulphide zone via the roots. The ability of the bacteria to efficiently oxidize the toxic H_2S may here be beneficial to the rice plants (Joshi & Hollis 1977). Mass developments of filamentous sulphur bacteria have also been observed on sediments in the upwelling zone along the coast of Chile. Oxygen-depleted sea-water here hits the bottom at 50–200 m depth and mats of *Thioploca* develop over vast areas (Gallardo 1976).

The thiobacilli are believed to be among the most important sulphide oxidizers in sediments and soils although they are less conspicuous than the gradient types. Thiobacilli can be isolated in large numbers from most marine and freshwater sediments. Viable counts of thiobacilli in, for example, coastal marine areas have yielded numbers of 10^4–10^6 g^{-1} sediment (e.g. Schröder *et al.* 1976). Their quantitative importance for the sulphide oxidation in such environments still remains to be demonstrated, however. The role of thiobacilli for the oxidation of pyrite and other metal sulphides in soils and sedimentary rocks is better established, as discussed later in this chapter. They may also be important for the oxidation of elemental sulphur and thiosulphate. The latter has been found in sediments in concentrations of 0–0.1 μmol cm^{-3} (Volkov *et al.* 1972, 1976).

In shallow lakes and coastal environments, phototrophic sulphur bacteria may sometimes grow abundantly at the sediment surface. The requirements of the bacteria for both light and H_2S restrict their distribution since the photic zone is usually oxygenated due to algal photosynthesis. Light only penetrates a few mm into sediments (Fenchel & Straarup 1971). For H_2S to reach that close to the sediment surface the rate of sulphate reduction must be high. This is often the case along protected beaches where algae and sea-grasses drift together and enrich the sediment with rapidly decomposable organic matter. Such environments, sulphureta, are visibly dominated by sulphide-oxidizing bacteria (Fenchel 1969) which may develop into a beautiful mosaic of purple, green, brown and white.

The phototrophic sulphur bacteria grow in a distinct vertical zonation according to the penetration of light and O_2 from above and H_2S from below (Blackburn *et al.* 1975). When both purple and green types such as *Chromatium* and *Chlorobium* are present, the purple ones will grow on top of the green bacteria (Jørgensen & Fenchel 1974). The purple sulphur bacteria are the most tolerant towards oxygen and they grow at higher light intensity and lower sulphide concentration than the green forms (Pfennig 1978). Furthermore, they store the produced sulphur inside the cells and can, therefore, be more independent of H_2S which may become depleted from their surroundings during the day (cf. Fig. 4.8). The sulphur bacteria have often been observed to grow underneath a mat of filamentous cyanobacteria. Although the blue-green layer reaches a thickness that does not permit penetration of light which can be absorbed by chlorophyll, the longer wavelengths may still reach the phototrophic sulphur bacteria. The bacteriochlorophylls have absorption peaks of 850–900 nm in the purples and 700–760 nm in the green bacteria as compared to an absorption peak around 650–690 nm for the chlorophyll of the blue-green bacteria. If the water depth exceeds a few metres, the longer wavelengths are physically absorbed and carotenoids (450–600 nm absorbance) become the important

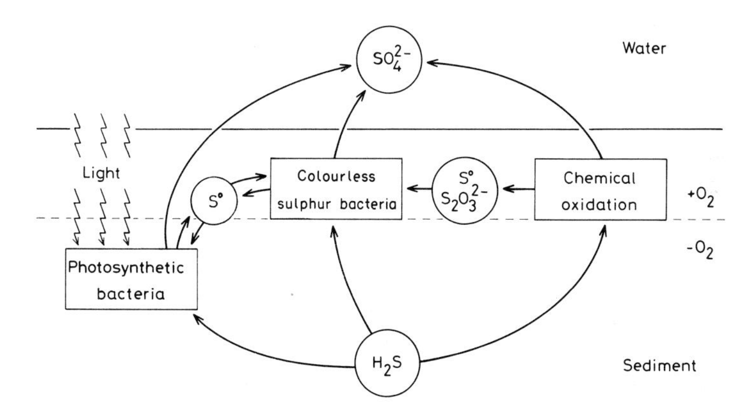

Fig. 4.11. Pathways of H_2S oxidation in sediments.

light-harvesting pigments of the sulphur bacteria. This complementarity in absorption spectra provides another important mechanism for the stratification of the phototrophic microorganisms.

The different types of sulphide oxidation in sediments are illustrated in Fig. 4.11. The phototrophic bacteria are only of local importance as they are restricted to water depths less than 10–20 m combined with high H_2S levels. Where they proliferate, they may, however, compete favourably with the colourless sulphur bacteria as they can utilize the H_2S before it reaches into the oxygenated zone (Jørgensen & Fenchel 1974). Furthermore, purple sulphur bacteria such as *Chromatium* may swarm up into the water column to follow an ascending H_2S zone during the night. In the early morning they rapidly deplete the H_2S from the water and again retreat into the sediment (Fenchel 1969).

The colourless sulphur bacteria which use O_2 to oxidize H_2S are in constant competition with the corresponding chemical process. It is not known how high a fraction of the H_2S oxidation can be performed by the bacteria. This is an important question as the bacterial metabolism may lead to chemosynthesis of new organic matter. Thereby some of the energy (10–15%) of the organic detritus, which orginally provided electrons for the H_2S formation, is conserved. In the case of chemical oxidation, all the energy of the H_2S is lost from the biological community. The dense populations of bivalves and tubeworms, which were recently discovered around H_2S-rich, thermal springs in the sea-bottom of the East Pacific, are an example of a benthic community living on the chemotrophic production of sulphur-oxidizing bacteria (Jannasch & Wirsen 1979).

The rapid bacterial conversion of H_2S to S^0 may be an important mechanism to conserve most of the energy for subsequent bacterial oxidation, as S^0 is not chemically oxidized. Elemental sulphur can be further oxidized to sulphate by colourless sulphur bacteria such as thiobacilli or by phototrophic sulphur bacteria, thereby completing the cycle. Or it can function as an electron acceptor for sulphur-reducing bacteria such as *Desulfuromonas*. These bacteria have been observed to grow in close syntrophy with green bacteria. In combination, the two forms can drive a sulphur cycle between H_2S and S^0 (Biebl & Pfennig 1978). The green bacteria even excrete large amounts of organic matter which serves as substrate for the sulphur reducers. Cyanobacteria and purple sulphur bacteria may also drive a H_2S-S^0 cycle, as they can oxidize H_2S phototrophically to S^0 in the light and reduce S^0 to H_2S in a dark respiratory metabolism (Gemerden 1967, Jørgensen *et al.* 1979a, Oren & Shilo 1979).

4.9 METAL SULPHIDE FORMATION AND OXIDATION

A part of the sulphide which is produced in sediments and soils is precipitated within the reducing environment by metal ions. As iron is by far the most abundant metal, iron sulphides constitute the major pool of sulphur in sediments. The fraction of the produced H_2S which is precipitated, depends upon the sediment type. If there is a high rate of sulphide production and a high H_2S level in the porewater, then most of the H_2S may be lost by diffusion to the surface sediment where it is oxidized. Bioturbation may contribute significantly to this loss since the animals ventilate zones of high sulphate reduction rate. This is the situation in coastal sediments where only a minor fraction of the produced H_2S goes into the iron sulphide pool (Jørgensen 1977a). In more offshore sediments, for example on the continental slopes, a much larger fraction of the sulphide is expectedly bound by iron (Goldhaber & Kaplan 1974).

The types of iron sulphide which are found in sediments or soil include amorphous ferrous sulphide (FeS), mackinawite ($FeS_{0.9}$), greigite (Fe_3S_4), and pyrite (FeS_2) (Goldhaber & Kaplan 1974). The amorphous ferrous sulphide is normally the first product to form. Ferrous ions are derived from iron oxide coatings on the mineral grains. Iron oxides such as geothite react chemically with H_2S present in the porewater:

$$2\ FeOOH + H_2S \rightarrow S^0 + 2\ Fe^{2+} + 4\ OH^-$$

and the free ferrous ions are then precipitated with additional H_2S:

$$Fe^{2+} + H_2S \rightarrow FeS + 2\ H^+.$$

The amorphous ferrous sulphide gradually crystallizes into mackinawite. This reacts with the produced elemental sulphur to form pyrite:

$$FeS + S^0 \rightarrow FeS_2.$$

The latter is a very slow process and FeS and S^0 are, therefore, able to coexist in sediments in significant concentrations. In the above reaction, greigite may be an intermediate which is also produced by reaction between FeS and S^0. Greigite subsequently reacts with additional S^0 to form FeS_2 (Goldhaber & Kaplan 1974).

At very low H_2S levels and under slightly acidic conditions ($pH < 6.5$) the concentrations of ferrous and sulphide ions may not exceed the solubility product of FeS. Instead of ferrous sulphide, pyrite may then form directly by reaction of the two ions with elemental sulphur:

$$Fe^{2+} + S^0 + H_2S \rightarrow FeS_2 + 2\ H^+.$$

This direct formation has been demonstrated in salt marshes as well as in other sediments (Howarth 1979). The exact mechanism of pyrite formation is, however, still not quite clear. It is also not clear to what extent bacteria may affect the rate and mechanism of pyrite formation. They may for example play a potential role in the initial reduction of ferric iron with H_2S or with S^0 (Brock & Gustafson 1976).

Pyrite is the stable form of iron sulphide under natural reducing conditions and it is the main sink of sulphide in sediments. In contrast to the other iron sulphides it dissolves only in oxidizing acids. The rate of pyrite formation is highest in the uppermost part of the sulphide zone where the sulphate reduction rate is also highest. In deeper strata, pyrite formation gradually ceases although both ferric oxides, H_2S and S^0 are present. The limiting reaction step here seems to be the reduction of ferric to ferrous iron.

An example of the distribution of H_2S, FeS, and FeS_2 in a marine sediment is shown in Fig. 4.12. There is an order of magnitude difference in maximum concentrations between H_2S and FeS as well as between FeS and FeS_2. A pyrite concentration of 100 μmol FeS_2-S cm^{-3} constitutes about 1% of the dry weight of the sediment. This is ten times more sulphur than

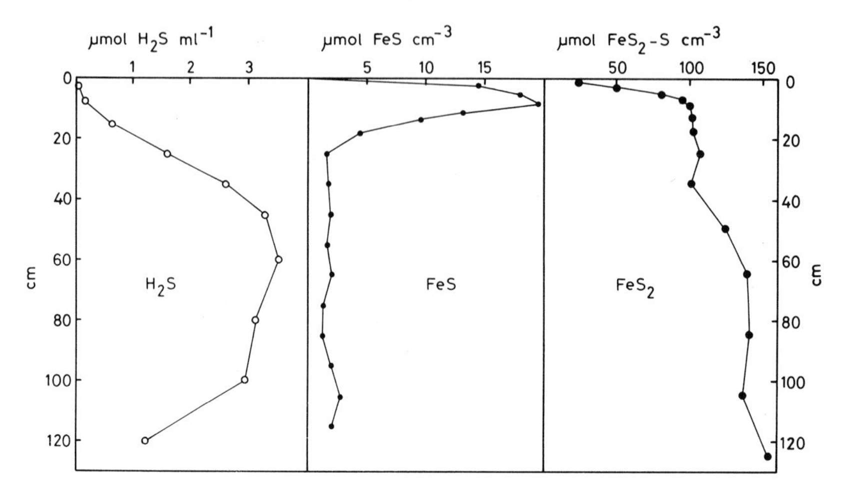

Fig. 4.12. Vertical distribution of free and iron-bound sulphide in a sediment of Limfjorden, Denmark (redrawn from Jørgensen 1978c).

was initially included in the porewater as sulphate. The additional sulphur was introduced later by diffusion of sulphate into the sediment and transformation into pyrite. By this mechanism, the sea-bottom acts as a sulphur trap, extracting and concentrating sulphur from the sea water into the sediments.

Metal sulphides other than those of iron have no significance for the sulphur balance itself. They occur at extremely low concentrations except for a few restricted locations. Two notable examples are the Red Sea hot brines and the hydrothermal areas of the East Pacific Rise (Degens & Ross 1969, Francheteau *et al.* 1979). In both of these areas, sulphide-containing water seeps out of the sea-bottom rich in heavy metals, and sulphides of, for example, Zn and Cu form massive deposits on the sediment.

Pyrite is stable only as long as the environment remains anoxic and reducing. It may become exposed to oxygen by erosion or bioturbation of sediments, by drainage of waterlogged soils and swamps, by mining, etc. In the presence of oxygen, pyrite oxidation immediately begins, catalysed by bacteria. The mechanism of oxidation in environments of neutral pH such as most sediments is only poorly understood. Bacteria such as *Thiobacillus ferrooxidans* may be directly involved in the oxidation of both the iron and the sulphur moiety of pyrite (Arkesteyn 1979). In many soils, mineral deposits, or in laboratory experiments, pyrite oxidation is, however, characterized by low pH due to intensive formation of sulphuric acid. Under suitable conditions, pH may drop down to 1.0–2.0. Only specialized thiobacilli, *T. ferrooxidans* and *T. thiooxidans*, tolerate this low pH.

Together with a few other chemolithotrophs such as the filamentous *Metallogenium*, they are active in the acid leaching of pyrites (e.g. Silverman 1967).

The suggested scheme for this process proceeds in three steps. First, the bacteria oxidize the sulphur atoms to sulphuric acid thereby releasing ferrous ions:

$$2\ FeS_2 + 2\ H_2O + 2\ O_2 \rightarrow 2\ Fe^{2+} + 4\ SO_4^{2-} + 4\ H^+.$$

At neutral pH, the ferrous iron reacts spontaneously with O_2 to form ferric iron which precipitates as $Fe(OH)_3$. The half-life of the ferrous iron is only a few minutes. Once the process has started, however, the environment may become acidic. When pH drops below 4.5 the ferrous ions become more stable and at pH 2.5 the chemical half-life in the presence of oxygen has increased to several years. Chemolithotrophic bacteria start to catalyse the oxidation and exploit the chemical energy for their autotrophic metabolism:

$$4\ Fe^{2+} + O_2 + 4\ H^+ \rightarrow 4\ Fe^{3+} + 2\ H_2O.$$

In the acid environment, the ferric ions remain in solution and initiate an oxidative attack on the pyrite:

$$FeS_2 + 14\ Fe^{3+} + 8\ H_2O \rightarrow 15\ Fe^{2+} + 2\ SO_4^{2-} + 16\ H^+.$$

The sulphur is now chemically oxidized to sulphuric acid by iron. Ferrous ions are regenerated and can again be oxidized by bacteria. A cyclic oxidation process is thereby established in which bacteria play a role by catalysing the oxidation of ferrous ions. The pyrite sulphur is then oxidized chemically by the iron. The process is strongly dependent upon the activity of the thiobacilli which speed up the ferrous ion oxidation by 10^5–10^6 times, relative to the chemical oxidation.

The cyclic pyrite oxidation can be exploited in the mining industry to leach and extract heavy metals from metal sulphide deposits. It has also an adverse effect in creating acid mine drainage as well as large precipitations of ferric hydroxides as the drainage water becomes neutralized.

4.10 THE SULPHUR CYCLE: AN EXAMPLE

The many different processes which have been discussed in this chapter combine in nature to form the complex sulphur cycles. Fig. 4.13 shows as an example the sulphur cycle of a coastal marine sediment. Sulphate reduction here proceeds at a rate of 9.5 mmol $SO_4^{2-}\ m^{-2}\ d^{-1}$ when integrated over the whole sediment column. Most of this reduction takes place in the upper 10 cm of the sediment and only 1–2% below 1 m depth. H_2S is also formed by the fermentative breakdown of organic sulphur compounds in the anoxic

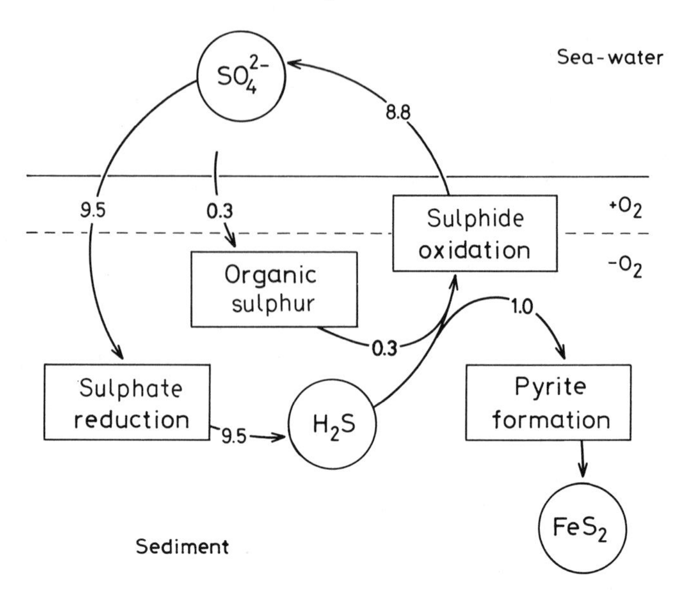

Fig. 4.13. Process rates of the sulphur cycle in a marine sediment. Rates are in mmol S m^{-2} d^{-1} integrated over the whole sediment column (redrawn from Jørgensen 1977a).

sediment. This, however, contributes only 3% of the total sulphide production. Assimilatory sulphate reduction is thus of little significance for the present sulphur cycle relative to dissimilatory reduction. The free sulphide constitutes a small but very dynamic pool. It has a turnover time of only one day in the upper 5 cm in contrast to sulphate which has a turnover time of 4–5 months.

About 10% of the produced sulphide is precipitated by iron and is ultimately converted into pyrite. As the sulphur is continuously cycling between sulphate and H_2S which flow into and out of the sediment, respectively, even this low degree of pyrite formation eventually leads to a pyrite concentration ten times higher than that of sulphate. Pyrite thus constitutes the largest pool of sulphur in the sediment and provides a continuous sink in the sulphur cycle. The remaining 90% of the H_2S is lost from the sediment by diffusion and bioconvection at the surface. Here it is reoxidized by oxygen, either chemically or by colourless sulphur bacteria. The sulphur has thus returned to the vast sulphate pool of the sea water and the cycle is completed.

Acknowledgements

I thank J.Gijs Kuenen for carefully reading the manuscript.

References

Abram J.W. & Nedwell D.B. (1978a) Inhibition of methanogenesis by sulphate reducing bacteria competing for transferred hydrogen. *Arch. Microbiol.* **117,** 89–92.

Abram J.W. & Nedwell D.B. (1978b) Hydrogen as a substrate for methanogenesis and sulphate reduction in anaerobic saltmarsh sediment. *Arch. Microbiol.* **117,** 93–7.

Almgren T. & Hagström I. (1974) The oxidation rate of sulphide in sea water. *Water Res.* **8,** 395–400.

Arkesteyn G.J.M.W. (1979) Pyrite oxidation by *Thiobacillus ferrooxidans* with special reference to the sulphur moiety of the mineral. *Antonie van Leeuwenhoek* **45,** 423–35.

Baalsrud K. & Baalsrud K.S. (1954) Studies on *Thiobacillus denitrificans. Arch. Mikrobiol.* **20,** 34–62.

Badziong W., Thauer R.K. & Zeikus J.G. (1978) Isolation and characterization of *Desulfovibrio* growing on hydrogen plus sulfate as the sole energy source. *Arch. Microbiol.* **116,** 41–9.

Barnes R.O. & Goldberg E.D. (1976) Methane production and consumption in anoxic marine sediments. *Geology* **4,** 297–300.

Berner R.A. (1964) An idealized model of dissolved sulfate distribution in recent sediments. *Geochim. Cosmochim. Acta* **28,** 1497–1503.

Berner R.A. (1970) Sedimentary pyrite formation. *Am. J. Sci.* **268,** 1–23.

Berner R.A. (1974) Kinetic models for the eary diagenesis of nitrogen, sulfur, phosphorus, and silicon in anoxic marine sediments. In *The Sea*, vol. 5 (Ed. E.D.Goldberg), pp. 427–50. John Wiley & Sons, New York.

Berner R.A. (1978) Sulfate reduction and the rate of deposition of marine sediments. *Earth Planet. Sci. Lett.* **37,** 492–8.

Biebl H. & Pfennig N. (1978) Growth yields of green sulfur bacteria in mixed cultures with sulfur and sulfate reducing bacteria. *Arch. Microbiol.* **117,** 9–16.

Blackburn T.H., Kleiber P. & Fenchel T. (1975) Photosynthetic sulfide oxidation in marine sediments. *Oikos* **26,** 103–8.

Brock T.D. (1978) *Thermophilic microorganisms and life at high temperatures.* Springer, New York.

Brock T.D. & J. Gustafson (1976) Ferric iron reduction by sulfur- and iron-oxidizing bacteria. *Appl. Environ. Microbiol.* **32,** 567–71.

Bryant M.P. (1979) Microbial methane production—theoretical aspects. *J. Animal Sci.* **48,** 193–201.

Buchanan R.E. & Gibbons N.E. (Eds) (1974) *Bergey's Manual of Determinative Bacteriology*, 8e. Williams & Wilkins, Baltimore.

Burton S.D. & Morita R.Y. (1964) Effects of catalase and cultural conditions on growth of *Beggiatoa. J. Bacteriol.* **88,** 1755–61.

Cappenberg T.E. (1974) Interrelations between sulfate-reducing and methane-producing bacteria in bottom deposits of a fresh-water lake. I. Field observations. *Antonie van Leeuwenhoek* **40,** 285–95.

Cappenberg T.E. (1975) Relationships between sulfate-reducing and methane-producing bacteria. *Plant Soil* **43,** 125–39.

Castenholz R.W. (1973) The possible photosynthetic use of sulphide by the filamentous phototrophic bacteria of hot springs. *Limnol. Oceanogr.* **18,** 863–76.

Chen K.Y. & Morris J.C. (1972) Kinetics of oxidation of aqueous sulfide by O_2. *Environ. Sci. Technol.* **6,** 529–37.

Chynoweth D.P. & Mah R.A. (1971) Volatile acid formation in sludge digestion. In *Anaerobic biological treatment processes* (Ed. F.G.Pohland), pp. 41–54. American Chemical Society, Washington.

Cline J.D. & Richards F.A. (1969) Oxygenation of hydrogen sulfide in seawater at constant salinity, temperature, and pH. *Environ. Sci. Technol.* **3,** 838–43.

Cohen Y., Padan E. & Shilo M. (1975) Facultative anoxygenic photosynthesis in the cyanobacterium *Oscillatoria limnetica*. *J. Bacteriol.* **123,** 855–61.

Davis J.B. & Yarbrough H.F. (1966) Anaerobic oxidation of hydrocarbons by *Desulfovibrio desulfuricans*. *Chem. Geol.* **1,** 137–44.

Degens E.T. & Ross D.A. (Eds) (1969) *Hot brines and recent heavy metal deposits in the Red Sea*. Springer Verlag, Berlin.

Fenchel T. (1969) The ecology of marine microbenthos. IV. Structure and function of the benthic ecosystem, its chemical and physical factors and the microfauna communities with special reference to the ciliated protozoa. *Ophelia* **6,** 1–182.

Fenchel T. & Blackburn T.H. (1979) *Bacteria and mineral cycling*. Academic Press, London.

Fenchel T.M. & Jørgensen B.B. (1977) Detritus food chains of aquatic ecosystems: The role of bacteria. In *Advances in Microbial Ecology*, vol. 1 (Ed. M.Alexander), pp. 1–58. Plenum Press, New York.

Fenchel T.M. & Riedl R.J. (1970) The sulfide system: a new biotic community underneath the oxidized layer of marine sand bottoms. *Mar. Biol.* **7,** 255–68.

Fenchel T. & Straarup B.J. (1971) Vertical distribution of photosynthetic pigments and the penetration of light in marine sediments. *Oikos* **22,** 172–82.

Francheteau J. *et al.* (1979) Massive deep-sea sulphide ore deposits discovered on the East Pacific Rise. *Nature* **277,** 523–8.

Friend J.P. (1973) The global sulfur cycle. In *Chemistry of the lower atmosphere* (Ed. S.I.Rasool), pp. 177–201. Plenum Press, New York.

Gallardo V.A. (1976) On a benthic sulfide system on the continental shelf of north and central Chile. In *International Symposium on Coastal Upwelling, Proceedings, Coquimbo, Chile, November 18–19, 1975*, pp. 113–18.

Garlick S., Oren A. & Padan E. (1977) Occurrence of facultative anoxygenic photosynthesis among filamentous and unicellular cyanobacteria. *J. Bacteriol.* **129,** 623–9.

Gemerden H. van (1967) On the bacterial sulphur cycle of inland waters. Ph.D. Thesis, Leiden.

Goldhaber M.B. & Kaplan I.R. (1974) The sulfur cycle. In *The Sea*, vol. 5 (Ed. E.D.Goldberg), pp. 569–655. John Wiley & Sons, New York.

Granat L., Rodhe H. & Hallberg R.O. (1976) The global sulphur cycle. *Ecol. Bull. (Stockholm)* **22,** 89–134.

Howarth R.W. (1979) Pyrite: Its rapid formation in a salt marsh and its importance to ecosystem metabolism. *Science* **203,** 49–51.

Howarth R.W. & Teal J.M. (1979) Sulfate reduction in a New England salt marsh. *Limnol. Oceanogr.* **24,** 999–1013.

Hungate R.E. (1966) *The rumen and its microbes*. Academic Press, New York.

Ivanov M.V. (1968) *Microbiological processes in the formation of sulfur deposits.* Israel Program for Scientific Translations, Jerusalem.

Jannasch H.W. & Wirsen C.O. (1979) Chemosynthetic primary production at East Pacific sea floor spreading center. *BioScience* **29,** 592–8.

Jørgensen B.B. (1977a) The sulfur cycle of a coastal marine sediment (Limfjorden, Denmark). *Limnol. Oceanogr.* **22,** 814–32.

Jørgensen B.B. (1977b) Bacterial sulfate reduction within reduced microniches of oxidized marine sediments. *Mar. Biol.* **41,** 7–17.

Jørgensen B.B. (1977c) Distribution of colorless sulfur bacteria (*Beggiatoa* spp.) in a coastal marine sediment. *Mar. Biol.* **41,** 19–28.

Jørgensen B.B. (1978a) A comparison of methods for the quantification of bacterial sulfate reduction in coastal marine sediments. I. Measurements with radiotracer techniques. *Geomicrobiol. J.* **1,** 11–28.

Jørgensen B.B. (1978b) A comparison of methods for the quantification of bacterial sulfate reduction in coastal marine sediments. II. Calculations from mathematical models. *Geomicrobiol. J.* **1,** 29–51.

Jørgensen B.B. (1978c) A comparison of methods for the quantification of bacterial sulfate reduction in coastal marine sediments. III. Estimation from chemical and bacteriological field data. *Geomicrobiol. J.* **1,** 53–68.

Jørgensen B.B. & Fenchel T. (1974) The sulfur cycle of a marine sediment model system. *Mar. Biol.* **24,** 189–201.

Jørgensen B.B., Kuenen J.G. & Cohen Y. (1979a) Microbial transformations of sulfur compounds in a stratified lake (Solar Lake, Sinai). *Limnol. Oceanogr.* **24,** 799–822.

Jørgensen B.B., Revsbech N.P., Blackburn T.H. & Cohen Y. (1979b) Diurnal cycle of oxygen and sulfide microgradients and microbial photosynthesis in a cyanobacterial mat sediment. *Appl. Environ. Microbiol.* **38,** 46–58.

Joshi M.M. & Hollis J.P. (1977) Interaction of *Beggiatoa* and rice plant: detoxification of hydrogen sulfide in the rice rhizosphere. *Science* **195,** 179–80.

Kaplan I.R., Emery K.O. & Rittenberg S.C. (1963) The distribution and isotopic abundance of sulfur in recent marine sediments off southern California. *Geochim. Cosmochim. Acta* **27,** 297–331.

Kellogg W.W., Cadle R.D., Allen E.R., Lazrus A.L. & Martell E.A. (1972) The sulfur cycle. *Science* **175,** 587–96.

Kondratjeva E.N. (1965) *Photosynthetic Bacteria.* Israel Program for Scientific Translations, Jerusalem.

Kosiur D.R. & Warford A.L. (1979) Methane production and oxidation in Santa Barbara Basin sediments. *Estuar. Coastal Mar. Sci.* **8,** 379–85.

Kuenen J.G. (1975) Colourless sulfur bacteria and their role in the sulfur cycle. *Plant Soil* **43,** 49–76.

Laanbroek H.J. & Pfennig N. (1981) Oxidation of short chain fatty acids by sulfate-reducing bacteria in freshwater and in marine sediments. *Arch. Microbiol.* **128,** 330–335.

Lackey J.B., Lackey E.W. & Morgan G.B. (1965) Taxonomy and ecology of the sulfur bacteria. *University of Florida, College of Engineering, Bull. Ser. no. 119,* vol. **19,** 1–23.

LeGall J. & Postgate J.R. (1973) The physiology of sulphate-reducing bacteria. *Adv. Microb. Physiol.* **10,** 81–133.

Lewis A.J. & Miller J.D.A. (1977) The tricarboxylic acid pathway in *Desulfovibrio*. *Can. J. Microbiol.* **23,** 916–21.

Martens C.S. & Berner R.A. (1974) Methane production in the inerstitial waters of sulfate depleted marine sediments. *Science* **185,** 1167–9.

Martens C.S. & Berner R.A. (1977) Interstitial water chemistry of anoxic Long Island Sound sediments. 1. Dissolved gases. *Limnol. Oceanogr.* **22,** 10–25.

Miller L.P. (1950) Tolerance of sulfate-reducing bacteria to hydrogen sulfide. *Contrib. Boyce Thompson Inst.* **16,** 73–83.

Murray J.W. & Grundmanis Y. (1980) Oxygen consumption in pelagic marine sediments. *Science* **209,** 1527–30.

Oppenheimer C.H. (1960) Bacterial activity in sediments of shallow marine bays. *Geochim. Cosmochim. Acta* **19,** 244–60.

Oremland R.S. & Silverman M.P. (1979) Microbial sulfate reduction measured by an automated electrical impedance technique. *Geomicrobiol. J.* **1,** 35–72.

Oremland R.S. & Silverman M.P. (1979) Microbial sulfate reduction measured by an automated electrical impedance technique. *Geomicrobiol. J.* **1,** 355–72.

Oren A. & Shilo M. (1979) Anaerobic heterotrophic dark metabolism in the cyanobacterium *Oscillatoria limnetica*: sulfur respiration and lactate fermentation. *Arch. Microbiol.* **122,** 77–84.

Panganiban A. & Hanson R.S. (1976) Isolation of a bacterium that oxidizes methane in the absence of oxygen. Annual Meeting of the American Society of Microbiology, Abstracts, p. 121.

Panganiban A.T., Patt T.E., Hart W. & Hanson R.S. (1979) Oxidation of methane in the absence of oxygen in lake water samples. *Appl. Environ. Microbiol.* **37,** 303–9.

Peeters T. & Aleem M.I.H. (1970) Oxidation of sulfur compounds and electron transport in *Thiobacillus denitrificans*. *Arch. Mikrobiol.* **71,** 319–30.

Pfennig N. (1967) Photosynthetic bacteria. *Ann. Rev. Microbiol.* **21,** 285–324.

Pfennig N. (1975) The phototrophic bacteria and their role in the sulfur cycle. *Plant Soil* **43,** 1–16.

Pfennig N. (1977) Phototrophic green and purple bacteria: a comparative systematic survey. *Ann. Rev. Microbiol.* **31,** 275–90.

Pfennig N. (1978) General physiology and ecology of photosynthetic bacteria. In *The Photosynthetic Bacteria* (Eds R.K.Clayton & W.R.Sistrom), pp. 3–18. Plenum Press, London.

Pfennig N. & Biebl H. (1976) *Desulfuromonas acetoxidans* gen. nov. and spec. nov., a new anaerobic, sulfur-reducing, acetate oxidizing bacterium. *Arch. Microbiol.* **110,** 3–12.

Postgate J.R. (1951) The reduction of sulphur compounds by *Desulphovibrio desulphuricans*. *J. Gen. Microbiol.* **5,** 725–38.

Postgate J.R. (1959) Sulphate reduction by bacteria. *Ann. Rev. Microbiol.* **13,** 505–20.

Postgate J.R. (1965) Recent advances in study of sulfate-reducing bacteria. *Bact. Rev.* **29,** 425–41.

Postgate J.R. (1979) *The Sulphate-reducing Bacteria*. Cambridge University Press, Cambridge.

Reeburgh W.S. & Heggie D.T. (1977) Microbial methane consumption reactions and their effects on methane distributions in freshwater and marine environments. *Limnol. Oceanogr.* **22,** 1–9.

Roy A.B. & Trudinger P.A. (1970) *The biochemistry of inorganic compounds of sulphur*. Cambridge University Press, London.

Sandkvist A. (1968) Microbiological investigation of modern Dutch tidal sediments. *Stockholm Contrib. Geol.* vol. **XV,** 67–123.

Schröder H.G.J., van Es F.B., Knol W.H., Kop A.J. & Wisser G.F. (1976) *Microbiological investigations of the Eems-Dollard Estuary* (in Dutch). Report, Eems-Dollard Project, University of Groningen.

Siegel L.M. (1975) Biochemistry of the sulfur cycle. In *Metabolic pathways*, vol. 7, *Metabolism of sulfur compounds* (Ed. D.M.Greenberg), pp. 217–86. Academic Press, New York.

Silverman M.P. (1967) Mechanisms of bacterial pyrite oxidation. *J. Bacteriol.* **94,** 1046–51.

Skyring G.W., Oshrain R.L. & Wiebe W.J. (1979) Sulfate reduction rates in Georgia marshland soils. *Geomicrobiol. J.* **1,** 389–400.

Sokolova G.A. & Karavaiko G.I. (1968) *Physiology and geochemical activity of Thiobacilli.* Israel Program for Scientific Translations, Jerusalem.

Sørensen J., Christensen D. & Jørgensen B.B. (1981) Volatile fatty acids and hydrogen as substrates for sulfate-reducing bacteria in anaerobic marine sediment. *Appl. Environ. Microbiol.* **42,** 5–11.

Sørensen J., Jørgensen B.B. & Revsbech N.P. (1979) A comparison of oxygen, nitrate, and sulfate respiration in coastal marine sediments. *Microb. Ecol.* **5,** 105–15.

Sorokin Yu.I. (1962) Experiment investigation of bacterial sulfate reduction in the Black Sea using ^{35}S. *Microbiology* **31,** 329–35.

Sorokin Y.I. (1966) Role of carbon dioxide and acetate in biosynthesis by sulphate-reducing bacteria. *Nature* **210,** 551–2.

Strohl W.R. & Larkin J.M. (1978) Enumeration, isolation, and characterization of *Beggiatoa* from freshwater sediments. *Appl. Environ. Microbiol.* **36,** 755–70.

Troelsen H. & Jørgensen B.B. (1982) Seasonal dynamics of elemental sulfur in two coastal sediments. *Estuar. Coast. Shelf Sci.* **15,** 255–60.

Tuttle J.H. & Jannasch H.W. (1973a) Dissimilatory reduction of inorganic sulfur by facultatively anaerobic marine bacteria. *J. Bacteriol.* **115,** 732–7.

Tuttle J.H. & Jannasch H.W. (1973b) Sulfide and thiosulfate-oxidizing bacteria in anoxic marine basins. *Mar. Biol.* **20,** 64–70.

Widdel F. (1980) Anaerober Abbau von Fettsaüren und Benzoesaüre durch neu isolierte Arten Sulfat-reduzierender Bakterien. Ph.D. Thesis, Göttingen.

Widdel F. & Pfennig N. (1977) A new anaerobic, sporing, acetate-oxidizing, sulfate-reducing bacterium, *Desulfotomaculum* (emend.) *acetoxidans. Arch. Microbiol.* **112,** 119–22.

Winfrey M.R. & Zeikus J.G. (1977) Effect of sulfate on carbon and electron flow during microbial methanogenesis in freshwater sediments. *Appl. Environ. Microbiol.* **33,** 275–81.

Volkov I.I., Rozanov A.G., Zjabina N.N. & Jagodinskaja T.A. (1972) Sulfur in sediments of the Pacific Ocean east of Japan (in Russian). *Lithology of Mineral Resources* **4,** 50–64.

Volkov I.I., Rozanov A.G., Zjabina N.N. & Fomina L.S. (1976) Sulfur compounds in sediments from the Gulf of California and adjacent parts of the Pacific Ocean (in Russian). In *Biogeochemistry of the diagenesis in marine sediments*, pp. 136–70. Nauka, Moscow.

Vosjan J.H. (1975) Respiration and fermentation of the sulphate-reducing bacterium *Desulfovibrio desulfuricans* in a continuous culture. *Plant Soil* **43,** 141–52.

ZoBell C.E. (1963) Organic geochemistry of sulfur. In *Organic Geochemistry* (Ed. I.A.Breger), pp. 543–78. Pergamon Press, Oxford.

CHAPTER 5
THE MICROBIAL SILICA CYCLE

WOLFGANG E. KRUMBEIN (SECTIONS 5.1–5.3) AND DIETRICH WERNER (SECTION 5.4)

> In the upper lithosphere, oxygen is the first and silicon the second most abundant element.
>
> Zajic (1969)

> Silicon constitutes 28% of the elemental composition of the earth's crust and is the most abundant crustal element after oxygen.
>
> Heinen and Oehler (1979)

> The element silicon is one of the most abundant in the earth's crust, ranking second only to oxygen.
>
> Ehrlich (personal communication)

> The most abundant elements of the lithosphere are silicon and aluminium; and the most abundant minerals are alumosilicates.
>
> Eckhardt (1979)

> Le quartz est un des minéraux les plus importants de l'écorce terrestre
>
> Krumbein (1972)

> Man ist heute allgemein der Meinung, daß das Silizium keine wesentliche Rolle in den Lebensprozessen der Organismen spielt.
>
> Voronkov *et al.* (1975)

5.1 INTRODUCTION

It is difficult to start a chapter on the microbial geochemistry of silicon with anything other than the above quotations. Reviewing the few textbooks on microbial cycling of minerals, on geomicrobiology or microbial ecology it appears that the most important element on earth besides the gaseous ones such as oxygen, hydrogen and nitrogen is often neglected and altogether not well understood in its relations to the living world, with the exception of one organism type, i.e. diatoms.

However, from its position in the periodical system of elements, from its abundance and its importance in many man-made and natural substances used by human society, silicon is extremely important. It is biologically transferred and altered and it influences microbial systems as well as macrophytes and animals. The position of silicon in the periodic system of elements brings it clearly into relation with carbon. Its valence states are the same (+2, +4 and −4).

Silicon forms tetrahedral configurations similar to phosphorus and double bonds similar to carbon are possible though perhaps under different temperature and pressure regimes. High molecular weight and optical active as well as ligand active compounds are possible though they cost more energy and are of different structural capacities. Chains, nets and helical structures may also be built with silicon.

Siloxane polymers may have structural capacities comparable to many complex carbon compounds. Consequently, there have been many theories about life on a silicon basis on planets with higher temperatures and pressures as well as about possibilities of life on a silicon basis under planetary conditions like those on earth, inasmuch as the character of the primordial earth was formed by silicon compounds rather than by carbon compounds. The fact that some minor organisms form siliceous, and other carbonate skeletal compounds has been forwarded as evidence for the theory of a competitive development of life on earth based either on carbon or on silicon. Our present knowledge of evolution does not yet allow the complete exclusion of such theories but the mathematical and physico-chemical extrapolations for the evolution of life relatively clearly exclude silicon as a real basis of life in so far as the development of organic compounds such as nucleic acids, membrane substances and enzymes indicates that carbon very early in evolution controlled the stability of living systems and thereby established its rule also over the physico-chemical environment of life (Lovelock 1979). Lovelock, however, also states that we know very little about the first two aeons (1 aeon represents 1000 million years) of life on earth. Therefore we may still expect many surprising findings and possibilities which may have been extinguished later by the ties which build up through the increasing control of 'carbonic life' on the geochemical cycles on earth.

Another quite important fact in the question of the relationship of silicon and carbon in the evolution of life is the theory of matrices necessary for duplication and copies. It has been suggested that silica gels as well as clay minerals and other silica surfaces are excellent matrices with 'memory' capacities (Matheja & Degeus 1971). In addition, optical characteristics such as left- or right-turning selectivity may occur with silicon compounds as well as with those of carbon. Some of the related experiments, have recently been criticized, however (Youatt & Brown 1981).

Thus on the one hand silicon is very similar to carbon in its capacities and potential and on the other hand it is one of the least reactive elements in terms of its compounds presently important for geochemical cycles. Quartz, the dioxide of silicon, is one of the most stable compounds on earth. Its hardness, its acid and base stability are matched by few other minerals. Among these, however, we also find a carbon compound, the diamond.

Speculations about the biological history of silicon are numerous and they have achieved recognition in science fiction novels.

5.2 GEOCHEMISTRY

Speaking of our planet of today and of the relations between silicon and carbon and silicon and life as they exist now we have little evidence to suspect findings related to evolutionary and extraterrestrial speculations. The element silicon is (possibly related to the biologically controlled cycle of oxygen) most stably bound in the earth's crust in one structural form of silicates, i.e. silicon dioxide or quartz. Silicates are minerals and compounds which contain silicon in tetrahedral configuration with oxygen, namely SiO_4. These may be correlated in the following forms.

(1) If a silicon tetrahedron is correlated only via cations and no direct forces are acting between the silica tetrahedra we speak of nesosilicates (e.g. zircon, olivine, garnets); biological degradation is fast.

(2) The tetrahedra may also be correlated via one of the four oxygen atoms. In this case isolated small units of several tetrahedra may form which are called sorosilicas. These occur rarely in nature.

(3) Since only binding on edges of tetrahedra occurs, these may be loosely linked to form a ring of often six tetrahedra bound via oxygen. These ring silicates are also called cyclosilicates (e.g. beryl). The degradation by biological attack is relatively slow.

(4) If the silica tetrahedra are combined into long chains or bands they are classified as inosilicates (e.g. chains = pyroxenes; bands = amphiboles). Both mineral types undergo biodegradation relatively rapidly.

(5) (a) If chains or bands are combined in one plane into large sheets, they are termed phyllosilicates (e.g. micas, biotite, muscovite or the clay mineral montmorillonite). In the latter case two layers are held together by octahedral coordinated cations. (b) In another important group only one tetrahedral and one octahedral layer exist and they are bound together and stacked. The main representative of this group is kaolinite. Biodegradation is very fast in cases of high concentrations of iron, potassium and/or calcium in the cationic groups.

(6) Finally the tectosilicates form an important group to which the most abundant minerals belong. In this case all tetrahedral corners share a common oxygen atom. All feldspars belong to this class of silicates. Also the modifications of SiO_2 (quartz, tridymite and cristobalite) when crystallized expose the tectosilicate type of lattice. Less-defined forms of silica in which a

crystal lattice is not clearly developed, micro- or subcrystalline or solidified from the molten stage in a very short time, embrace siliceous compounds as glasses, opal, chert, silica gel, petrified wood, etc. Many of the less well-defined crystalline forms of silica are biogenic or related biological processes in the natural environment. Biodegradation of tectosilicates is clearly cation related. Potassium- and calcium-rich silicates undergo biodegradation and biotransformation.

Another important factor which combines biological and physico-chemical parameters of the cycle of material is the solubility of silicon, silica and silicates. In contrast to many other chemical compounds of the earth's crust below a pH of 9.0 silicon is present in solution only as silicic acid in the monomeric stage (H_4SiO_4). At pH above 9.0 it dissociates to $H_3SiO_4^-$ and $H_2SiO_4^{2-}$. Since the pH of most natural environments is below pH 9.0, most of the reactions involving silicon are ruled by the chemistry of silicic acid. The concentration in rivers and lakes is between 10^{-3} and 10^{-5}M and the average is about 2.2×10^{-4}M in rivers. In the ocean the range is lower (2×10^{-4} to 10^{-6}M). The physico-chemical solubility of silica is ruled by the following reaction:

$$SiO_2 \text{ (amorph)} + 2\,H_2O \rightleftharpoons H_4SiO_4 \text{ (aqueous)} \qquad (5.1)$$

The constant at 25 °C being $K = 2 \times 10^{-3}$.

Table 5.1 gives an impression of the solubilities at different temperatures.

The polymerization of dissolved silica can be explained biologically, but also physico-chemical reactions may lead to polymerized silica. Generally hydrolytic reactions can easily be catalysed by several groups.

$$2\left(\begin{matrix} & OH & \\ & | & \\ OH- & Si & -OH \\ & | & \\ & OH & \end{matrix}\right) \rightarrow \left(\begin{matrix} & OH & & OH & \\ & | & & | & \\ OH- & Si & -O- & Si & -OH \\ & | & & | & \\ & OH & & OH & \end{matrix}\right) + H_2O. \qquad (5.2)$$

The occurrence of siloxanes is thus a very probable reaction in the natural environment. Silanes and silicones are apparently compounds which only

Table 5.1. Solubility of silica in water at pH lower than 9.0.

Temperature	SiO_2 (p/10^6	$^mH_4SiO_4 \times 10^{-3}$)
0 °C	60–80	1.0–1.3
25 °C	100–140	1.7–2.3
90 °C	300–380	5.0–6.3

occur in technical products or in specific biological reactions. The same seems to be the case with its ethers and esters and other organic silicon compounds.

Due to the fact that silicates and silicon dioxide (as quartz) are the most stable minerals in terms of abrasion, corrosion and solubility in the natural environment geoscientists have tended for very long periods to regard silicon as an abiogenic element. However, they always made one exception, i.e. the behaviour of silicon in the marine environment where its presence and metabolism are closely related to and absolutely ruled by mainly two classes of microorganisms, namely the radiolarians and the diatoms, named in order of their apparent evolutionary arrival in the oceanic sediments. Textbooks on the exogenic cycle of material (e.g. Carroll 1970) usually stress the inorganic chemistry of corrosion rather than the numerous processes of a biological nature involved (see Chapter 8). Thus the silicon cycle in nature is still encrusted in classical approaches. The correct approach of biological control of the exogenic cycle, long established for elements such as sulphur, nitrogen, carbon, iron, calcium and manganese, has to be better established.

5.3 MICROORGANISMS AND THE SILICA CYCLE

As with many other elements a purely physico-chemical view of the silicon cycle is impossible on the earth's surface. All steps and stages of low-tem-

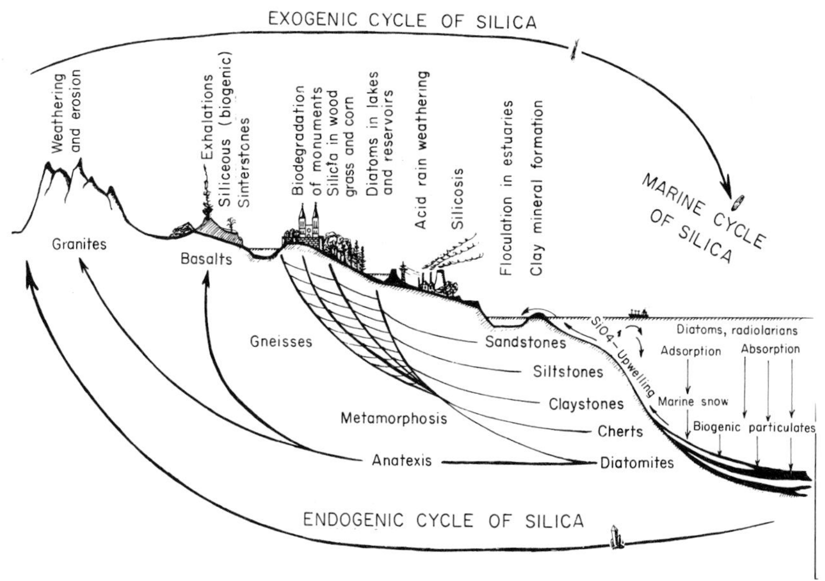

Fig. 5.1. The silica cycle in nature and some of the associated microbial and biogeochemical processes.

perature reactions of silicon in nature are controlled more or less to completeness by biological and mainly microbial pathways and interactions.

Figure 5.1 gives an impression of the interactions of microorganisms with the silicon cycle. Lauwers and Heinen (1974), Berthelin (1976; see Chapter 8), Heinen and Oehler (1979) and Silverman (1979) have summarized much of the recent work on the microbial cycle of silicon in nature. It appears that almost any mineral encountered at the earth's surface will be attacked by, for example, bacteria, fungi or lichens to deliver cations and anions embedded in the mineral lattices (Jackson & Keller 1970). Thus the destruction and degradation of silicates are often directly controlled by microbial activity. Indirect control is exerted by the general activity of organisms yielding, for example, organic acids or complexing compounds which attack rocks and soil minerals. The presence or absence of oxygen, nitrate, sulphate and other radicals of importance for the surface cycle of elements including silicon in the atmosphere and their equilibrium with the atmosphere in aqueous solution is clearly a combined effect of the activities of microorganisms and macroorganisms.

It is evident, therefore, that the silicon cycle too is ruled by biological activities. In respect to the detailed study of such influences and interactions we encounter, however, more complications than in the case of other important elements and their minerals. Some of the specific reactions involved remain to be studied in detail.

5.3.1 Methods of study

In this section we deal exclusively with prokaryotic microorganisms and the eukaryotic group of fungi and lichens. The important group of diatoms is treated separately (cf. Chapter 5, section 5.4).

The interaction of minerals with microorganisms in respect to silicon is relatively difficult to study inasmuch as silicon is a very abundant element and in wide use in laboratory equipment. In order to define quantitatively interactions of silicon and microorganisms either protective measures have to be taken for work with glassware or laboratory equipment for microbiological work has to be made up completely of plastic materials, including equipment used for the storage and preparation of culture media and analytical solutions. Silicone rubber has to be avoided. Classical methods of soil microbiology and geomicrobiology did not ensure against this problem and often results were biased because of this.

Several fields of study can be defined in respect of silicon and silicates. The intimate contact of microorganisms with the mineral is usually studied by colonization experiments which then are controlled by fixation embedding methods and transmission electron microscopy (TEM) of peels (e.g.

Pohlmann & Oberlies 1960) or direct examination of the surfaces and the microorganisms attached to surfaces (Rades-Rohkohl 1975, Rades-Rohkohl *et al.* 1979) using scanning electron microscopy (SEM).

Cultural work includes the isolation of microorganisms able to furnish their mineral nutrition from silica minerals (Smyk 1970). Plate techniques where direct contact is made between the mineral and the microorganism are in wide use (Webley *et al.* 1960). Indirect influences by exoenzymes and/or acids and chelating compounds excreted into the environment are studied by using dialysis bags, for example (Eno & Reuszer 1951, 1955). Another approach to the question of weathering and mineral formation is directly related to the corrosion stability and solution equilibrium of many minerals. Therefore, many experiments were done using soil or mineral perfusion columns or percolators in which the medium of culture is circulated through glass or plastic columns filled with the mineral being studied or with soil and rock particles. Much attention has been given in such experiments to questions of: (1) reactive surface area as related to particle size; (2) to mobilization processes modified by immediate immobilization by chelate formation or precipitation of fresh minerals, and also (3) by the immediate transformation of one mineral into another one (Zajic 1969, Berthelin 1976; see also Chapter 8).

Another possible method of studying microbial influences on the silicon cycle is the analytical approach in which the amount and form of silicon found in organisms is determined (e.g. Holzapfel & Engel 1954, Heinen 1967).

The interaction between clay minerals and microorganisms in soil is a sorption/solution and a nutritional phenomenon. It is also related to the question of reduced metabolic activity and survival of vegetative cells inasmuch as attached microorganisms seem to have better growth yields than motile microorganisms and that microbial populations in the water column are always higher in the presence of particles than without (Filip 1975, Marshall 1976). Therefore many studies have been concentrating on the effect of solid surfaces on the survival and activity of microorganisms. In this type of study it is very difficult to include consideration of the complicated chemistry of sorption of nutrients to reactive surfaces as compared with pure solutions (e.g. Lorenz *et al.* 1981).

Filip (1975), for example, has shown clearly that the degradation of sugars by soil microorganisms is enhanced in the presence of many clay minerals. Van der Waal's forces are certainly active in this process and the structure of clay minerals with the possibility of forming organic mixed layers may largely enhance the interaction of microorganisms, nutrients and mineral surfaces. In another case, it was demonstrated that the adsorption of DNA to quartz grains protects it from the decomposition by nucleases

(Lorenz *et al.* 1981). These processes may be studied by using different silicon mineral suspensions or columns and glass beads and nylon or activated charcoal particles as a reference system. Microbial methods also involve the exchange of silicon when it is offered instead of phosphorus (Holzapfel & Engel 1954, Heinen 1960, Lauwers & Heinen 1974). In this case it was shown that under artificially produced stress conditions silicon can replace phosphorus in the cell and can be exchanged for phosphorus. This process, however, leads to the inability of the bacterial cell to divide.

Several experiments have been done to show the formation of organosilicic compounds similar to carbon and phosphorus compounds. Fractionate centrifugation and analysis of cell fractions may demonstrate that silicon can be enriched in the cytoplasm fraction, the membrane fraction and the cell wall and extracellular capsular material fractions. The compounds which are formed in the different fractions are not always clearly defined or analysed. By deduction and analytical methods it was concluded that microorganisms have the capability of incorporating silicon in various organic compounds forming C—O—Si linkages as well as more complicated structures. Some of these must play an important role in silification of micro- and macroorganisms.

Another method of studying the interaction of siliceous compounds and microorganisms is by direct microscopy using SEM methods combined with energy dispersive X-ray analysis or STEM-EDX combinations. These methods enable the direct observation of metal and cation transfer from silicates to microorganisms and related solubilization processes.

Enzymes specifically active and reactive in respect to silica could be enriched and purified, but so far no reports on such work seem to exist at least for bacteria.

5.3.2 Biodegradation of silicon compounds

Most of the problems in the degradation of silicon compounds are considered in Chapter 8. One of the most fascinating and yet not completely solved problems in the interaction of silicates and microorganisms is the question whether direct microbial attack can cause the breakdown of quartz and glass. Some progress has been made in this field by Oberlies and Pohlmann (1958) who studied the microbial attack on optical glasses. Berthelin (1976) and Silverman (1979) have summarized much of the literature on the microbial attack on minerals and the effect of weathering on materials, especially important in the case of works of art (e.g. Krumbein 1972, Eckhardt 1979).

Glass in the form of medieval stained glass windows or precious glass objects which were embedded in soil at archaeological sites is a special case of such destructive processes inasmuch as glass itself is unstable and has the physical tendency to crystallize and then to decompose. Recent work in France (Bauer, personal communication) and work in our laboratory (unpublished data) have produced strong evidence that antique glasses in medieval windows as well as in Phoenician and Roman glass objects are destroyed much faster by microbial attack than by mere physical and chemical processes. It is, however, difficult to define exactly the biological part and the inorganic part in these weathering processes. Figures 5.2 to 5.6 give an impression of the activity of microorganisms in the destruction of glass. This destruction can be classified into mechanical damage (Figs 5.3

Fig. 5.2. Part of an medieval glass window with typical lesions caused by biodeterioration through lichens.

and 5.4), chemical damage (Figs 5.5 and 5.6) and combinations of the two processes. Rades-Rohkohl (1975) has shown that microorganisms attach themselves relatively closely to quartz surfaces but she was not able to demonstrate microbial solubilization of quartz (Fig. 5.7). Silverman (1979) gives several references on the microbial attack on quartz but none of these give clear evidence of microbial solubilization. Another set of data on the solubilization of quartz was collected by Krumbein (1969) and Krumbein and Jens (1981) in a study on the dynamic equilibrium of microbial weathering and rock quality. A pattern of response was found in which 'microbial solution fronts' are penetrating into solid rock according to its mineralogical, chemical and petrographic parameters. The softest rocks

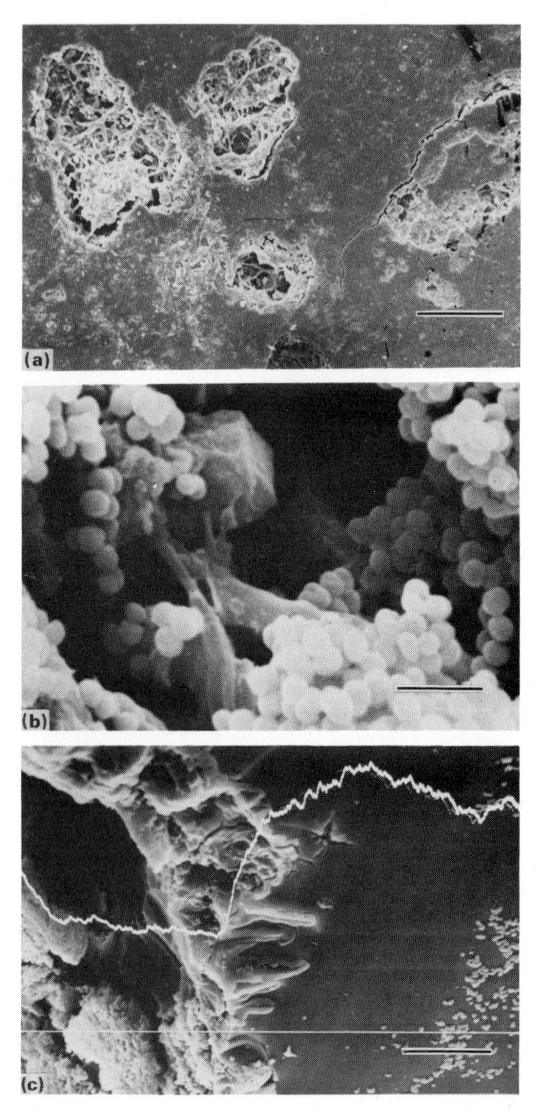

Fig. 5.3. SEM-micrographs and energy dispersive X-ray analysis of lesions on antique glass caused by lichens. (a) erosion pits filled with fungal mycelium (bar = 100 μm); (b) coccoid bacteria growing up to thick carpets after 24 h incubation of glass chips in YP-medium. The strain was isolated and was found to enrich for silver and to be extremely silver tolerant (bar = 1 μm; (c) erosional pit filled with fungal mycelium and cocci after 24 h incubation analysed for potassium. Potassium, iron and manganese and in some cases phosphorus and silver concentrations usually decrease considerably within the biogenic pits. The bacteria mobilize more of these elements than of the silica. The white baseline is the place of analysis, the curve gives relative abundance of potassium. Potassium increases after decrease towards the centre of microbial growth (K-uptake).

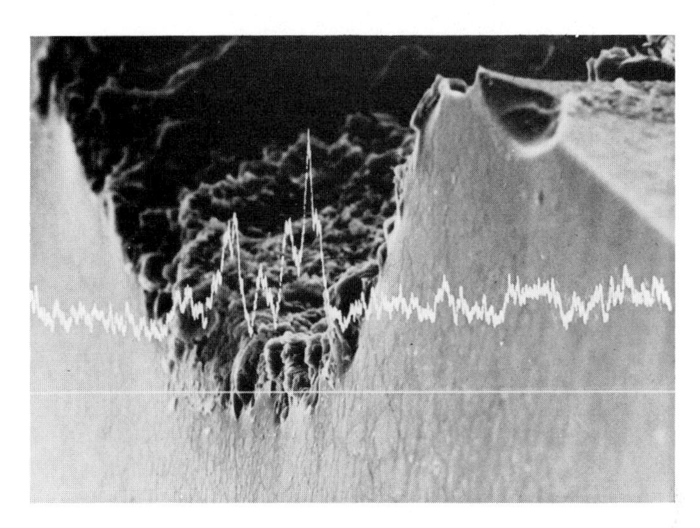

Fig. 5.4. Si-linescan of lichen-produced pits in antique glass. The ancient glass was a mixture of several different compounds with silica concentrations of up to 50%. The biological degradation selectively mobilizes the associated compounds such as iron, manganese, potassium and phosphorus. Silica remains and thus is enriched in the decaying glass remainders.

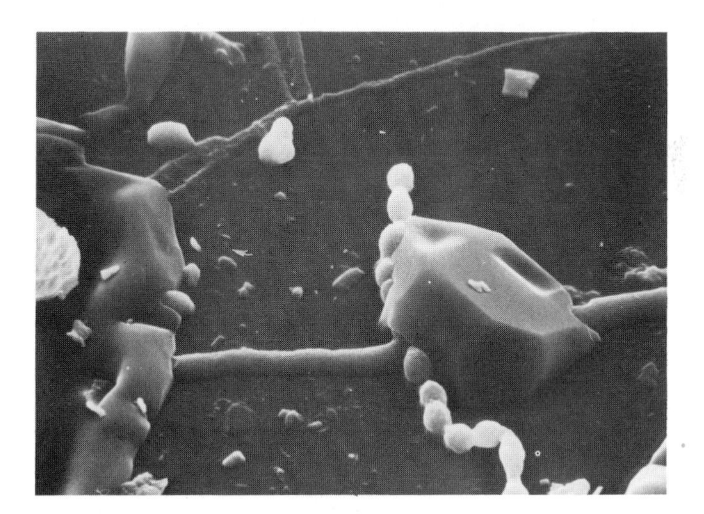

Fig. 5.5. Mechanical and chemical damage on a phoenician glass bottle from an excavation in Israel. The bottle was embedded in soil and the soil flora attacked the glass. Fungal mycelia covered the glass, penetrated between layers and impurities. The glass surface is etched (e.g. lower left) and mechanical forces destroyed the glass (chlamydospores).

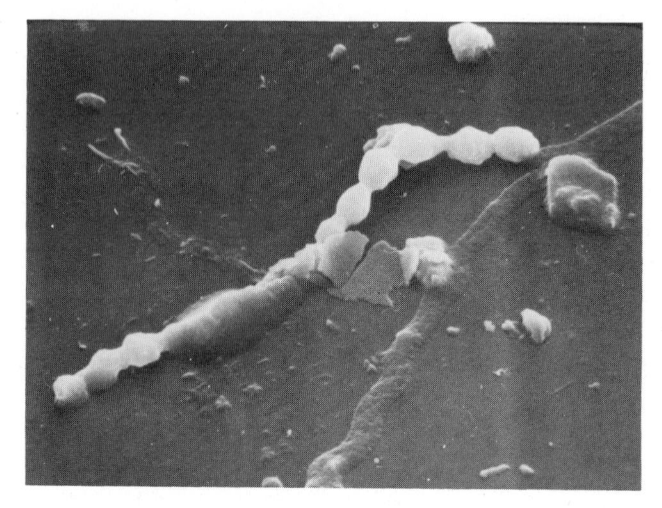

Fig. 5.6. Fungal mycelium with vegetative parts and chlamydospores. The chlamydospores have penetrated underneath the surface and formed cracks and mechanical damage. The vegetative mycelium has floated apart. This is not caused by fixation. The samples have been rewetted and fixed in glutaraldehyde. Probably the mycelium was already decaying at the time of sampling.

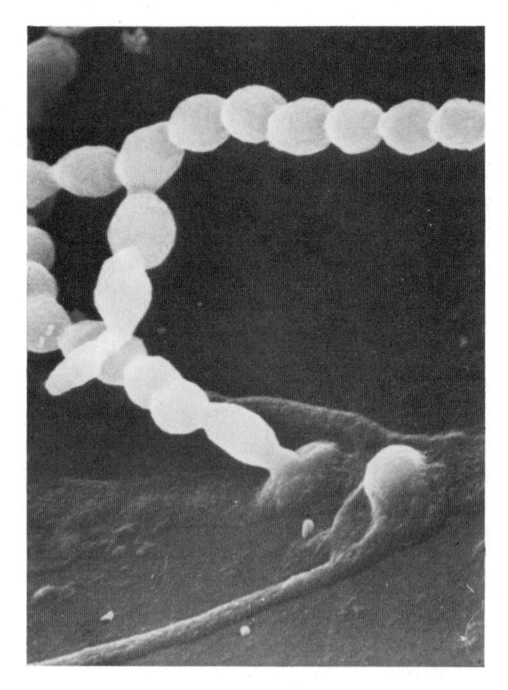

Fig. 5.7. Attachment of fungal mycelium to a glass surface, which had previously been pitted by former colonization or by physico-chemical deterioration.

attacked were limestones, the hardest ones cryptocrystalline quartz in the form of cherts. The study was done in desert areas for two main reasons.

(1) Desert and mountainous regions rarely develop modifying soil covers and the interaction of microorganisms and rock is not biased by the complex soil structure phenomena.

(2) Stress conditions under high evaporation rates and high sun irradiation force biological communities to develop protective patterns. These may be:

(a) biological crusts (epilithic lichens, mosses, etc.);

(b) mineralogical crusts (biogenic desert varnishes or mineral dust aggregation by fungal mycelia and/or cyanobacterial slimes);

(c) rock penetration and invasion by chasmolithic and/or endolithic boring or penetrating microflores (Krumbein 1969, 1971, Friedmann & Galun 1974).

Figures 5.8 and 5.9 show endolithic boring patterns of cyanobacteria, while Figs 5.10 and 5.11 show endolithic lichen boring patterns. Special cases of crust development partially producing pits, partially exfoliation patterns, are shown in Figs 5.12 and 5.13. These dynamic processes in desert rocks have been demonstrated in the Negev Desert close to the ancient site of Avdat where Eocene limestones are gradually silicified to silica concentrations of up to 99%. In the case of totally silicified rocks microcrystalline quartz is the mineral which is exposed to the attack of lichens and independent cyanobacteria, fungi and chemoorganotrophs. The boring pattern by endolithic lichens (Fig. 5.10 in chert; Fig. 5.13 in partially silicified carbonate) by epi-endolithic lichens (Fig. 5.12) and epilithic lichens (Fig. 5.11) is different. It may however occur on one and the same kind of rock in relation to micromorphology, sun exposition, light penetration into the rock and micro-climate conditions. Krumbein and Jens (1981) have demonstrated that the pattern of mineral solution and rock decomposition is associated with mineral deposition and rock-varnish formation. The factors regulating these quartz- and chert-attacking and crust-forming microbial ecosystems have not yet been totally understood.

Thus, at least for microcrystalline quartz it can clearly be shown that silica is mobilized and dissolved under natural conditions by a variety of microorganisms.

The diagram in Fig. 5.14 illustrates the dynamic equilibrium between boring, crust formation and protecting 'desert varnish' or better 'rock varnish' layers forming under the influence of microbial populations on weathering siliceous and silicate rocks. The equilibrium and/or crust formation can be envisaged as follows. Soft rocks are attacked easily by microorganisms and the weathering speed is considerable. In these cases the rock is decaying fast, often exposing a smooth surface or mineral grains loosely attached to the surface which are washed away by desert floods or

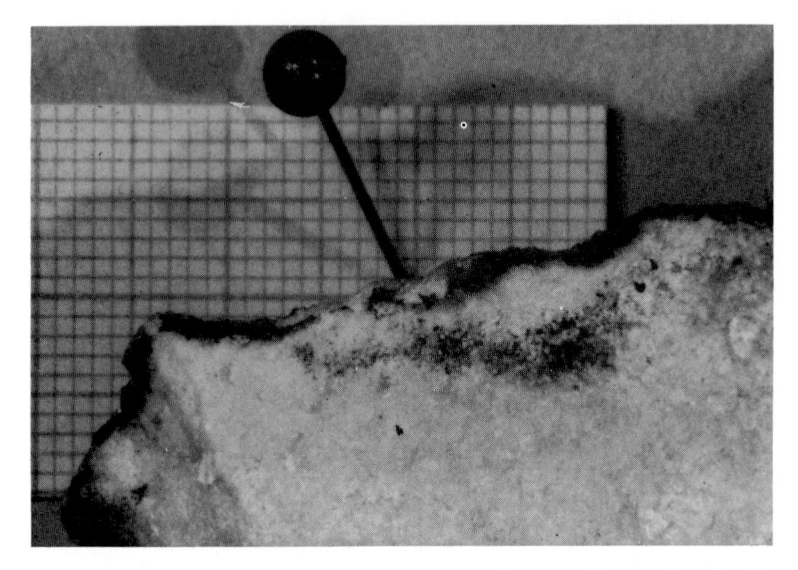

Fig. 5.8. Partially silicified Eocene limestone reef of Avdat (Negev, Israel). The silica concentration is 67%. The outer parts of the specimen are coated with biogenic desert varnish (Krumbein & Jens 1981). The dark cavernous zone indicates the site of a colonization by endolithic coccoid cyanobacteria producing large cavities. The dark colour is caused by green photosynthetic pigment. In black and white photographs black Mn-Fe coatings cannot be distinguished from green chlorophyll.

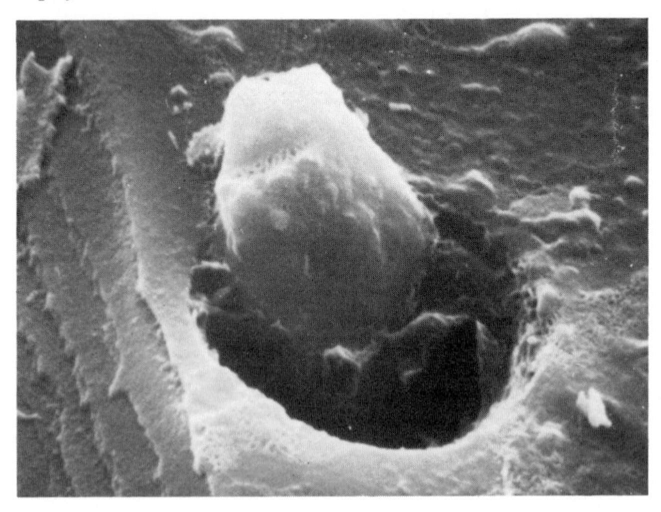

Fig. 5.9. A totally silicified mollusc shell (see Fig. 5.14) is penetrated by a filamentous cyanobacterium with its slime adhering to the cavity which extends about 250 μm into the rock. The lamination is caused by the preserved laminated carbonate of the initial shell which later was dolomitized and then completely silicified during dedolomitization.

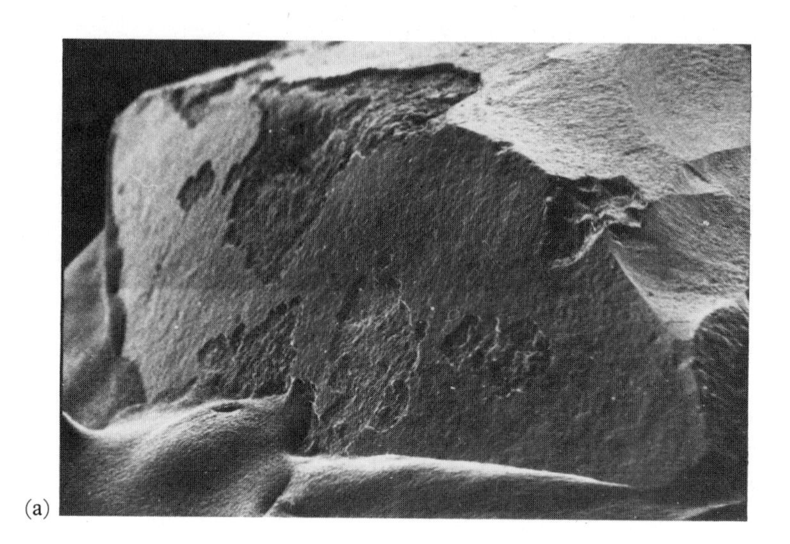

(a)

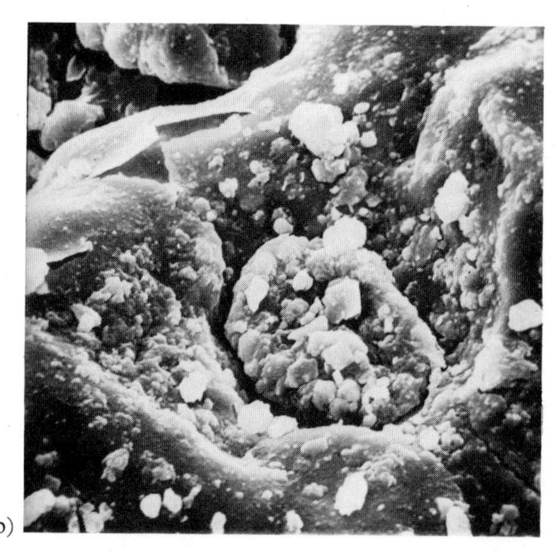

(b)

Fig. 5.10. (a) Neolithic tool found in the desert plains of the Negev near the Nabbatean city of Avdat. The tool consists of flint (chert) of the Mishash flint of Cretaceous age. It is bored by an endolithic lichen with the typical colony-sized rounded pits occurring also sometimes on limestones. (b) Section of (a) enlarged considerably to expose one of the deep holes filled by a fruiting body of the endolithic lichen. These tiny black spots are the most conspicuous indications of biological activity of endolithic lichens (see Fig. 5.12b, c).

Fig. 5.11. Epilithic lichen (one of the cases of brush-type ramified lichens) on dark-banded chert (i.e. microcystalline quartz). The epilith has formed a disc-shaped pit of 1 cm diameter and 3–4 mm depth in the hard chert material.

Fig. 5.12. (a) Partially lithified Eocene limestone-reef of Avdat (Negev). The nodular black material is part of a coralline algal colony which was selectively silicified to 100%, while the surrounding detrital material was silicified only to 35% silica concentration (see Fig. 5.14). The carbonate-containing material is covered by a dense carpet of epi-endolithic lichens. The totally silicified material is covered by a rock varnish coat of 200 μm of Mn-Fe-hydroxides of biogenic nature.

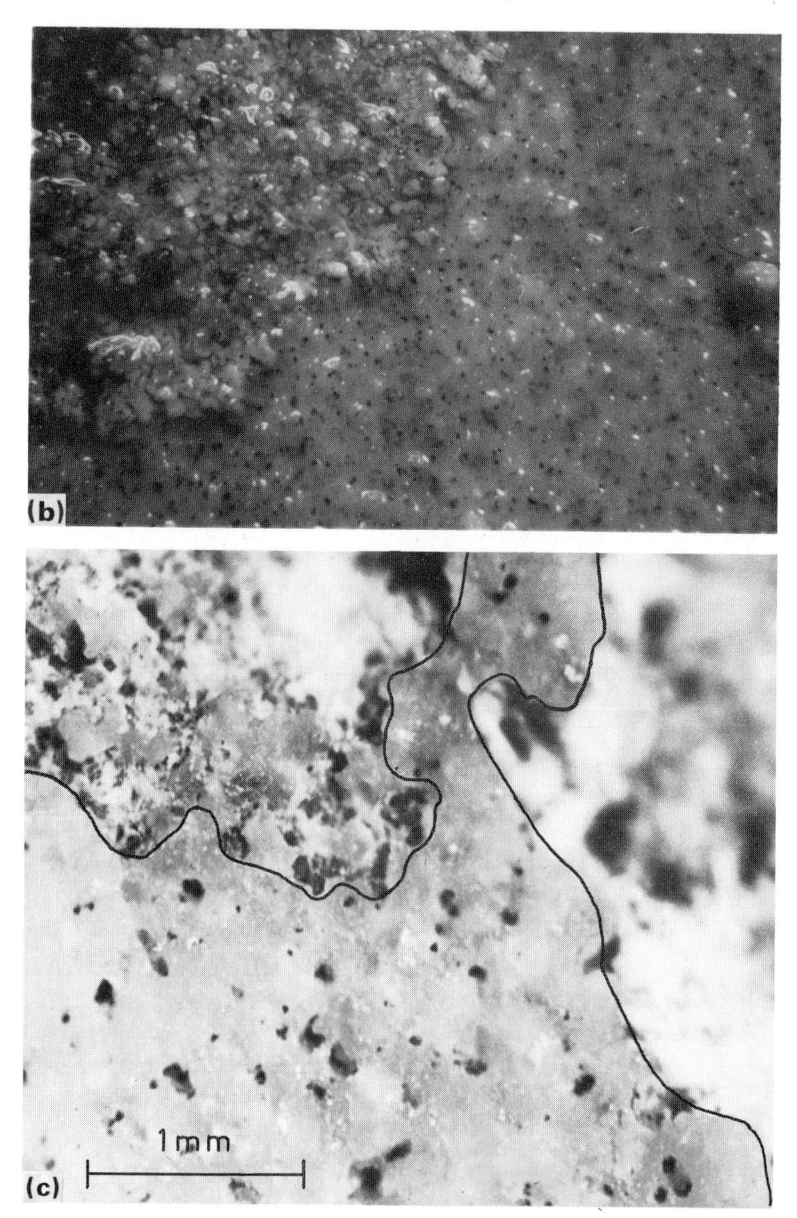

It protrudes from the biological solution front produced by the epi-endolith. (b) Enlargement of a section of the lichen, exhibiting the typical crustose pattern with Apothecia and the black spots of the endolithic part showing the boring activity. (c) In still higher magnification the pocket-like bore-holes of the associated alga and fungi are exhibited. The dark shadows in the holes are produced by photosynthetic pigments. In this case an almost horizontal surface area is worn down by boring activity almost into a levelled plain.

Fig. 5.13. In contrast to the epi-endolith of Fig. 5.12 this crustose endolithic lichen is distributed in a patchy way over the entire rock and produces boreholes almost as deep as the diameter of each individual colony. This is the first initial step to the pitting action by endoliths which produces a very sharp-edged undulating surface with an extremely fast and deep destruction. The eroding front of this kind of biodegradation of partially silicified rock advances very fast until it hits completely silicified parts where it comes to a standstill for some time. Eventually the endolithic flora breaks through the chert surface as described in Fig. 5.14 or as shown in Fig. 5.10 (a).

carried away by wind action. Harder rocks will be bored, drilled or pitted and a very rough, often darkish surface pattern develops which exhibits many small borings on closer view. Sometimes iron and manganese are immediately redeposited in the pores and cavities created by the penetrating microflora, thus yielding an endolithic crust formation. Still harder (often silicified) rocks are rarely bored and if so, not very deeply. In these cases the microflora cannot penetrate deeper into the rock and a protective layer of desert varnish is formed which often is made up of redeposited iron and manganese oxides which have been mobilized and fractionated by micro-organisms. Under such extreme conditions it can be shown that micro-organisms, among them many endolithic fungi more or less loosely associated with cyanobacteria, dissolve quartz and create niches in which they may survive the extreme desert conditions.

Colloidal forms of silica seem to play an important role in these processes. Often such processes are associated with colour changes of the silicified rocks. The chert itself is often changing in colour from grey or black into white and in cases where enough iron and manganese are present

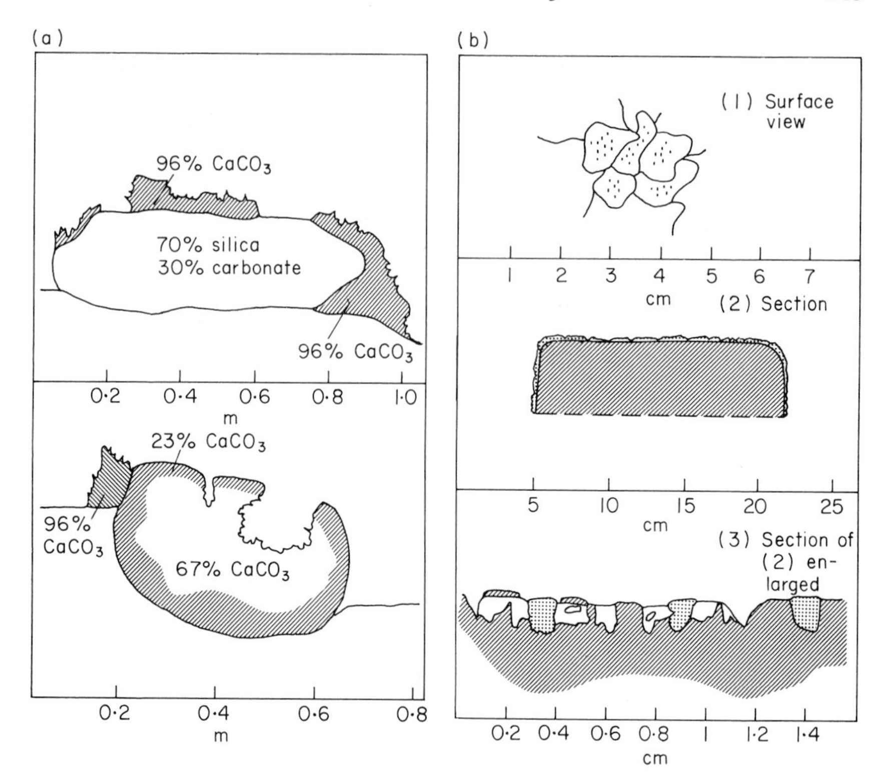

Fig. 5.14. Diagram of some of the observed boring and destruction activities on partially and totally silicified limestone of the eocene reefs of Avdat (Negev). (a) Pitting endolithic lichen are practically eating up the 96% limestone material until they reach the 70% silica part. A smooth surface is produced by the difference in hardness and by biogenic desert varnish. The activity of the endoliths after some time can break through partially silicified rock and form deep cavities corresponding to the silica concentration of the rock. Sections through the clear, white non-eroded rock material do not show any difference in material. Only chemical analyses and the lichen activity reveals the whole story. (b) Demonstrates the erosive action of epi-endoliths as shown in the micrographs of Fig. 5.12. The speed of erosion is very fast when the rock is only slightly silicified. It decreases as the silica content increases and the crustose lichen is replaced by an endolith when silica concentration or other conditions (climate, weather, precipitation, night thaw conditions) exclude epilithic crustose forms (redrawn from Krumbein 1969).

or enriched by the microflora, fresh brown or black thin layers and coatings may redeposit on the 'bleached' chert. It has been suggested by several geologists that desert opal in the Australian, African and Asian deserts is also produced by biological solubilization and precipitation of silica. So far no cultural or biochemical experiments have been reported regarding this question.

5.3.3 Formation of silicates and siliceous minerals

It was already suggested that some microorganisms may have the potential to enrich and precipitate silicon or silica to an extent that silicate minerals are formed. Heinen and Oehler (1979) have reported several mechanisms in microorganisms and plants by which silica is enriched and finally precipitated as colloidalamorphous silica. Possibly organic intermediate forms of silica are produced by incorporation of dissolved silicic acid or by direct uptake from or dissolution of silica.

Theoretical reactions involved have been presented by Voronkov *et al.* (1975) and by Heinen and Oehler (1979). They include Si—Si, Si—C, Si—O—C and Si—O—Si linkages. Polymeric siloxanes are theoretically deducible and some of them have been identified in organic compounds. Carboxyl and hydroxyl groups may be involved leading to patterns such as those in equations (5.3), (5.4) and (5.5).

$$\text{(adenine, }NH_2\text{)–(ribose, OH OH)–}CH_2\text{—O—}\overset{O}{\underset{O}{P}}\text{—O—}\overset{O}{\underset{O}{P}}\text{—OH} + Si(OH)_4 \longrightarrow \text{(adenine, }NH_2\text{)–(ribose)–}CH_2\text{—O—}\overset{O}{\underset{O}{P}}\text{—O—}\overset{O}{\underset{O}{P}}\text{=O—}\overset{O}{\underset{O}{Si}}\text{—OH} + H_2O \qquad (5.3)$$

Carboxylic acids which are important in rock and silica weathering (e.g. formic acid, malonic acid, oxalic acid, gluconic acid) may also react with silica:

$$R\text{—}C\overset{/\!/O}{\underset{\backslash OH}{}} + Si(OH)_4 \longrightarrow R\text{—}C\overset{/\!/O}{\underset{\backslash O\text{—}\overset{OH}{\underset{OH}{Si}}\text{—OH}}{}} + H_2O \qquad (5.4)$$

In addition the hydroxyl group of many organic compounds can possibly react with silica:

$$R\text{—}\overset{H}{\underset{H}{C}}\text{—OH} + Si(OH)_4 \longrightarrow R\text{—}\overset{O}{\underset{O}{C}}\text{—O—}\overset{OH}{\underset{OH}{Si}}\text{—OH} + H_2O \qquad (5.5)$$

Several of these reactions may be responsible for toxic effects of organic siliceous compounds in use in chemistry and medication. The enrichment of silica by cyanobacteria, fungi and chemoorganotrophic bacteria, which has been demonstrated by Glazovskaya (1950), Holzapfel and Engel (1954), Heinen (1967), Jackson and Keller (1970) and Berthelin (1976), can easily

lead to the precipitation of noteworthy amounts of amorphous minerals. Phytoliths, plant opal and desert opal are classical examples of such processes. Reports on such mineral precipitates are often troublesome to interpret since the analysis of the siliceous minerals is frequently done after ignition of the organic material which may lead to crystallization via heat. Wilding and Drees (1974), however, seem to have clear evidence of the formation of quartz and cristobalite at low temperatures, at least for macrophytes. In most cases silicon is associated with other elements as is the case with many silica minerals formed under volcanic or magmatic conditions (Oehler 1979).

Berthelin (1976) points out that microbial processes lead to a sequence of transformation of minerals in which the silica texture is only partially transformed by microbial attack. Recrystallization takes place without passing through a truly soluble form. It is assumed that much of the transformation of feldspars and micas under tropical and temperate climate microbial weathering leads to 'biogenic' minerals by leaching of cations and, therefore, enriching of aluminium and silicon in relation to potassium, sodium, magnesium, iron, manganese, etc. These processes yield bauxitic alumo-silicates from granite and other silica rocks.

$$\text{Biotitie} \; \frac{+\,\text{bacteria}}{+\,\text{organic energy source}} \rightarrow \text{vermiculite} \rightarrow \text{kaolinite.} \quad (5.6)$$

An example of direct biological precipitation of quartz from solution studied by us is the silicification of cyanobacteria in a hydrothermal environment. During a visit to the Lake View hot spring deposits in Oregon, samples were collected of a white mineral aggregate which initially was smooth and slimy and was intimately mixed with cyanobacteria. During accumulation it hardened considerably and finally formed greenish-white, extremely hard layers. Figs 5.15, 5.16, 5.17 and 5.18 give an impression of the crust which formed on the borders of the spring. It turned out to be a siliceous biogenic stromatolite. X-ray diffraction analyses on a Philips diffractometer, carried out in our laboratory, yielded quartz signals. Walter (1976) has described inorganic siliceous stromatolites from the Yellowstone area. In the case of the Lake View samples, SEM analyses combined with EDX clearly indicate that the initially low silicon concentration in the spring water was enriched in the cell material and envelopes of the cyanobacteria growing in the hot water. Silicon concentrations were initially higher in the cell walls than in the surrounding soft deposit. It was concluded, therefore, that the cyanobacteria were playing an active role in the deposition of the resulting laminated siliceous 'stromatolite'. Other examples of biological silicification are derived from soil crusts in which chert or opal is produced around mycorrhiza-like fossils in pre-Pleistocene soil profiles.

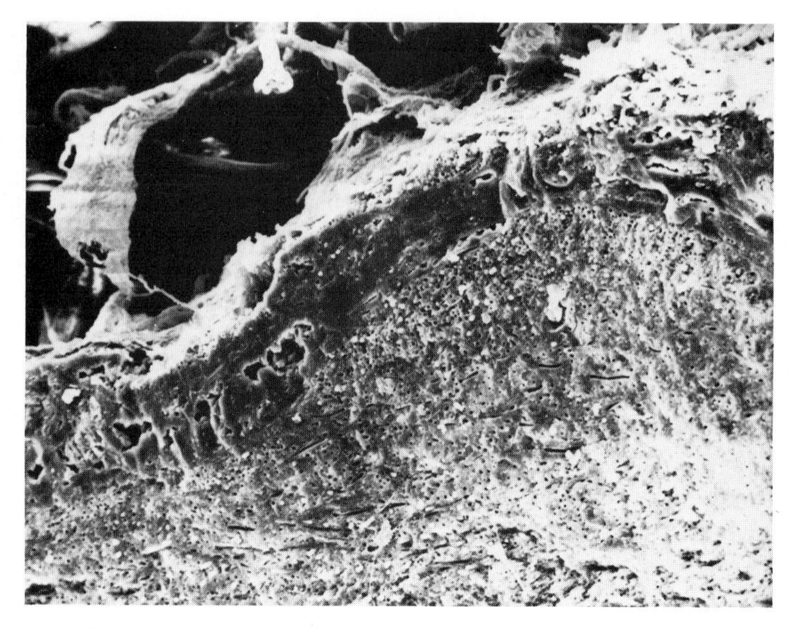

Fig. 5.15. Partially laminated siliceous stromatolitic rock formed at a hot spring Lake View, Oregon. Three different species of cyanobacteria are embedded within the silica matrix. (The site was introduced to the author by R.Castenholz.)

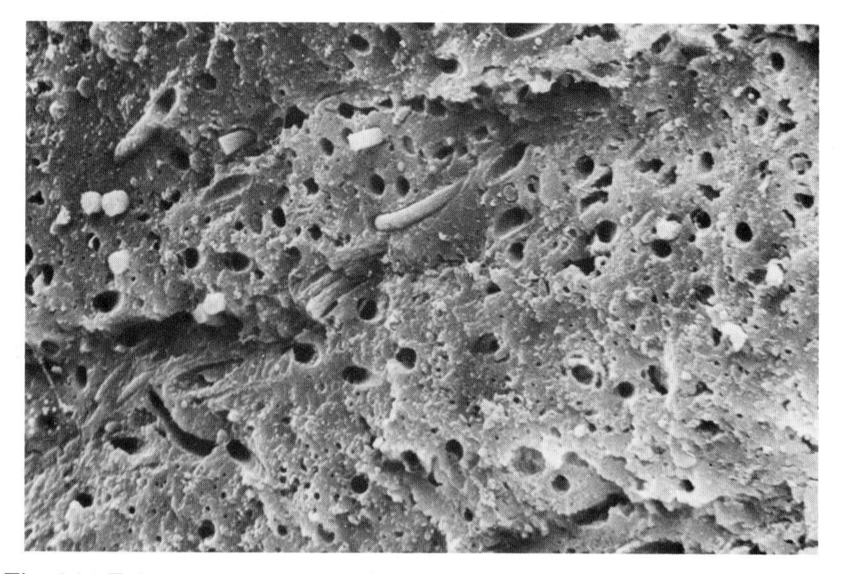

Fig. 5.16. Enlarged section of 5.15. Two different cyanobacteria, one with sheath, another without sheath, both belonging to the LPP-group are embedded in siliceous material.

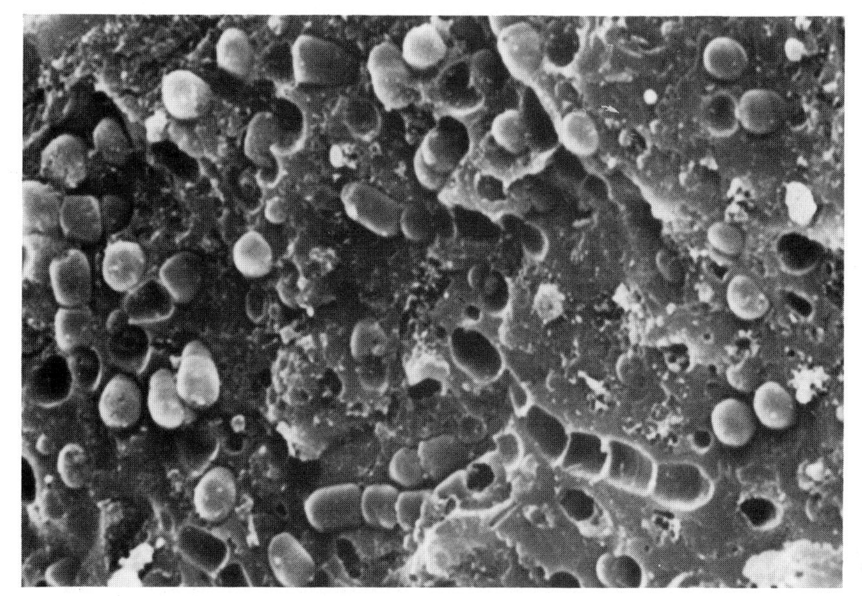

Fig. 5.17. *Anabaena* or *Nostoc* embedded in silica carcass at the same site as 5.15.

Thus, it can be concluded that not only the mobilization of elements including silicon from silicate minerals, but also the deposition of new siliceous minerals are often related to microbial activities. More spectacularly and under defined genetic control silicification is occurring with eukaryotic microorganisms, for example diatoms and radiolarians.

5.4 DIATOMS AND THE AQUATIC SILICA CYCLE

5.4.1 Introduction

Diatoms (Bacillariophyta) contribute about 20 to 25% of biological world (land and ocean) primary production, measured as carbon assimilation (Werner 1977). Due to their abundance in the most productive areas of the oceans, they are by far the most important organisms in the aquatic silica cycle. Foraminifera, silicoflagellates and some sponges are other groups contributing some additional silica structures and deposits.

5.4.2 Physiology and biochemistry of silica in diatoms, in culture and in aquatic habitats

Silicon in the nutrient silicic acid [$Si(OH)_4$] is an essential macroelement for the growth of diatoms. An elementary analysis of the large marine diatom

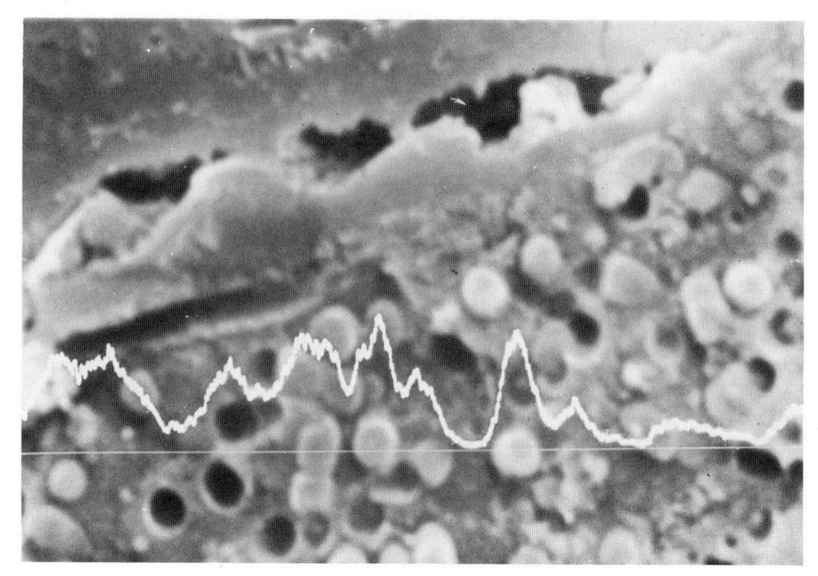

Fig. 5.18. EDX-line-scan of silica over cyanobacteria embedded in silica precipitate. The line-scan demonstrates that the cyanobacteria, though embedded in siliceous matrix, enrich even more silica on their sheaths. The analytical method has been carefully checked to exclude artefacts by relief 'shadows'. No explanation could be found by physiological experiments with the associated cyanobacteria. A reasonable explanation is the embedding of silica in slime as it has been observed by Heinen with slime-excreting soil bacteria (Heinen, personal communication). The polysaccharide material of the glycocalyx apparently absorbs or incorporates Si instead of carbon in the complex macromolecules lipopolysacclaniloes.

Coscinodiscus asteromphalus (stage 135 μm valve diameter) cells from a log phase culture in the light under defined conditions (Werner 1978b), demonstrates an equal silicon and carbon content in the cells with 170 μg silicon and carbon per mg dry weight (Table 5.2). The silicon content is 60 to 80 times larger than the nitrogen and phosphorus percentage of these cells. This is equal to a silica content of about 40% SiO_2 per mg dry weight. In diatoms with very thin shell components this value can be as low as 3% SiO_2 per mg dry weight, as found in *Bellerochea yucatanensis* (Werner 1978a). The absolute silicon content per cell depends primarily on the size of the cells (Table 5.3). The small cells of *Skeletonema costatum* have a minimum silicon content of 0.6 pg silicon per cell, a large *Coscinodiscus asteromphalus* cell 40,000 pg Si or more. The doubling time of vegetative cells of these two species differs only by a factor of five (5 h against 24 h). Thus in one hour the *Coscinodiscus* cell assimilates 8000 times more silica than a cell of *Skeletonema costatum.* Per μm^2 cell surface and per hour, the two species still differ by a factor of 250 in silica incorporation. These few

Table 5.2. Element analysis of cells of *Coscinodiscus asteromphalus* (135 μm valve diameter) determined by proton-induced X-ray emission spectroscopy.

Element	Atomic no.	ng mg^{-1} cell dry weight
H*	1	27 600
C*	6	170 500
N*	7	3 100
O	8	450 000–500 000
Si†	14	170 000
P	15	2 515
S	16	5 764
Cl	17	72 905
K	19	34 189
Ca	20	973
Ti	22	8
Cr	24	8
Mn	25	24
Fe	26	309
Ni	28	22
Cu	29	62
Zn	30	49
Br	35	435
Rb	37	10
Sr	38	18
Ba	56	27

* Determined by CHN-analysis.
† Determined additionally by chemical methods and used as reference value.

Table 5.3. Silicon content of diatoms.

Species from cultures	pg Si $cell^{-1}$	Size (μm)	References
Skeletonema costatum	0·56	5 × 4–12	Harrison (1974)
Skeletonema costatum	1–2		Paasche (1973)
Navicula pelliculosa	1.3–1.4	8–11 × 4–5	Busby & Lewin (1967)
Nitzschia alba	9		Lewin & Chen (1968)
Cyclotella cryptica	7.9–11.9	9 × 8–10	Werner (1966) Werner & Pirson (1967)
Nitzschia palea	50–190	20–65 × 2.5–5	Jørgensen (1953)
Thalassiosira decipiens	90–330	17–28 × 8–18	Paasche (1973)
Ditylum brightwellii	200–900	13–52 × 80–130	Paasche (1973)
Coscinodiscus asteromphalus	40 000	60 × 135	Werner (1978a)

examples are an illustration of the difficulty in generalizing about diatoms and their contribution to the aquatic silica cycle.

Uptake of $Si(OH)_4$ is energy dependent (Lewin 1955), potassium dependent (Roth & Werner 1978) or depends on other monovalent ions or potassium (Sullivan 1976). Silicoborates, made by fusion of silica with boric acid, are utilized in addition to silicic acid and boric acid by some diatom species (Werner 1969). Thus silica metabolism in diatoms is directly connected to other nutrients and elements. However, for diatoms as well as for all other organisms there is no experimental evidence for stable organic silicic acid compounds with, for example, sugars or ATP. The lowest possible concentration utilized by diatoms (K_0 values) differs significantly. Data of 3×10^{-7}M for *Skeletonema costatum* and *Ditylum brightwellii* and 1.3×10^{-6}M for *Licmophora* sp. and *Thalassiosira decipiens* are reported (Paasche 1973). Under $Si(OH)_4$-limitations the mean uptake ratio of N:Si:P in *Skeletonema costatum* was 10:2:1 (Harrison 1974).

Diatoms react to silicic acid deficiency with several typical and rapid metabolic shifts, inhibitions and enhancements: for example changes in pools of organic acids and amino acids, an early stop of protein synthesis, an enhancement of fatty acid synthesis, and of a rather later effect on photosynthesis. These reactions cannot be mimicked by any other nutrient deficiency (Werner 1977).

Concentrations of orthosilicic acid in freshwater lakes can vary by almost 3 orders of magnitude, between 20 μg and almost 20 mg Si l^{-1} (Jørgensen 1955, Tessenow 1966). In the marine environment at the surface of the oceans and shores the variation in the silica content, due to biological activity, seasons, currents and upwelling is almost 4 orders of magnitude, between 5 mg Si l^{-1} and less than 1 μg Si l^{-1} (Armstrong 1965).

5.4.3 Formation and dissolution of silica structures

Diatom frustules with an intricate ornamentation in the silica structures belong to the most studied types of cell walls. More than 1000 species have been studied so far by transmission and scanning electron microscopy (Helmcke *et al.* 1963–1975, Hasle 1977). The various parts of the frustules (valves, girdle bands, copulae) have been given a special terminology (Ross *et al.* 1979). Two examples of SEM-pictures of diatom valves are given in Fig. 5.19. Formation of these structures is under strict genetic control; however, we have so far no definite answer which organic template and which sequence in these templates finally determines the silica ultrastructure of the shells (Werner 1977). Dissolution of silica structures is much easier to study than formation since most of these experiments are concerned with dead cells or purified frustule fragments. Between 90 and

99% of the average diatom valves dissolve during the sedimentation to the ocean floor (Lisitzin 1971). The non-dissolving portion of the frustules accumulate under areas with a medium or high primary phytoplankton production at a rate of 1 to 2 g cm^{-2} in 1000 years. In these areas with high primary production biogenous silica (opal A) can be more than 50% of the whole sediment.

Radiolarians contribute less than 10% of the biogenous opal, diatoms almost the whole of the rest (Calvert 1966). The identification of the frustule components is complicated by silica dissolution in some species and

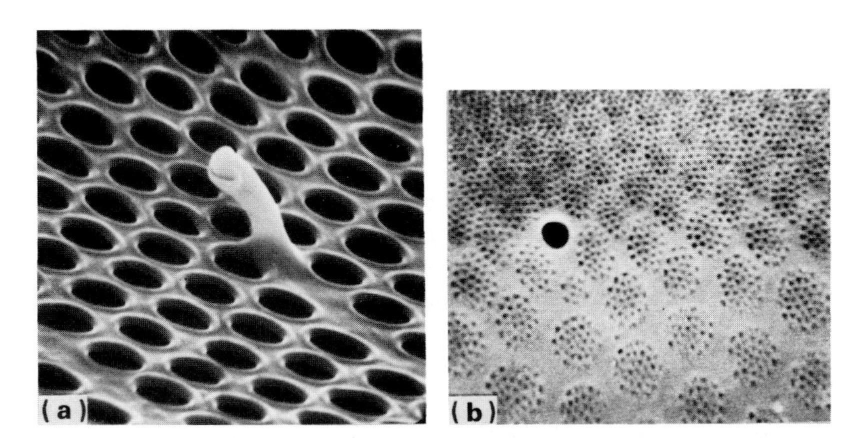

Fig. 5.19. (a) *Coscinodiscus concinnus*: valve, view from the inside with a process, × 9000; (SEM picture). (b) *Actinocyclus ehrenbergii*: valve, view from the outside with the opening of a process, × 10 350 (SEM picture). (By courtesy of K.Fecher, Marburg.)

overgrowth in others (Mikkelsen 1977). In natural sea-water within 30 min at 100 °C 50 to 75% of diatom silica is solubilized from *Chaetoceros gracilis* and *Skeletonema costatum* but only about 30% in *Ditylum brightwellii* and *Nitzschia closterium*. With the latter species more than 90% is dissolved after separation of the organic components of the cell valves by treatment with nitric acid. In general, recent diatoms resemble silica gel, fossil diatoms are more like opaline silica (Kamatani 1971).

For the rate of dissolution of diatom skeletons the following equation has been proposed (Wollast 1974):

$$\frac{dq}{dt} = -k_1\Omega_r = -4\pi\alpha r^2 k_1 \tag{5.7}$$

where q = molar amount of silica dissolved; k_1 = a constant depending on pH and chlorinity; r = Stokes' radius of the particle; α = shape factor relating the Stokes' radius of the particle to the apparent radius deduced from measurements of specific surface area; Ω_r = surface area of the particle. This equation, however, does not consider in detail the additional biological factors that can influence the structure and composition and thereby also the dissolution of the skeletons, for example:

(1) growth rate of the cells;
(2) stage in the mitotic life cycle;
(3) stage in the meiotic life cycle;
(4) temperature at which the cells have been grown;
(5) silica concentration at which the cells have been grown.

Valves of three stages of the meiotic (sexual) life cycle in *Coscinodiscus asteromphalus* are shown in Fig. 5.20. Due to a highly specialized type of cell division, the average diameter of the valves is reduced by 1.5 μm per division. The number of radial rows and the size of the polygonal loculi is reduced. In the larger cells, the pattern of areolation in the centre is also interrupted (Werner 1978a). These structural changes significantly affect the dissolution rate, although systematic studies have not been performed. Several other important factors have been discussed in more detail elsewhere (Werner 1977).

The parts of the shells finally reaching the ocean floor undergo further transformation. Most deep-sea siliceous sediments are biogenic with the following diagenetic maturation sequence:

opal-A (siliceous ooze)→opal-CT (porcelanite)→chalcedony or cryptocrystalline quartz.

The transformation of opal-A to opal-CT is much higher in carbonate- than in clay-rich minerals. Competition of clay minerals and opal-CT formation explains the present alkalinity in the solvent sea-water (Kastner *et al.* 1977). Transformation of opal-A to opal-CT in Bering Sea deposits was found at temperatures between 35 and 50°C. At least 500 m of diatomaceous material was necessary to reach this temperature. In fact, an average of 600 m diatomaceous ooze and diatomaceous mudstone was found in the deep-sea drilling project at Leg 19 (Hein *et al.* 1978).

Formation of diagenetic cristobalite is assumed to proceed through the dissolution of diatomite (diatom frustules) and a following precipitation of cryptocrystalline cristobalite. The density of the layers changes from 0.5 g ml^{-1} to 2.0 g ml^{-1} in compact cristobalite chert. The d (101) spacing of cristobalite changes from 0.4115 to 0.4040 nm (Murata & Nakata 1974). Interesting differences in the trace metal composition of diatomite and layer chert have been found, the largest in Al_2O_3, Fe_2O_3 and MgO (Murata &

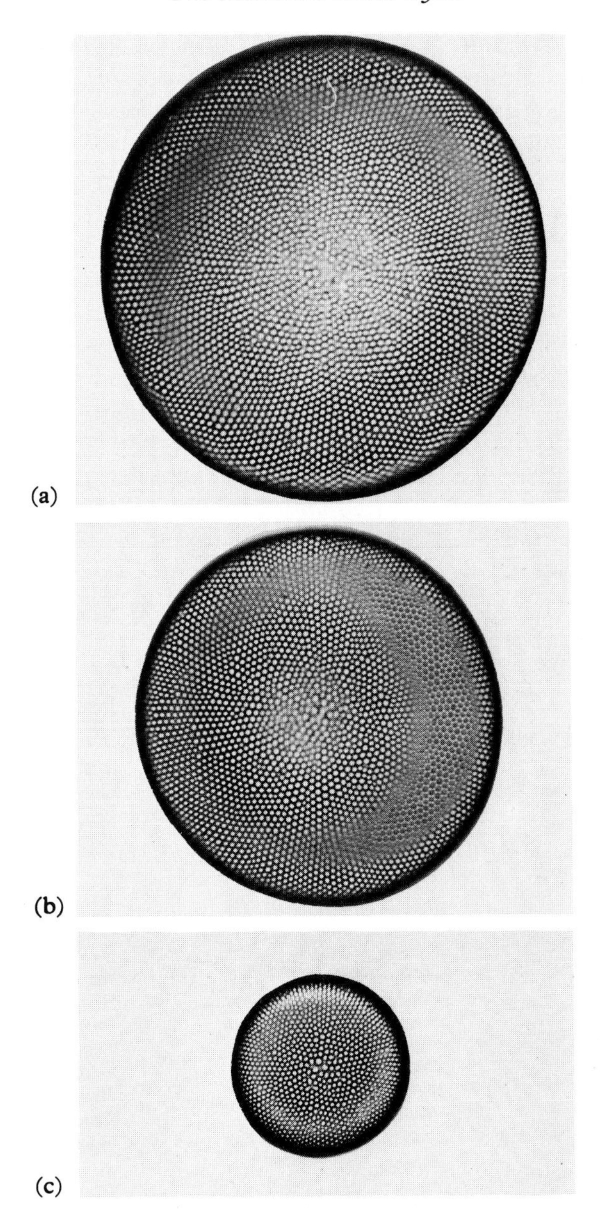

Fig. 5.20. *Coscinodiscus asteromphalus*. Valves: (a) 125 μm, (b) 100 μm, (c) 50 μm diameter, stages in the meiotic life cycle. (Modified from Werner 1978a.)

Larson 1975). Oxygen isotope ratios from three diagenetic zones of Miocene Monterey Shale (California) shows a significant decrease from 37.4 $^{\delta 18}O_2$ per ml standard mean ocean water in biogenetic opal to 29.4 ± 1.5 in cristobalite and 23.8 ± 0.3 in microquartz (Murata *et al.* 1977).

A very interesting relation between the evolution of radiolarians and diatoms by competition for silica has been discussed by Harper and Knoll (1975). The large evolutionary success of diatoms since the Jurassic time could be responsible for the decrease of the average radiolarian test weight from 0.2 μg to less than 0.08 μg Si within 60 million years. This example illustrates very nicely that the aquatic silica cycle is not only very important for biological primary production and for sedimentation processes in the oceans but also for the mutual interaction of the evolution of organisms, depending on large quantities of silicon. This aspect has to be separated completely from the studies on silicon as an essential trace element in diatoms, higher plants and vertebrates, as discussed in another review (Werner 1981).

References

Armstrong F.A.J. (1965) Silicon. In *Chemical Oceanography* (Eds J.P.Riley & G.Skirrow), pp. 409–32. Academic Press, London.

Berthelin J. (1976) Etudes expérimentales de mécanismes d'altération des minéraux par des microorganismes hétérotrophes. Thèse de Doctorat, Université de Nancy I.

Brightman F.H. & Seaward M.R.D. (1977) Lichens of man-made substrates. In *Lichen Ecology* (Ed. M.R.D.Seaward), pp. 253–94. Academic Press, London.

Busby W.F. & Lewin J. (1967) Silicate uptake and silica shell formation by synchronously dividing cells of the diatom *Navicula pelliculosa* (Bréb) Hilse. *J. Phycol.* **3,** 127–31.

Calvert S.E. (1966) Accumulation of diatomaceous silica in the sediments of the Gulf of California. *Geol. Soc. Am. Bull.* **77,** 569–96.

Carroll D. (1970) *Rock weathering*. Plenum Press, New York.

Eckhardt F.E.W. (1979) Über die Einwirkung heterotropher Mikroorganismen auf die Zersetzung silikatischer Minerale. *Z. Pflanzenernaehr. Bodenkd.* **142,** 434–45.

Eno C.F. & Reuszer H.W. (1951) The availability of potassium in certain minerals to *Aspergillus niger*. *Proc. Soil Sci. Soc. Am.* **15,** 155–9.

Eno C.F. & Reuszer H.W. (1955) Potassium availability from biotite, muscovite, greens and and microcline, as determined by growth of *Aspergillus niger*. *Soil Sci.* **80,** 199–209.

Filip Z. (1975) Wechselbeziehungen zwischen Mikroorganismen und Tonmineralen und ihre Auswirkung auf die Bodendynamik. Habilitationsschrift, Giessen.

Flaig W. (1978) Effect of interactions of silicious components and organic substances on life in soil—a contribution to plant production. In *Biochemistry of silicon and related problems* (Eds G.Bendz & I.Lindqvist), pp. 93–108. Plenum Press, New York.

Friedmann E.I. & Galun M. (1974) Desert algae, lichens and fungi. In *Desert biology 2* (Ed. G.W.Brown Jr.), pp. 166–213. Academic Press, New York.

Glazovskaya M.A. (1950) Vyvetrivanie gornykh porod v nival'nom poyase tsentral'-nogo Tyan-Shanya. (Rock weathering in the arable belt of Central Tyan-Shan). *Tr. Pochv. Inst., Akad. Nauk SSSR* **34,** 28–48.

Harper H.E. & Knoll A.H. (1975) Silica, diatoms, and cenozoic radiolarian evolution. *Geology* **3,** 175–7.

Harrison P.J. (1974) Continuous Culture of the Marine Diatom *Skeletonema costatum* (Grev.) Cleve under Silicate Limitation. Ph.D. Dissertation. University of Washington, Seattle.

Hasle G.R. (1977) Morphology and taxonomy of *Actinocyclus normanii* f. *subsalsa* (Bacillariophyceae). *Phycologia* **16,** 321–8.

Hein J.R., Scholl D.W., Barron J.A., Jones M.G. & Miller J. (1978) Diagenesis of late cenozoic diatomaceous deposits and formation of the bottom simulating reflector in the southern Bering Sea. *Sedimentology* **25,** 155–81.

Heinen W. (1960) Silicium-Stoffwechsel bei Mikroorganismen. *Archiv Mikrobiologie* **37,** 199–210.

Heinen W. (1967) Ion-accumulation in bacterial system III. Respiration-dependent accumulation of silicate by a particulate fraction from Proteus mirabilis cell-free extracts. *Arch. Biochem. Biophys.* **120,** 101–7.

Heinen W. & Oehler J.H. (1979) Evolutionary aspects of biological involvement in the cycling of silica. In *Biogeochemical cycling of mineral-forming elements* (Eds P.A.Trudinger & D.J.Swaine), pp. 431–43. Elsevier, Amsterdam.

Helmcke J.G., Krieger W. & Gerloff J. (1963–75) Diatomeenschalen im Elektronenmikroskopischen Bild. Vol. I–X, 1023 Plates. J.Cramer, Lehre.

Holzapfel L. & Engel W. (1954) Der Einfluß organischer Kieselsäure-Verbindungen auf das Wachstum von *Aspergillus niger* und Triticum. *Z. Naturforsch.* **93,** 602–6.

Irion G. (1978) Soil infertility in the Amazonian rain forest. *Naturwissenschaften* **65,** 515–19.

Jackson T.A. & Keller W.D. (1970) A comparative study of the role of lichens and 'inorganic' processes in the chemical weathering of recent Hawaiian lava flows. *Am. J. Sci.* **269,** 446–66.

Jørgensen E.G. (1953) Silicate assimilation by diatoms. *Physiologia Pl.* **6,** 301–15.

Jørgensen E.G. (1955) Variations in the silica content of diatoms. *Physiologia Pl.* **8,** 840–5.

Kamatani A. (1971) Physical and chemical characteristics of biogenous silica. *Mar. Biol.* **8,** 89–95.

Kastner M., Keene J.B. & Gieskes J.M. (1977) Diagenesis of siliceous oozes—I. Chemical controls on the rate of opal-A to opal-CT transformation—an experimental study. *Geochim. Cosmochim. Acta* **41,** 1041–59.

Krumbein W.E. (1969) Über den Einfluß der Mikroflora auf die exogene Dynamik (Verwitterung und Krustenbilding). *Geol. Rundschau* **58,** 333–65.

Krumbein W.E. (1971) Sedimentmikrobiologie und ihre geologischen Aspekte. *Geol. Rundschau* **60,** 438–71.

Krumbein W.E. (1972) Rôle des microorganismes dans la genèse, la diagenèse et la dégradation des roches en place. *Rev. Ecol. Biol. Sol* **9,** 283–319.

Krumbein W.E. & Jens K. (1981) Biogenic rock varnishes of the Negev Desert (Israel)—an ecological study of iron and manganese transformation by cyanobacteria and fungi. *Oecologia* **50,** 25–38.

Lauwers A.M. & Heinen W. (1974) Biodegradation and utilization of silica and quartz. *Arch. Microbiol.* **95,** 67–78.

Lewin J.C. (1955) Silicon metabolism in diatoms. III. Respiration and silicon uptake in *Navicula pelliculosa. J. Gen. Physiol.* **39,** 1–10.

Lewin J.C. & Chen C. (1968) Silicon metabolism in diatom VI. Silicic acid uptake by a colorless marine diatom, *Nitzschia alba* Lewin and Lewin. *J. Phycol.* **4,** 161–6.

Lisitzin A.P. (1971) Distribution of siliceous microfossils in suspension and in bottom sediments. In The *Micropaleontology of Oceans* (Eds B.M.Funnel & W.R.Riedel), pp. 173–96. Cambridge University Press, Cambridge.

Lorenz M.G., Aardema B.W. & Krumbein W.E. (1981) Interaction of marine sediment with DNA and DNA availability to nucleases. *Mar. Biol.* **64,** 225–230.

Lovelock J.E. (1979) *Gaia. A new look at life on earth.* Oxford University Press, Oxford.

Marshall K.C. (1976) *Interfaces in microbial ecology.* Harvard University Press, Cambridge.

Matheja J. & Degens E.T. (1971) *Structural Molecular Biology of Phosphates.* Fischer, Stuttgart.

Mikkelsen N. (1977) Silica dissolution and overgrowth of fossil diatoms. *Micropaleontology* **23,** 223–6.

Murata K.J., Friedman I. & Gleason J.D. (1977) Oxygen isotope relations between diagenetic silica minerals in Monterey Shale, Temblor Range, California. *Am. J. Sci.* **277,** 259–72.

Murata K.J. Murata K.J. & Larson R.R. (1975) Diagenesis of miocene siliceous shales, Temblor Range, California. *J. Research U.S. Geol. Survey* **3,** 553–66.

Murata K.J. & Nakata J.K. (1974) Cristobalitic stage in the diagenesis of diatomaceous shale. *Science* **184,** 567–8.

Oberlies F. & Pohlmann G. (1958) Einwirkung von Mikroorganismen auf Glas. *Naturwissenschaften* **45,** 487.

Oehler J.H. (1979) Deposition and diagenesis of biogenic silica. In *Biogeochemical cycling of mineral-forming elements* (Eds P.A.Trudinger & D.J.Swaine), pp. 467–84. Elsevier, Amsterdam.

Paasche E. (1973) Silicon and ecology of marine plankton diatoms. II. Silicate uptake kinetics in five diatom species. *Mar. Biol.* **19,** 262–9.

Pohlmann G. & Oberlies F. (1960) Angriff von Glasoberflächen durch tierisches Gewebe. *Naturwissenschaften* **47,** 58.

Rades-Rohkohl E. (1975) Verhalten, Aktivitäten und Effekte von Bakterien auf Oberflächen von Quarzeinkristallen. Diplomarbeit, Universität Kiel.

Rades-Rohkohl E., Franzle O. & Hirsch P. (1979) Behavior, activities, and effects of bacteria on synthetic quartz monocrystal surfaces. *Microb. Ecol.* **4,** 189–205.

Ross R., Cox E.J., Karayeva, N.I., Mann D.G., Paddock T.B.B., Simonsen R. & Sims P.A. (1979) An amended terminology for the siliceous components of the diatom cell. In *Fifth Symposium on Recent and Fossil Diatoms* (Ed. R.Simonsen) Beih. Nova Hedwigia **64,** 513–30. J.Cramer, Vaduz.

Roth R. & Werner D. (1978) Kaliumabhängige Kieselsäureaufnahme im Dunkeln bei *Cyclotella cryptica*. *Z. Pflanzenphysiol.* **89,** 239–49.

Silverman M.P. (1979) Biological and organic chemical decomposition of silicates. In *Biogeochemical Cycling of Mineral-forming Elements* (Eds P.A.Trudinger & D.J.Swaine), pp. 445–66. Elsevier, Amsterdam.

Smyk B. (1970) The microbial degradation of silicates and aluminium silicates. *Postepy Mikrobiologii* **9,** 121–35.

Sullivan C.W. (1976) Diatom mineralization of silicic acid. I. $Si(OH)_4$ transport characteristics in *Navicula pelliculosa*. *J. Phycol.* **12,** 390–6.

Tessenow U. (1966) Untersuchungen über den Kieselsäurehaushalt der Binnengewässer. *Arch. Hydrobiol. Suppl.* **32,** 1–136.

Voronkov M.G., Zelchan G.I. & Lukevitz E. (1975) *Silizium und Leben*. Akademie-Verlag, Berlin.

Walter M.R. (1976) *Stromatolites. Development in sedimentology 20*. Elsevier, Amsterdam.

Webley D.M., Duff R.B. & Mitchell W.A. (1960) A plate method for studying the breakdown of synthetic and natural silicates by soil bacteria. *Nature* **188,** 766–7.

Werner D. (1966) Die Kieselsäure im Stoffwechsel von *Cyclotella cryptica*, Reimann, Lewin und Guillard. *Arch. Mikrobiol.* **55,** 278–308.

Werner D. (1969) Silicoborate als erste nicht C-haltige Wachstumsfaktoren. *Arch. Mikrobiol.* **65,** 258–74.

Werner D. (1977) Silicate metabolism. In *The Biology of Diatoms* (Ed. D. Werner), pp. 110–49. Blackwell Scientific Publications, Oxford.

Werner D. (1978a) Regulation of metabolism by silicate in diatoms. In *Nobel-Symposium: Biochemistry of Silicon and Related Problems* Stockholm 1977 (Ed. G.Bendz & L.Lindquist), pp. 149–75. Plenum Press, New York.

Werner D. (1978b) Silicon and iodine metabolites from marine plants. In *Drugs and Food from the Sea* (Eds P.K.Kaul & C.J.Sinderman) pp. 171–80. University of Oklahoma Press, Norman, Oklahoma.

Werner D. (1983) Silica metabolism. In *Encyclopedia of Plant Physiology*, New Series, Vol. 15. Inorganic Plant Nutrition (Eds A.Läuchli & R.L.Bieleski). Springer, Berlin (in press).

Werner D. & Pirson A. (1967) Über reversible Speicherung von Kieselsäure in *Cyclotella cryptica*. *Arch. Mikrobiol.* **57,** 43–50.

Wilding L.P. & Drees L.R. (1974) Contributions of forest opal and associated crystalline phases to fine silt and clay fractions of soils. *Clays Clay Miner.* **22,** 295–306.

Wollast R. (1974) The silica problem. In *The Sea* (Ed. E.D.Goldberg) Vol. 5, pp. 359–92. John Wiley & Sons, New York.

Youatt J.B. & Brown R.D. (1981) Origins of chirality in nature: A reassessment of the postulated role of bentonite. *Science*, **212,** 1145–6.

Zajic J.E. (1969) *Microbial biogeochemistry*. Academic Press, New York.

CHAPTER 6
THE MICROBIAL IRON CYCLE

KENNETH H. NEALSON

6.1 INTRODUCTION

Iron is the fourth most abundant element in the Earth's crust; it is a component of virtually all types of rocks, and represents about 5% (by weight) of the crust (Wedepohl 1971, Lepp 1975). Due to the prevailing oxidizing conditions on earth, the distribution of iron is not uniform. Having no abundant soluble or volatile phases that are chemically stable under present conditions, iron is primarily confined to oxidized phases in the lithosphere. This is in marked contrast to what iron distribution may have been in the Precambrian, before the evolution of oxygenic photosynthesis. At that time soluble ferrous iron should have been abundant and stable, representing the source of iron for many iron deposits found in Precambrian rock formations. In fact, the relationship between iron deposition, biological evolution and atmospheric evolution provides a fascinating puzzle (Cloud 1973, 1974, Walker 1980), the answer to which may provide insight into the importance of biological–geological feedback controls during the evolution of the earth.

Although iron can form many different minerals with a variety of other elements (Table 6.1), the largest iron accumulations in the crust are the Precambrian oxidized deposits known as banded iron formations (BIFs) (James 1966, Lepp 1975, Lundgren & Dean 1979). These BIFs, which average about 28% iron, illustrate the capacity of iron to accumulate to anomalously high concentrations, which are economically valuable as ore deposits. Other crustal iron reservoirs include bog ores, red beds, hydrothermal deposits generated by submarine volcanism or magmatic activity, and ferromanganese deposits of both lakes and oceans, but all of these are quantitatively small in comparison to the BIF deposits. Presumably because large amounts of soluble iron are unavailable, the large-scale accumulation of oxidized iron such as may have occurred in the Precambrian is not possible in present-day environments. Figure 6.1 is a representation of a global iron cycle, based on that of Lepp (1975). The lack of soluble or volatile phases under present surface conditions means the iron cycle is largely uncoupled from both the atmospheric and hydrospheric cycles except for its relation to oxygen (Deevey 1970), so that its global cycle is a geological one, with burial, metamorphosis, volcanism, uplifting, weather-

Table 6.1. Iron-containing minerals. (Based on data from Murray 1979 and Wedepohl 1971.)

Compounds	Mineral name(s)	Crystal form
Oxides		
$\alpha FeO(OH)$*	goethite (α authosiderite)	orthorhombic
$\gamma FeO(OH)$*	lepidocrocite	orthorhombic
$\beta FeO(OH)$*	akaganeite	tetragonal
$\delta' FeO(OH)$*	feroxyhyte	hexagonal
αFe_2O_3	hematite	hexagonal
$5Fe_2O_39H_2O$	ferrihydrite	hexagonal
γFe_2O_3	maghemite (oxymagnetite)	cubic or tetragonal
Fe_3O_4 *or* $(Fe,Mg)Fe_2O_4$	magnetite	cubic
$(Fe,Mg)(OH)_2$	amakinite	hexagonal
Sulphides		
FeS $(Fe,Ni)_{1.1}S$	mackinawite (kansite)	tetragonal
FeS	troilite (pyrrhotite)	hexagonal
FeS_2	pyrite	cubic
FeS_2	marcasite	orthorhombic
Fe_3S_4	greigite (melnikovite)	orthorhombic
$FeCu_2SnS_4$	stannite (stannine)	tetragonal
Fe_2CuS_3	cubanite (chalmersite)	orthorhombic
Others		
$(Fe,Mg,Mn)CO_3$	siderite (chalybite)	rhomohedral
(K,Fe^{3+},Al) silicate	glauconite	(mica)-group
(Mg,Fe^{2+}) silicate	chamosite	chlorite-group
$(Mg,Fe)_2$ silicate	hypersthene	pyroxene-group
$(Mg,Fe,Ca)_2$ silicate	pigeonite	chiropyroxene-group

* A mixture of all hydroxides and hematite is also named limonite.

ing and other slow geological processes acting as transformation and transport mechanisms.

Nevertheless, there are environments in which iron cycles rapidly. A prerequisite for such environments is the presence of a reduced or semianaerobic zone, where iron reduction and mobilization can occur. Such environments include the bottoms of eutrophic and stratified lakes, seasonally anaerobic ponds or lakes, anaerobic seepages, flooded bogs and anaerobic or semianaerobic sedimentary environments of both marine and freshwater origin.

Iron is an essential element of virtually all organisms, performing a variety of essential functions; thus, it is not surprising that there are several ways in which the biota interact with iron. These interactions, while they may be quantitatively insignificant to the iron cycle on a global scale, are, nevertheless, extremely important to the biota. It is the purpose of this

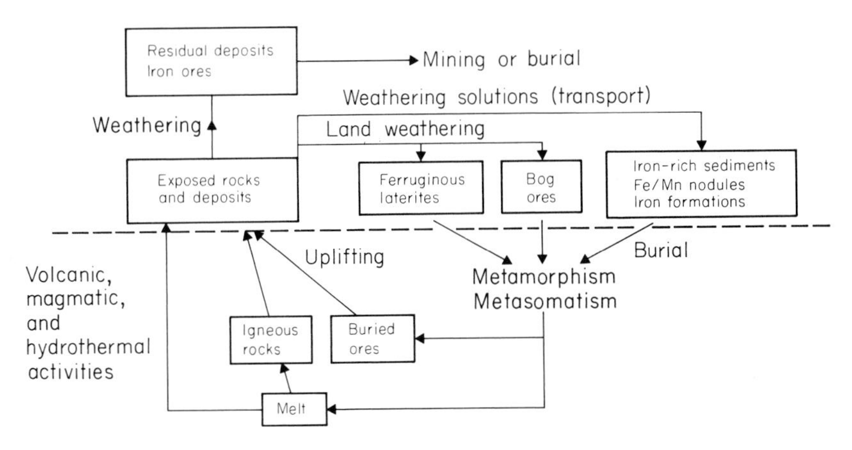

Fig. 6.1. Global iron cycle. This diagram represents the major sources of iron that cycle at or near the crust (see also Figs 2.1 and 2.3). Above the dotted line weathering represents the major transport phenomenon (Lepp 1975).

chapter to review the interactions of microbes with iron, discussing the roles and significances of these interactions both to the biota and to the geochemistry of iron.

6.2 IRON CHEMISTRY

Certain aspects of iron chemistry must be kept in mind in any consideration of the biological iron cycle. These include the common valences taken by iron, the relative stabilities of iron compounds, both organic and inorganic, formed in conjuction with these valence states, and the energetics and kinetics of iron interconversions.

It is possible to predict on thermodynamic grounds (from the free energies of formation of the components) the stability of various inorganic iron phases in solution (Garrels & Christ 1965, Stumm & Morgan 1970). Iron has two common oxidation states, Fe(II) and Fe(III), so that iron phase diagrams usually have the electron potential, either pE ($-\log(e^-)$) or Eh as one parameter (Fig. 6.2). While these diagrams are valid for the systems and concentrations used for their construction, they must be interpreted cautiously, as they are almost never representative of the complexities found in nature. They do, however, form a starting point for predicting iron distribution in the environment.

In most oxygen-containing waters, for instance, it is predicted that iron will be oxidized and maintained at very low levels. In fact, natural waters commonly contain from 0.1 ppb to 0.7 ppm iron (see Wedepohl 1971),

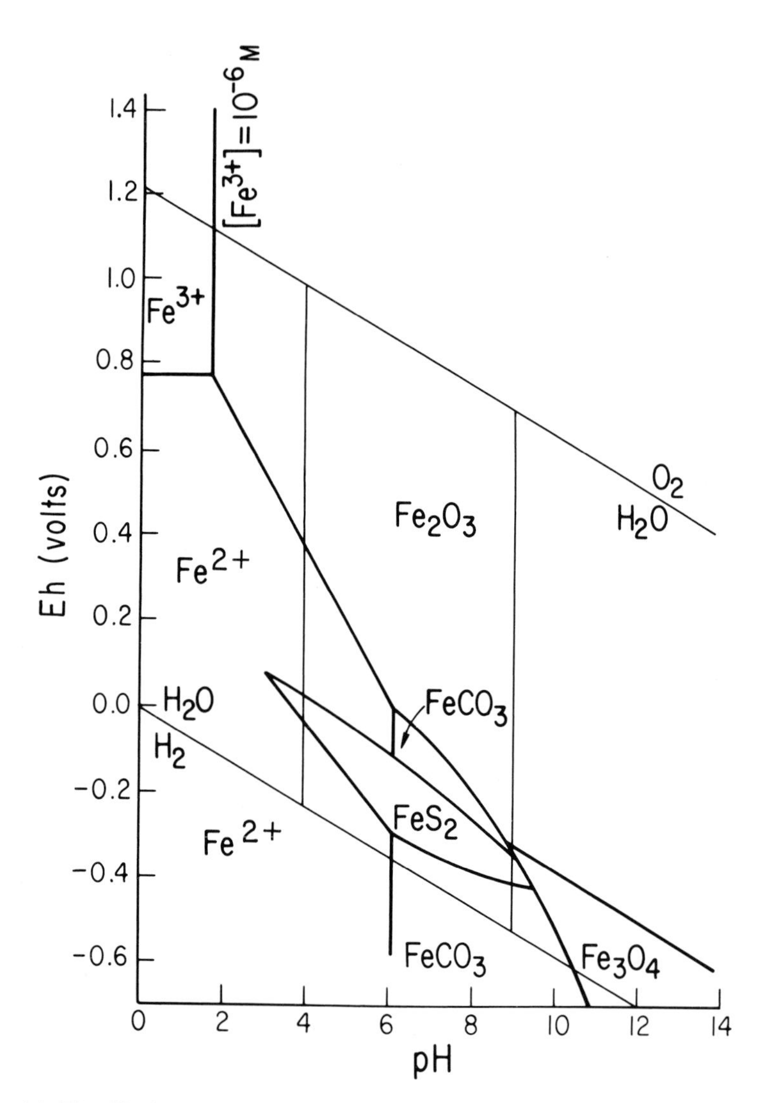

Fig. 6.2. Eh–pH phase stability diagram for several common iron minerals. Total activity of dissolved carbonate 1M, of dissolved sulphur, 10^{-6}M, dissolved iron 10^{-6}M (after Garrels & Christ 1965).

significantly above the level predicted from thermodynamic calculations (10^{-5} to 10^{-6} ppm).

Iron oxidation is usually rapid, and is sensitive to pH and oxygen concentration as described by the following equation (Stumm & Lee 1960, Singer & Stumm 1970, Stumm & Morgan 1970):

$$\frac{d\,\mathrm{Fe(II)}}{dt} = k\,\mathrm{Fe(II)}\,(\mathrm{OH}^-)^2\,p\mathrm{O}_2$$

where $k = 8.0 \pm 2.5 \times 10^{13}$ min^{-1} atm^{-1} mole^{-2} at 20 °C.

The rate constant, k, is affected by both temperature (increasing with increasing temperature) and by ionic strength. As the concentration of anions in solution is raised, the rate constant decreases, with certain anions being more effective in retarding oxidation than others (e.g. $SO_4{=} > Br^- > Cl^- > NO_3^-$). Thus, in sea water, iron precipitation will be considerably slower than in fresh water due to the effect of both sulphate and chloride ions on the rate constant. Even in sea water, however, the half-life ($T_{1/2}$) of added iron is of the order of one to three minutes (Dr J.Murray, personal communication).

Once oxidized to the ferric state, iron has a strong tendency to hydrolyse in solution, so that ferric iron is maintained at very low levels, of the order of 10^{-17}M (Spiro & Saltman 1969, Neilands 1977, Raymond & Carrano 1979). Thus, upon addition of ferric iron to aqueous solutions, a variety of hydrated and oxidized phases are formed. The order and speed of formation and interconversion is dependent on environmental conditions. The interactions occur as the ferric iron in solution hydrolyses, interacts with the various ions in solution, forming spherical polycationic complexes and eventually crystalline forms (Fig. 6.3; Murray 1979). The kind of minerals

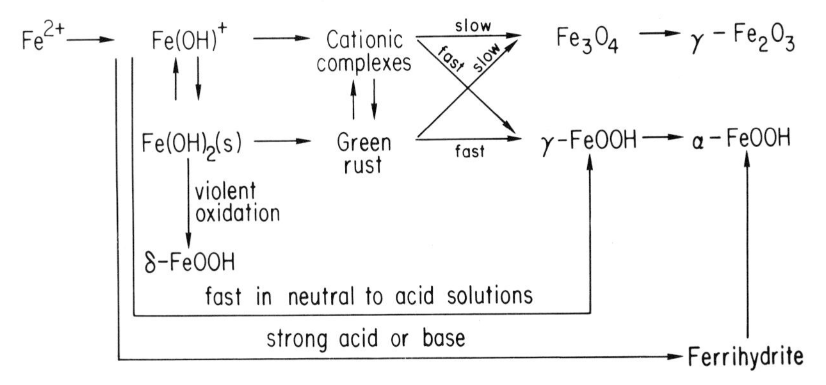

Fig. 6.3. Schematic pathway for oxidation of Fe(II) and interconversions of various oxidized forms (after Murray 1979).

formed will depend on the environment of formation, and many of the forms are interconvertible if the environment is subsequently changed. Nevertheless, the presence of some mineral types can be of use in the extrapolation to the environment of formation, and may be of general use as more is learned about mineral formation and interconversion (Murray 1979).

Iron is readily chelated by a variety of organic molecules including haem, oxalate, citrate, humic acids, tannins and siderophores (or siderochromes as they are also called). In natural environments, such chelation reactions may stabilize iron (either in the ferrous or ferric state) as soluble chelates thereby increasing the amount of iron in solution far above that expected from thermodynamic calculations. This is, at least in part, a problem of definition; aquatic chemists have traditionally defined particulate iron as that retained on a 0.45 μm filter, and soluble iron as that which passes through. Thus the term 'soluble' does not necessarily imply free ferrous or ferric iron in solution, but only iron that passes through the filter even though this may be primarily in the form of organic complexes. The existence of such excess soluble iron has been noted by several workers including Lewis and Goldberg (1954), Sugimura *et al.* (1978), and Eaton (1979).

6.3 INTERACTIONS OF MICROORGANISMS WITH IRON

Microbial iron interactions can be grouped into three major categories: (a) iron scavenging and uptake; (b) iron oxidation or precipitation; and (c) iron reduction or solubilization; although in some cases the various interactions are not easily separated or categorized. The first class of reactions, those involved with the uptake and scavenging of iron, presumably arose for the purpose of supplying iron to microbes in aerobic environments where it is in such low concentrations that it would otherwise be a limiting element. Such needs would seem to provide a sufficient evolutionary selective force to account for the existence of the specific, high-affinity transport systems now known. The selective advantage of the other iron interactions, with the exception of iron oxidation by the *Thiobacilli*, where iron is used as a source of energy, is much less clear.

The uptake of iron is an essential function of many microbes; with the exception of some lactobacilli, virtually all organisms require iron as an essential trace nutrient, where it is used primarily in electron transfer reactions. Several properties of iron make it ideal as a cofactor in cellular electron transfer reactions: (a) it is readily chelated by organic compounds; (b) in the bound state it can change its valence, thus acting as either an electron donor or acceptor; and (c) as a bound cofactor it can assume a wide range of electron potentials (Table 6.2). In addition to its role in electron

Table 6.2. Iron-containing cellular constituents (after Yoch & Carithers 1979).

Component	Type of iron	E_m(mV)*	Function
Cytochrome aa_3	haem	+290	O_2 reduction
Cytochrome c	haem	+220	electron transport
Cytochrome b	haem	+50	electron transport
Catalase	haem	−207	H_2O_2 removal
Ferredoxins:			
Pseudomonas	Fe-S	−235	camphor hydroxylation
Clostridium	Fe-S	−300	pyruvate oxidation
Halobacterium	Fe-S	−345	α-keto acid oxidation
Sulphite Reductase	Fe-S/haem		SO_3^{-2} reduction
Nitrogenase	Fe-S-Mo Fe-Co-Mo		N_2 reduction
Nirate reductase	Mo,Fe/S, haem		NO_3 reduction
Formate dehydrogenase	Se, Mo, Fe/S		Formate oxidation

* E_m indicates the midpoint of the electron potential (E_0') of the Fe^{2+}/Fe^{3+} couple in the particular association. Values from various bacteria may vary substantially; the numbers are meant only to illustrate the wide range of E_m values possible when iron is combined with proteins.

transfer, iron particulates as a cofactor for several other enzymes including catalase, peroxidase and some superoxide dismutases. Due to its many biological uses, iron is concentrated by the biota, ranging from 50 to 200 ppm dry weight. Thus, in certain iron-poor environments, obtaining sufficient iron for metabolism, growth and cell division may be a major problem faced by aerobic microbes.

In this chapter, the mechanisms of iron scavenging and uptake are reviewed, as are the various interactions that microbes undergo in both the oxidation and reduction of iron. These interactions are then discussed in terms of a microbiological iron cycle, and this cycle is then related to iron chemistry on a larger (global) scale, both past and present.

6.3.1 Uptake, solubilization and stabilization of iron

Without some organisms with high-affinity iron uptake systems, aquatic food chains as we know them might not be possible in the world's lakes and oceans. In the laboratory, bacteria are commonly supplied with iron in the growth medium, usually at levels of 0.4 to 4.0 μM and assimilate iron via a low-affinity transport system. In nature, however, iron is almost always at concentrations below the lower of these levels, and under these conditions microorganisms use high-affinity specific systems for iron uptake. Siderochromes or siderophores as they are sometimes called (Neilands 1973, 1974,

1977) are classes of compounds synthesized by bacteria, fungi and some algae; they are excreted into the growth medium to facilitate iron uptake. 'Probably all aerobic and facultative anaerobic microbial cells produce or require siderochromes' (Neilands 1973), which are usually produced only during iron-limited growth. They are excreted into the growth medium where they bind ferric iron with extremely high affinity (formation constants of the order of 10^{30}) and specificity. Once the iron is bound to the siderophore as a ligand, the complex can be taken up by the cell that excreted the siderophore, or by other cells, and the iron can be released intracellularly by any one of several mechanisms (Byers & Arceneaux 1977, Hartmann & Braun 1980). Some of the great variety of known siderophores are shown in Table 6.3.

To illustrate the point of iron scavenging, we can consider the enteric bacterium *Escherichia coli*, which has at least three high-affinity iron transport systems, each coded for by chromosomal genes (Rosenberg & Young 1974, Hantke & Braun 1975, Woodrow *et al.* 1978). These include the enterochelin system in which a siderophore (enterochelin) is synthesized, excreted and taken up as described above. There are also two systems for the uptake of iron-binding ligands (citrate and ferrichrome) that are not produced by *E. coli*. Thus, compounds produced by other organisms can be used by *E. coli* as specific sources of iron for its growth. Recently another high-affinity system has been described for *E. coli*, one whose synthesis is controlled by transmissible genetic elements (plasmids) of the ColV type (Stuart *et al.* 1980, Williams & Warner 1980). The ColV plasmids apparently carry information that controls the synthesis of cell-associated hydroxamate-containing compounds. The existence of such multiple high-affinity systems as well as one carried on a transmissible genetic element serve to emphasize that iron transport is a major problem of aquatic microbes, the solution to which has been addressed in several ways during evolution. The plasmid-mediated iron transport system may have other implications as well, since many higher organisms use iron limitation as one line of defence against bacterial infection (Weinberg 1978, Aisen & Listowsky 1980, Elwell & Shipley 1980); for instance, the pathogenicity of a marine fish pathogen, *Vibrio anguillarum*, has been correlated with the presence of a plasmid-mediated iron uptake system (Crosa 1980).

The solubilization and stabilization of iron in solution is thus greatly affected by siderophores, but not by them alone. Microbes produce a variety of organic compounds with the capacity to combine with iron thus stabilizing it from oxidation or precipitation (Table 6.3). Many of these compounds are not exclusively microbial in origin or nature. The tetrapyrrole, haem, for instance, is a biologically ubiquitous iron-binding compound, associated with cytochromes, haemoglobin and other proteins.

Table 6.3. Iron-binding and transport compounds.

Non-specific iron chelators:
- citrate
- oxalate
- EDTA
- dicarboxylic acids
- humic acids
- tannins

Specific iron chelators:
- (a) haem and haem derivatives—used for electron transport; all bacteria can take up
- (b) transferrins—used by vertebrates to sequester iron and inhibit bacterial growth[1]
- (c) ferritin—iron storage compounds; widely distributed in nature
- (d) siderophores—iron transport compounds; produced by bacteria and fungi[2]

Compound	Organism
1. hydroxamates:	
mycobactin	*Mycobacteria*
schizokinin	*Arthrobacter*
ferrichromes	many fungal genera
rhodotorulic acid	*Rhodotorula*
aerobactin	*Aerobacter*
2. phenolates:	
enterobactin	*Salmonella*
enterochelin	*Escherichia*
itoic acid	*Bacillus*

[1] Aisen & Listkowsky (1980)
[2] Neilands (1973, 1974, 1977), Byers & Arceneaux (1977)

Virtually all bacteria take up haem, and haem iron from any source can be used to supply iron to most bacteria in culture, although the haem transport systems are not understood. Whether or not such chelators are important in nature is not known, and this question may be worthy of more study. The importance, however, of iron uptake and scavenging mechanisms in general probably cannot be overestimated. With many organisms competing for limiting iron in aerobic aquatic environments, the availability of mechanisms to stabilize iron, keeping it 'soluble' and available to the biota are probably of great significance. In fact, it seems likely that the production of iron-chelating compounds may be of geochemical significance as well. There are several examples in which the amount of soluble iron measured has been found to be significantly above that expected by calculation (Lewis & Goldberg 1954, Wedepohl 1971, Sugimura *et al.* 1978). Recent studies in Chesapeake Bay (Eaton 1979) demonstrated similar excess soluble iron, and showed that the release of iron from solution was bacterially mediated. If poisons were added the iron remained soluble, while in unpoisoned samples, iron was precipitated, presumably due to the activity of bacteria consuming the organic iron chelates. Thus iron mobility may be, at least in part, mediated by production and consumption of organic chelators in aquatic systems.

6.3.2 Oxidation and accumulation of iron by bacteria

There are many bacteria that have at one time or another been called 'iron bacteria' including: (a) those that oxidize iron; (b) those that accumulate iron oxides irrespective of whether they oxidize it themselves; and (c) those that reduce iron (Pringsheim 1949, Lundgren *et al.* 1974, Lundgren & Dean 1979). Those that oxidize iron and/or accumulate iron oxides are, in general, placed into one of two groups: (a) the acidophilic (acid-tolerant) forms; and (b) the neutral pH bacteria. The acidophilic forms are the only group unequivocally shown to oxidize ferrous to ferric iron. Among the acidophiles, the most thoroughly studied organism is *Thiobacillus ferrooxidans*, considered synonymous with *Ferrobacillus ferrooxidans* and *Ferrobacillus sulfooxidans* (for discussion, see Lundgren & Dean 1979). The different names are the result of the capacity of *T. ferrooxidans* to oxidize not only ferrous iron, but inorganic sulphur and metal sulphides as well (Silverman & Ehrlich, 1964).

In low pH environments such as acid mine waters or acid hot springs, it is not uncommon to find iron-oxidizing bacteria, and several others besides *Thiobacillus ferrooxidans* have been identified (Table 6.3). *Sulfolobus* (Brock *et al.* 1972) and *Sulfolobus*-like organisms (Brierley & Brierley 1973, Brierley & Murr 1973) are thermophilic acidophiles, and like *T. ferrooxidans*, can oxidize sulphur compounds as well as ferrous iron. *Leptospirillum ferrooxidans* (Balashova *et al.* 1974) is a spirillum-type organism that oxidizes iron only. An acid-tolerant iron-oxidizing *Metallogenium* has recently been characterized on the basis of branching filament morphology which is similar to other identified *Metallogenium* types (Walsh & Mitchell 1972). As will be noted below, the taxonomy and even the validity of this genus has been disputed, so some caution must be exercised in the discussion of this organism.

Virtually all knowledge of the enzymology and bioenergetics of microbial iron oxidation has come from the study of *Thiobacillus ferrooxidans*. This Gram-negative acidophilic rod grows optimally over a pH range of 2.0 to 3.5. While the details of iron oxidation have not been entirely elucidated, the overall reaction has a stoichiometry of four Fe(II) oxidized per O_2 consumed, and is thought to proceed as shown below, with iron oxidation (reaction (a)) being biologically catalysed, and the hydrolysis (b) being spontaneous (Lundgren & Dean 1979).

$$4\ FeSO_4 + 2\ H_2SO_4 + O_2 \rightarrow 2\ Fe_2(SO_4)_3 + 2\ H_2O \qquad (a)$$
$$2\ Fe_2(SO_4)_3 + 12\ H_2O \rightarrow 4\ Fe(OH)_3 + 6\ H_2SO_4. \qquad (b)$$

At the pH range at which *Thiobacillus ferrooxidans* grows, ferrous iron is stable, and the oxidative reaction is sufficiently exergonic to allow the organism to use iron as the sole source of energy; about 25–29 kJ mol^{-1} are available. Some caution must be used in extrapolation of solution thermodynamic theory to cell energetics, however; for instance, the chelation state of iron may be very important. Iron forms an inorganic complex with sulphate that may significantly lower its electron potential (Ingledew *et al.* 1977), and sulphate is required for the growth of *T. ferrooxidans* (Dugan & Lundgren 1965). Tuovinen and Kelly (1972) have pointed out that the oxidation of one iron may not supply sufficient energy for the synthesis of an ATP, and have suggested that two electrons may be required (two ferrous ions oxidized) per ATP.

The relation between electron flow and ATP generation has recently been posed in terms of the chemi-osmotic theory (Ingledew *et al.* 1977). This model suggests that the internal pH of *Thiobacillus ferrooxidans* is maintained at a high value (between 5.0 and 6.0) in comparison to that of the external milieu by virtue of external iron oxidation supplying electrons to the interior. The respiratory chain then is viewed simply as a proton sink for the interior, where molecular oxygen is reduced to water by the terminal cytochrome. The organism thus uses the acidic environment to its advantage, generating a transmembrane pH gradient by internal proton consumption, and using this gradient to drive ATP synthesis via a proton-linked ATPase. The net effect is the same as if protons were pumped outward, as is done in other systems, but the mechanism of obtaining the gradient is unique. This model is consistent both with modern ideas of bioenergetics and ATP generation, and with the ecology of the organism.

The details of the respiratory chain are still under study, but several components appear to be involved. Both a copper-containing protein called rusticyanin (Cobley & Haddock 1975, Ingledew *et al.* 1977) and cytochrome c (Blaylock & Nason 1963, Din *et al.* 1967a) have been implicated as interacting directly with iron in the initial oxidation. The latter authors have presented enzymatic evidence for an iron cytochrome c reductase. Cytochrome oxidase (an a-type cytochrome) is involved as the terminal component in oxygen reduction to water. When the components of the respiratory chain are all known, it should be possible to interface that knowledge with a model similar to that proposed by Ingledew *et al.* and understand more fully the mechanism of ATP synthesis in these bacteria.

Thiobacillus ferrooxidans is an iron chemoautotroph, using CO_2 as the sole source of carbon for growth. It is well established that carbon assimilation is via the Calvin–Benson cycle, using the enzyme RuBP carboxylase (Maciag & Lundgren 1964, Din *et al.* 1967b, Gale & Beck 1967). Since the Fe(II)/Fe(III) couple is so electropositive, about 0.77 V, it is not possible to directly reduce NADP, and the mechanism of NADPH production is assumed to be via a reverse electron transport mechanism (Aleem *et al.* 1963, Broda & Peschek 1979). The components involved have not been identified or purified.

The overall energetics of bacterial growth on iron are one of the most interesting and complicated aspects of the system. Silverman and Lundgren (1959) calculated that 504 kJ are needed for the fixation of 1 mole of carbon at 100% growth efficiency. Thus, if the energy obtained from iron oxidation and the efficiency of growth are known, the amount of ferrous iron oxidized as a function of growth can be estimated. Reported growth efficiencies range from 3.2% (Temple & Colmer 1951) to 20.6% (Silverman & Lundgren 1959). The latter authors predicted that about 90 moles of Fe(II) should be consumed per mole of carbon fixed. Experiments by them and others have shown values from 50 to 100 moles iron consumed per mole of carbon fixed. Clearly, for a small amount of growth very large environmental effects on iron occur if the organisms grow on iron alone. Such hypothetical calculations must be tempered by the knowledge that in the environments where iron oxidation occurs, reduced sulphur is almost always available. If the organisms can also harvest energy from sulphur oxidation, their abundance in many acidic environments is indeed to be expected. In fact, it is hard to construct a significant ecological role for these bacteria based on iron oxidation alone. If they could oxidize only iron, they would be restricted to acidic environments containing high iron concentrations, and would be dependent on other organisms for pH maintenance and iron stabilization. Iron oxidation itself produces soluble ferric iron which is stable in the acidic environment in which the bacteria thrive, but as the dissolved iron moves downstream or into higher pH environments, precipitation occurs, presumably as described above (Murray 1979). In other words, acidophilic bacterial iron oxidation is usually dependent on sulphur oxidation (environmental acidification), whether the oxidation is accomplished by enzymes from the iron oxidizing organism itself, or by an accompanying organism. In terms of ecological significance then, the sulphur metabolism of the *Thiobacilli* must be regarded as more significant than iron oxidation, the acidification of the environment through the production of sulphuric acid and the resulting stabilization of ferrous or ferric iron in solution for periods long enough to allow substantial mobility and movement.

The neutral pH iron bacteria are a controversial group; whether or not they actually oxidize iron, or merely accumulate it is still disputed. The problem is, of course, that they catalyse a process that occurs spontaneously under the conditions in which they grow. Thus in no case has it been possible to unequivocally demonstrate autotrophic growth (or even that iron oxidation is enzymatic), although there are many examples of organisms that catalyse iron oxidation and/or accumulation at rates significantly faster than occurs in the absence of the organisms (Table 6.4). It is curious that many of these bacteria are structurally complex, with sheathed, budding or stellate forms predominating. Both laboratory experiments in which iron oxidation is enhanced by the presence of bacteria, and ecological studies of bacterial distribution in various environments have suggested that these various bacteria (Table 6.4) are active in the catalysis of iron oxidation and/or accumulation in nature (Harder 1919, Silverman & Ehrlich 1964; Perfil'ev *et al.* 1965, Drabkova & Stravinskaya 1969, Kuznetsov 1970, Ghiorse & Hirsch 1982).

It has long been hypothesized and argued that the neutral pH iron bacteria might obtain energy for growth from iron oxidation. Autotrophic metabolism has been proposed for sheathed forms (*Sphaerotilus–Leptothrix* group) and for *Gallionella*, but unequivocal proof has not been presented. Wolfe (1963) hypothesized that some of the neutral pH iron bacteria are gradient organisms, capable of establishing themselves at anaerobic–aerobic interfaces where they can successfully compete with autooxidation as it occurs via molecular oxygen. This would allow the bacteria to harvest energy from iron oxidation in microenvironments where oxygen gradients are established. Recent work on the ecology and distribution of iron bacteria indicate that they are indeed found in specialized local environments, but whether oxygen is the determining factor in this distribution is not known (Ghiorse *et al.* 1980).

Many iron bacteria probably do not oxidize iron but are, nevertheless, commonly associated with accumulated iron oxides. It is common for bacteria to take up chelated iron compounds leaving the iron deposited on the cell surface (Kuznetsov 1970, Hirsch 1974). In many cases, the iron is undoubtedly already oxidized, and the bacterial activity is one of precipitation or accumulation.

The sheathed bacteria, the *Sphaerotilus–Leptothrix* group (vanVeen *et al.* 1978), are small Gram-negative motile rods that can form filaments surrounded by sheaths. The sheaths can be heavily encrusted with metal oxides, some species accumulating either iron or manganese and others accumulating both (Ghiorse *et al.* 1980). Autotrophic growth has been proposed but not proven, and it seems likely that iron oxidation proceeds spontaneously while organic compounds are utilized for energy. Sheathed bacteria are often abundant in well-preserved natural iron deposits as discussed by Harder (1919), who favoured the view that bacteria played a

Table 6.4. Iron oxidizing and depositing microorganisms.[1]

I. Acidophilic iron oxidizers:	
	Thiobacillus ferrooxidans[1]
	Ferrobacillus ferrooxidans
	Ferrobacillus sulfooxidans
	Sulfolobus[2]
	Leptospirillum[3]
	Metallogenium[4]
II. Neutral pH iron bacteria[5]	
(a) Deposition apparently directly on surfaces	
1. cocci—	*Siderococcus*
	Planctomyces
	Caueococcus
	Naumaniella
2. Trichomes—	*Peloploca*
3. Hyphae	*Hyphomicrobium*
	Pedomicrobium
4. Branched filaments—	*Actinomyces* spp.
5. Open coat—	*Ochrobium tactum*
(b) Deposition on visible polymer layers	
	Siderocapsa (Arthrobacter)[6]
	Thiopedia
	Leptothrix (Sphaerotilus)
	Clonothrix
	Crenothrix
(c) Deposition on erected stalks	
	Planktomyces
	Gallionella
	Toxothrix
(d) Bacteria without rigid cell walls	
	Metallogenium[7]
	Acholeplasma
III. Fungi	
	Papilospora[8]
	Aureobasidium[9]
	Coniothyrium[9]
	Cryptococcus[9]

[1] Lundgren & Dean 1979.
[2] Brock *et al.* 1972, Brierley & Brierley 1973.
[3] Balashova *et al.* 1974.
[4] Walsh & Mitchell 1972.
[5] Hirsch 1974.
[6] Dubinina & Zhdanov 1975.
[7] Dubinina 1969.
[8] Beijerink 1913.
[9] Mulder 1972.

positive role in metal (ore) deposition. More recent ecological studies (Caldwell & Caldwell 1980, Ghiorse *et al.* 1980, Wieczorek *et al.* 1980), have investigated both the distribution and ultrastructure of lake and cave iron deposits. There seems little doubt from these studies that members of the

Leptothrix group are involved with the accumulation and precipitation of iron oxides in these environments, but the nature of the involvement remains unknown.

The budding bacteria form new growth processes (buds) that are extensions of the previous cell, thus assuming a variety of morphologies. *Hyphomicrobium* and *Pedomicrobium* species have been found associated with iron and manganese deposits in soils, lakes and shallow marine basins (Tyler & Marshall 1967a, Gebers & Hirsch 1978, Ghiorse & Hirsch 1978, 1979, 1982, Ghiorse 1980). These forms tend to be quite pleomorphic (T'ieng Ha-Mung 1967, Tyler & Marshall 1967b), with some of their morphologies being similar to fossilized bacterial remains seen in sediments located in and around ancient iron formations (Barghoorn & Tyler 1965). Like sheathed bacteria, the budding forms may oxidize both iron and manganese, or be restricted to one or the other metals (Ghiorse & Hirsch 1978). External acidic polysaccharides appear to be involved with the chelation and subsequent oxidation of the metals (Ghiorse & Hirsch 1979), and there is no evidence that iron oxidation is coupled to the metabolism in any direct way.

Gallionella ferruginea, one of the first iron bacteria ever recognized, is sometimes grouped with the budding bacteria (Hirsch 1974). The small, bean-shaped cells of this organism produce long cell-associated stalks of $Fe(OH)_3$, presumably the excretion product of growth on ferrous iron. It is microaerophilic, preferring 0.1 to 1.0 mg oxygen ml^{-1} (Wolfe 1963). It has been hypothesized to be a chemolithotroph or autotroph because it: (a) grows in a mineral salts medium in the absence of added organic carbon; (b) will not grow without ferrous iron; and (c) is capable of CO_2 assimilation (Kucera & Wolfe 1957, Hanert 1968). However, unequivocal evidence of autotrophy has not been presented and, since the studies cited above, little has been done with this organism to complete the understanding of its physiology and biochemistry.

Perhaps the most controversial iron bacteria are members of the group *Metallogenium*. Several species in this genus have been described, including *M. invisum*, *M. personatum* and *M. symbioticum* (Hirsch 1974), but little success has been achieved in obtaining pure cultures in the laboratory. The extant species are associated primarily with manganese oxides, although in nature, up to 30% of the *Metallogenium*-associated deposits may contain iron (Perfil'ev *et al.* 1965). Some workers have successfully cultured the symbiotic species, *Metallogenium symbioticum*. *Metallogenium personatum*, however, has never been cultivated (Zavarzin 1982). Dubinina (1969) has placed them among the group *Mycoplasmatales*, while others have been unable to reproduce these results and have challenged the validity of the genus *Metallogenium* (Schweisfurth *et al.* 1978). On the basis of field studies,

Metallogenium has been implicated in the accumulation of iron deposits (see Kuznetsov 1970, for discussion), and recent studies (Gregory *et al.* 1980) have shown that *Metallogenium* morphytes are associated with newly formed particulate manganese in Lake Washington (Fig. 6.4). However, the latter workers also concluded that the *Metallogenium*-like microcolonies seen in the lake were not the living form of an organism. The controversy is clearly not settled.

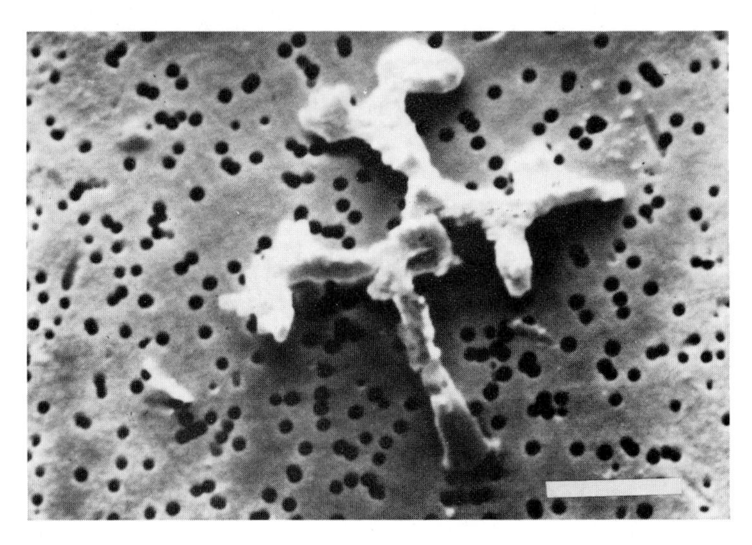

Fig. 6.4. These '*Metallogenium*' structures have been found in many lakes, usually at the borderline, at which oxygen pressure decreases and before anoxic conditions of the hypolimnion prevail, i.e. where oxygen still is available and there are large amounts of reduced iron and manganese. The structures usually do not show clear internal membrane organization in TEM micrographs. They have not been cultivated successfully; bar = 1 μm (photograph courtesy of B.Hickel).

Krumbein (1969, 1971) and Krumbein and Jens (1981) have isolated numerous Fe- and Mn-precipitating microorganisms from recent sediments and from rock varnishes. Some of the isolates, with which artificial rock varnishes were produced in the laboratory had epiphytes, which closely resembled the structures described as *Metallogenium symbioticum* (Figs 6.5–6.9).

The interest in *Metallogenium* springs not only from its abundance in present-day deposits and areas of active metal deposition, but also from its similarity to fossil organisms. In studies of the lower Gunflint iron formations, Barghoorn and Tyler (1965) reported abundant microfossils called *Eoastrion simplex*, which were morphologically similar to modern *Metallogenium*. Similar microfossils have been seen in many other iron

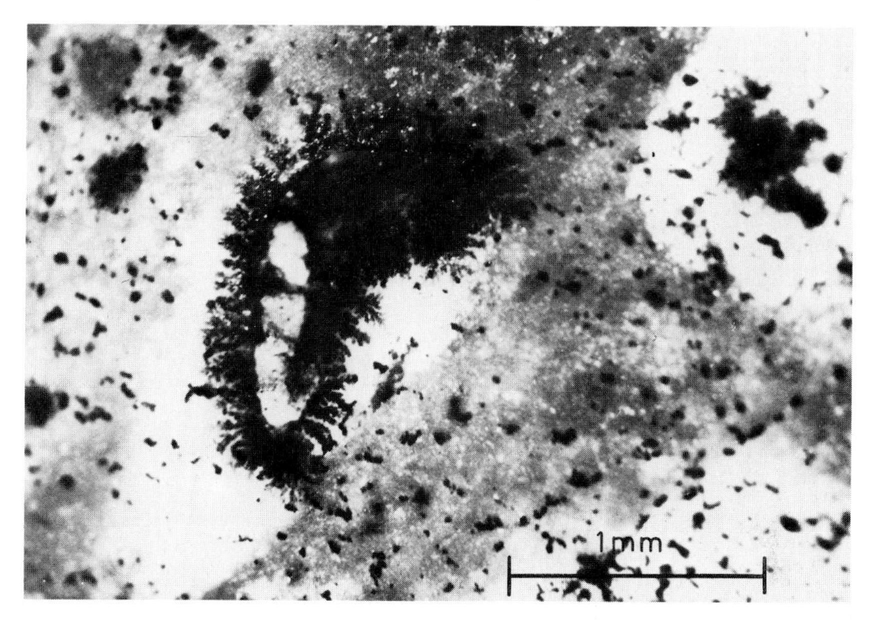

Fig. 6.5. Biogenic dendritic rock or desert varnish of the Negev (Israel). From the centre of the black coating fungi have been isolated, which apparently are associated with *Metallogenium symbioticum* (Krumbein & Jens 1981).

deposits, lending strength to the hypothesis that *Metallogenium*-like organisms were important in the formation of these early deposits.

Krumbein (1983) and Krumbein *et al.* (1983) have recently compared cretaceous iron stromatolites, which were produced by fungal activity with similar morphologies produced in the laboratory during experiments for laboratory-made biogenic rock varnishes or rock coatings. These corresponded very well with the fossil material (Fig. 6.10). They strongly suggest not to interpret these spherical agglomerates as fossil *Metallogenium*. The culture experiments with fungi yielded such structures by several different physiological and growth patterns. The advantage of iron and manganese deposition in the case of the rock varnish seems to be that a protective coating is formed, which protects the organisms from dessication and damaging irradiation. Physiological experiments on possible energy gains by the oxidation of iron in the case of the fungi have not yet been reported.

Considering the number of different types of iron bacteria (Table 6.4), and their apparent abundance in nature, it is interesting to speculate as to what, if any, the advantages of iron oxidation might be. First, bacteria may derive energy from iron oxidation, an obvious advantage in nutrient-limited environments; for the acid-tolerant forms this is so, but in no other case has it been shown to be true. Iron oxidation may have developed as a mechanism of oxygen detoxification (removal). The presence of a simple cell-associated oxygen removal system such as a sheath or extracellular component might

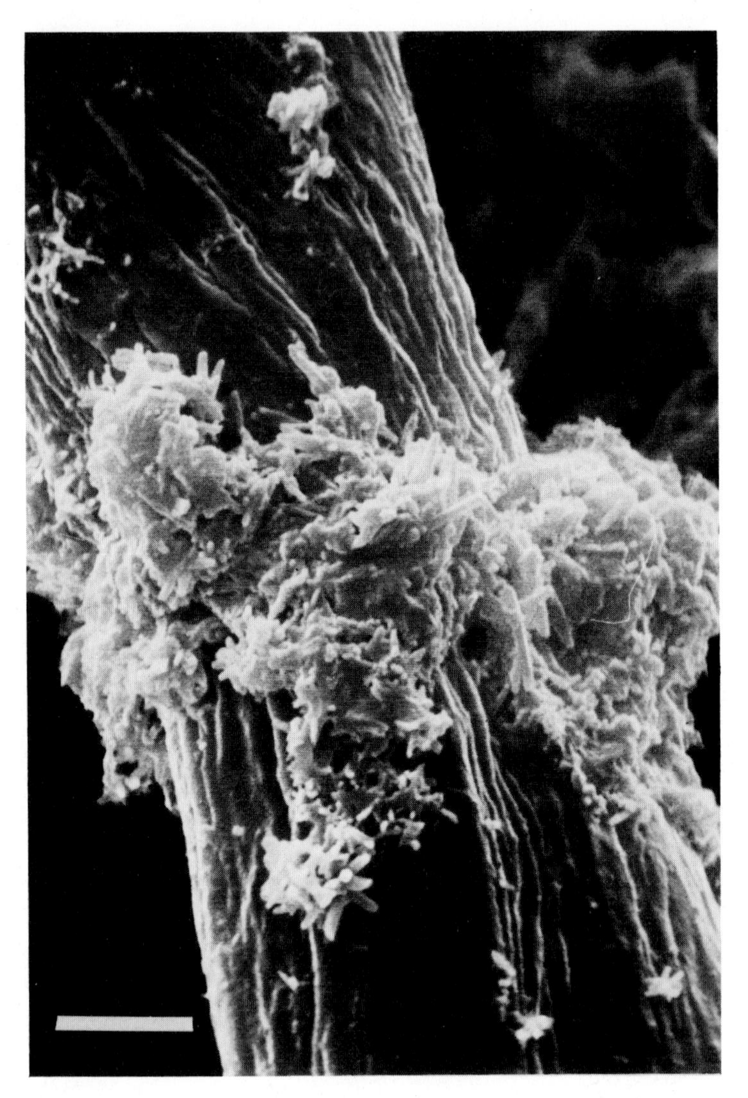

Fig. 6.6. Fungal hyphae from an isolate of the dendritic desert varnish shown in Fig. 6.5. An epiphyte, or an actinomycete *Metallogenium symbioticum*, grows on the fungal filament in structures comparable to those described by Dubinina (1969). These are heavily incrusted with Fe- and Mn-oxides. In the lower right and upper left, the budding (branching) epiphyte is visible, while the central part is mostly crystallized; bar = 2 μm (Krumbein & Jens 1981).

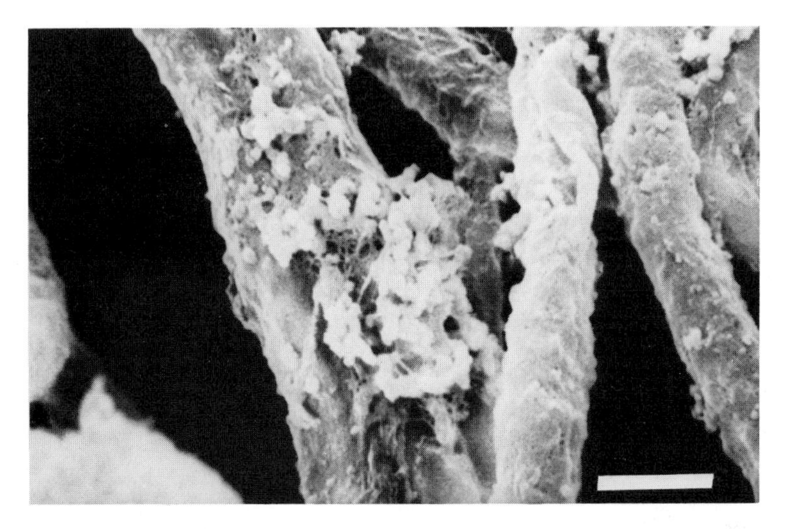

Fig. 6.7. Granular Fe-Mn-oxides with Mn predominating from a fungus, isolated from rock varnish. The structure of this deposit corresponds to pictures given for *Metallogenium* or possibly for budding bacteria or actinomycetes; bar = 2 μm (photograph courtesy of W.E.Krumbein).

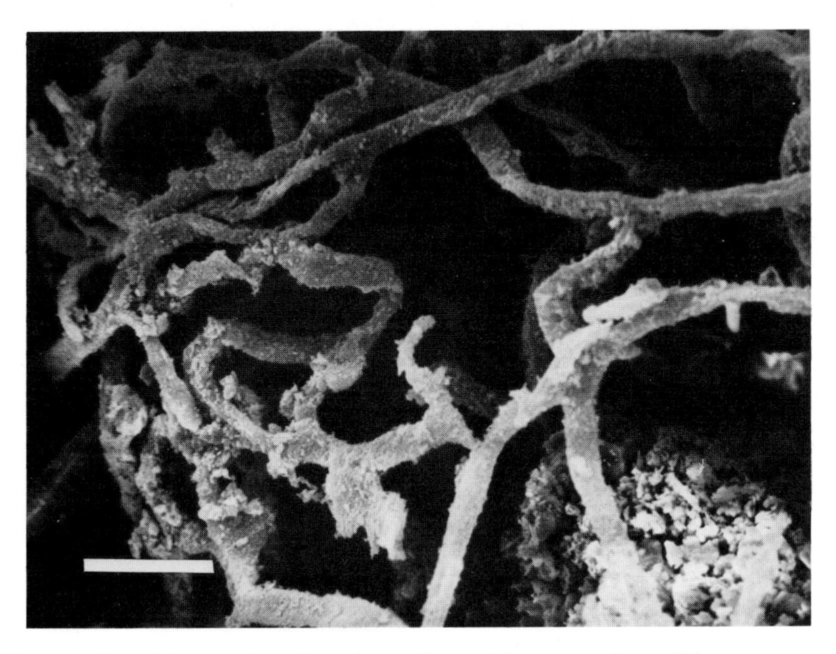

Fig. 6.8. Fe-oxide deposit on fungal mycelium without any discernible structures of associated *Metallogenium* or budding bacteria. This culture was used to produce rock varnishes in the laboratory on cleaned rock surfaces; bar = 20 μm (photograph courtesy of W.E.Krumbein).

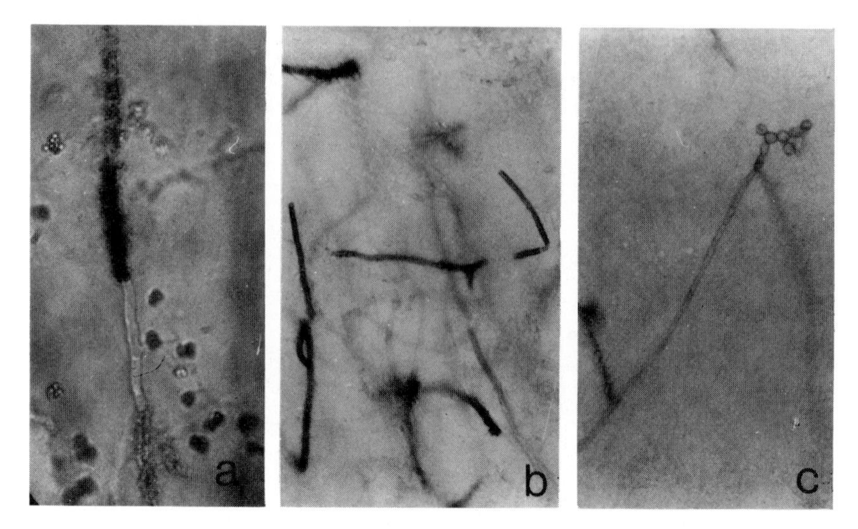

Fig. 6.9. Fe and Mn deposits on mycelia of two different Fe- and Mn-precipitating fungi, isolated from rock varnish (Krumbein & Jens 1981). (a) Very small particles are produced at and near the cell wall. (b) Fe deposit representing closely cell wall-related incrustations on strain N-17. No explanation has been found so far for the abrupt intermissions of precipitates on fungal hyphae. (c) Chlamydospores of strain N-17 with Mn-oxide deposits. The fractionate staining was achieved by using Berbelin-blue for Mn(IV) and Berlin-blue for Fe(III) selective staining (Krumbein & Altmann 1973); magnification × 260 (photograph courtesy of W.E.Krumbein).

have been of great advantage during the periods before the evolution of enzymes for protection against oxygen. This hypothesis views iron oxidation as an extracellular oxygen sink of no particular advantage in terms of energy production but of great advantage as an oxygen-protective mechanism. It might be noted that protection from oxygen toxicity is a postulated function of iron in several proteins, including haemoglobin, peroxidases, catalase and some superoxide dismutases, so the hypothesis that iron might have served as a primitive extracellular oxygen-protection mechanism is a tempting one.

The deposition of insoluble iron or metal oxides around bacteria might of course be a significant disadvantage in some situations, and it may not be surprising that organisms commonly associated with iron oxidation are sheathed or budding forms, able to escape from their precipitates by a variety of mechanisms. It may also be possible that the precipitates themselves confer some advantage on the bacteria, such as stabilization of cell walls or holdfasts or the formation of protective cell coverings ((Figs 5.7 and 6.9; Krumbein 1969, 1980, Caldwell & Caldwell 1980).

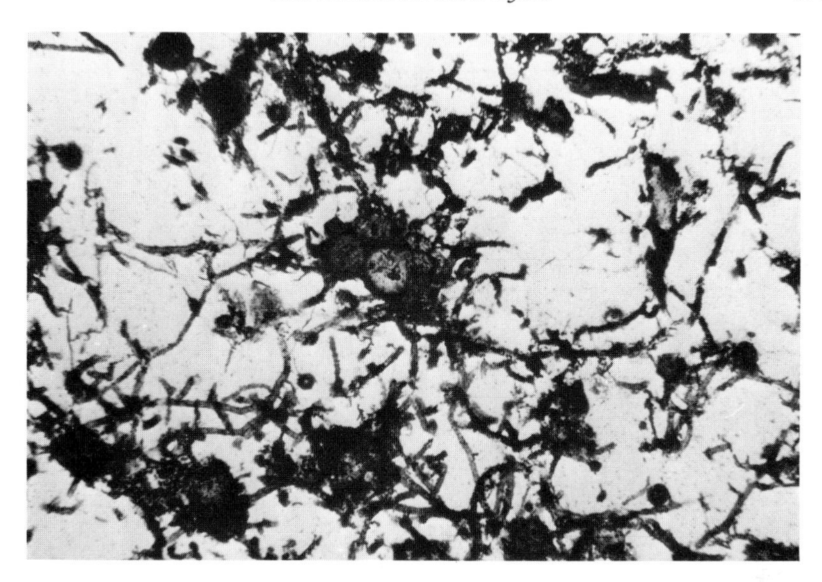

Fig. 6.10. Encrusted fungal mycelia from Cretaceous iron ore deposit. These deposits show a variety of fungal Fe-encrusted fossils, identical in their morphologies to laboratory-produced fungal iron deposits (Krumbein 1983).

6.3.3 Iron reduction by microorganisms

Although photoreduction of iron can lead to Fe(II) accumulation in acidic lakes (Colliene 1983), in the aerobic environment that characterizes most of the present-day earth, reduced iron is rare. However, in many anaerobic zones iron reduction and mobilization occurs; environments such as anaerobic marine sediments (Froelich *et al.* 1979), bogs (Crerar *et al.* 1979) and almost any stratified environment in which organic input is sufficient to allow oxygen depletion (Kuznetsov 1970). Almost without exception, such zones are established as the result of microbial activities. In experiments where microbial growth was inhibited with poisons, the reduction and restratification of iron in homogenized sediment samples did not occur (Troshanov 1965). Several workers have studied the distribution of bacteria that can reduce iron, and all concur that iron reducers are diverse and widely distributed (Bromfield 1954, Pochon *et al.* 1964, Troshanov 1965, 1968, 1969, Ottow 1968, 1969). *Bacillus* species are particularly widespread, abundant and active in iron reduction, although many other bacterial genera also have this capacity (Table 6.5). Recent studies of marine sediments (Sorensen 1982) and a stratified lake (Jones *et al.* 1983) both concluded that iron reduction was strongly influenced by bacteria.

Table 6.5. Iron-reducing bacteria.

Genus	Species	Reference	Comments
Bacillus	*circulans*	Bromfield 1954, Troshanov 1969	
	megaterium	Roberts 1947, Bromfield 1954	
	mesentericus	Troshanov 1969	
	polymyxa	Ottow 1968	DeCastro and Ehrlich showed that *Bacillus* 29A reduced iron in limonite best, then goethite and hematite
	spp.	Ottow 1969, DeCastro & Ehrlich 1970	
Aerobacter	*aerogenes*	Ottow 1969	
Pseudomonas	*aeruginosa*	Ottow 1968, Troshanov 1968, Cox 1980	Cox showed that only complexed iron could be reduced by cell-free extracts
	liquefaciens	Troshanov 1969	
	cloacae	Ottow 1968	
	spp.	Ottow 1969	
Escherichia	*coli*	Halvorson & Starkey 1927	
	spp.	Ottow 1969	
Proteus	*vulgaris*	Troshanov 1968	
	spp.	Ottow 1969	
Clostridia	spp.	Ottow 1969	
Achromobacter	spp.	Ottow 1969	
Staphylococcus	spp.	Lascelles & Burke 1978	active cell-free extracts obtained

What are the mechanisms of iron reduction? Ferric iron can be chemically reduced by many compounds, some of which are characteristic excretion products of microbes grown anaerobically (e.g. formate or sulphide) as well as by lowering the Eh and pH of the environment and establishing conditions that favour reduced over oxidized phases of iron (Fig. 6.2). Such indirect mechanisms were proposed by some early workers (Halvorson & Starkey 1927) and undoubtedly occur to some extent in nature. In addition, however, many bacteria have the capacity to reduce iron under conditions in which it would not spontaneously be reduced (Bromfield 1954, Troshanov 1968, 1969, Ottow 1968, 1969, 1970, Lascelles & Burke 1969).

The exact mechanisms of iron reduction remain obscure, and it is likely that there may be several. Nitrate inhibits bacterial iron reduction in strains possessing nitrate reductase (Ottow 1968), leading to the speculation that the electron transport chain to nitrate is involved. In fact nitrate reductase has been suggested as a possible electron donor to iron in several genera including *Aerobacter*, *Bacillus*, *Serratia* and *Pseudomonas* (Ottow 1970). Several nit$^-$ mutants (lacking the capacity to reduce nitrate) were also deficient in iron reduction; the exact nature of the mutants was, however, not determined. Furthermore, in some nit$^-$ mutants, the amount of iron reduction was equal to or greater than that of the wild type (nit$^+$). Other studies using cell-free extracts of *Staphylococcus aureus* have shown that electrons are donated to iron from a component located before cytochrome b in the respiratory chain to nitrate reductase, and not from the nitrate reductase itself (Lascelles & Burke 1978). It may well be that many iron-sulphur proteins, including the one in the nitrate reductase complex have the capacity to reduce iron (Lascelles, pers. comm.), and many mechanisms will thus be possible. Field studies of iron reduction (Sørensen 1982; Jones *et al.* 1983) also support the idea that faculative nitrate reducers are involved in iron reduction. Alternatively, Obuekwe *et al.* (1981) have proposed that the inhibitory effect of NO_3 is due to the chemical oxidation of FE(II) by NO_2.

Roberts (1974) reported alteration of the fermentation balance of *Bacillus polymyxa* by ferric iron; this presumably occurred because the ferric iron was being used as an alternate electron acceptor. If so, it suggests that there should be organisms that grow anaerobically using ferric iron as the only electron acceptor (Ottow 1969). Jones (1983) has reported a faculative chemolithotroph that meets these criteria.

The preservation potential of oxidized iron is reasonably good, and the accumulation of iron into permanent deposits is well known. However, the weathering of iron, both chemical and biological, is an important component of the global cycle (Fig. 6.1), and thus the study of bacteria involved with the reduction and mobilization of iron may lend considerable insight into the understanding of the role of microbes in the weathering component of the iron cycle (Pochon *et al.* 1964, Krumbein 1972). As can be seen from the above discussion, these studies are only in their infancy.

6.3.4 Magnetite formation by bacteria

The subject of iron interactions is incomplete without mentioning the recently discovered magnetite-containing bacteria (Blakemore 1975). These bacteria, called magnetotactic because they respond to magnetic lines of force, contain intracellular 'organelles' (Balkwill *et al.* 1980) composed of crystalline magnetite (Frankel *et al.* 1979). The magnetite is formed *in vivo* by the bacteria grown on ferrous iron supplied in the growth medium (Blakemore *et al.* 1979). Magnetotactic bacteria are abundant in a variety of sedimentary environments (Moench & Konetzka 1978, Blakemore *et al.*

1979), and presumably all are rendered magnetotactic by the presence of intracellular magnetite. In contrast to most biota, where iron rarely exceeds 0.02% of the total dry weight, iron can represent as much as 3–4% of the dry weight of the magnetotactic bacteria (Moench & Konetzka 1978, Frankel *et al.* 1979). The ecological role of magnetotaxis for bacteria is not yet clear, nor are any of the mechanisms for either iron uptake or magnetite synthesis. It should be noted, however, that magnetite functions as a magnetic sensor in many organisms, from bacteria to birds (Frankel *et al.* 1979), and whether or not bacteria are involved in the transfer of magnetite through the food web will be an interesting and important question. One might also speculate that if populations of magnetotactic bacteria reach high levels under natural conditions, then the deposition of sediments containing large numbers of magnetotactic bacteria could result in significant iron (magnetite) accumulation due simply to bacterial uptake oxidation, and concentration of iron.

6.3.5 Indirect effects of microbes on iron

In addition to direct effects of microbes on iron there are many important indirect effects which are the result of biological activity (Silverman & Ehrlich 1964). First and foremost are those of the cyanobacteria and photosynthetic eukaryotes which, by virtue of possessing photosystem II, evolve molecular oxygen. All autotrophs consume CO_2, often increasing environmental pH and stimulating iron precipitation. Many fungi and bacteria, when growing on protein or amino-acid-rich substrates excrete basic by-products such as ammonia; through pH effects these products can stimulate iron oxidation. Several of the reactions already mentioned above are indirect or non-specific effects: uptake of iron-containing organics; local pH/Eh changes; and excretion of strong oxidants or reductants.

Iron is also reduced or solubilized by indirect mechanisms. The sulphur-oxidizing *Thiobacilli* are true autotrophs, which gain energy from the oxidation of sulphide in many environments. Since iron sulphide is a commonly precipitated form, the oxidation of the sulphide results in the release of iron into solution (Lundgren & Silver 1980). The dissolved iron is often stable in local environments, as the product of sulphur oxidation (sulphuric acid) makes the environment ideal for iron solubilization and mobilization. At least at the local level, this form of microbiological weathering may be quite important, and it has been exploited economically as a valuable method for the leaching of many metals, including iron, from ores (Kelly *et al.* 1979, Lundgren & Silver 1980, Trudinger *et al.* 1980).

A final type of indirect interaction is that of precipitation of iron by the production of components that combine with and precipitate iron. Iron can combine with a variety of counter ions (Table 6.1), and presumably any

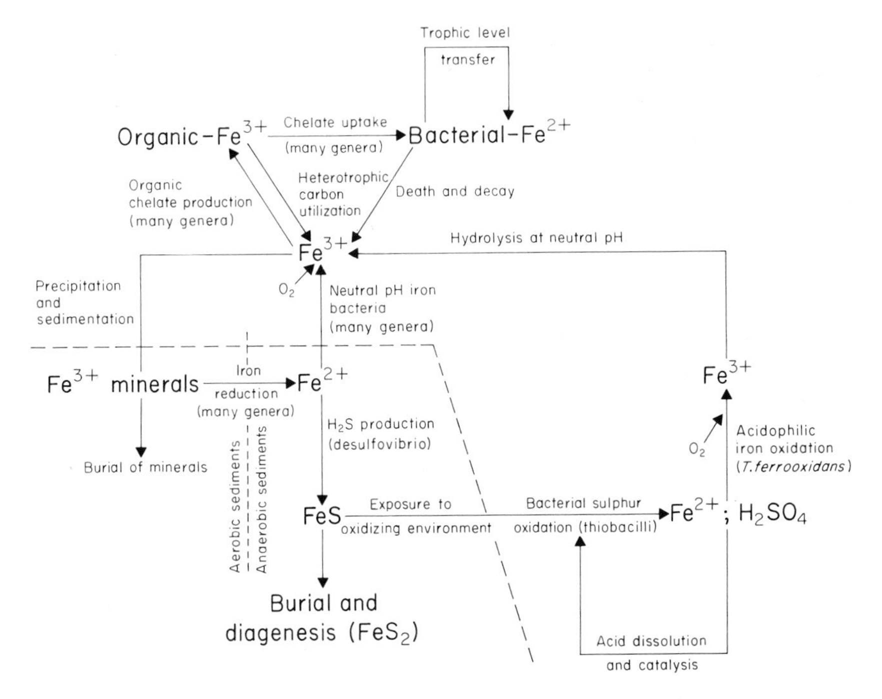

Fig. 6.11. The microbiological iron cycle. The known activities of microbes are arranged into a cycle to show the various ways in which they affect iron chemistry and distribution. These reactions are probably quantitatively important only in selected environments. The heavy dashed line indicates the sediment–water interface, the lighter dashed line represents a redox interface in the sedimentary environment. The aerobic part of the cycle is dominated in acidic environments by iron oxidation, in neutral environments by siderophore production and activity. The anaerobic portion is strongly influenced by sulphide production and/or iron reduction and precipitation as Fe-sulphides.

organisms capable of producing or concentrating such ions could stimulate the formation of iron minerals or precipitates. Especially with regard to marine black muds and other iron sulphide deposits, the role of *Desulfovibrio* species and other sulphate reducers may be significant. Similarly, organisms that can accumulate carbonate or phosphate may cause the precipitation of iron minerals at interfaces where the counter ions and iron can meet and interact. Examples of such processes for environments with redox interfaces have been discussed by Hallberg (1978), who has presented a model for the formation of precipitates in modern sediments.

6.4 MICROBIOLOGICAL IRON CYCLE

Figure 6.11 presents a diagram of the microbial iron cycle. It includes the known direct and indirect effects of microbes: oxidative, reductive and chelating reactions. Such a cycle must be viewed with the proper perspective; it is not to be taken as an indication of a global iron cycle or of any particular significance quantitatively on a planetary scale. Rather, it sets out those reactions and processes known to occur in various microbes and which may be of importance for maintenance of iron to the biota. In general, iron bacteria are important in the accumulation and deposition of iron in local environments, although whether or not they are actively involved in oxidation is often enigmatic. The iron-reducing microbes may act specifically or non-specifically, and whether iron oxides can be utilized directly as electron acceptors in anaerobic environments is not proven. It seems beyond question that iron acts externally both as an electron donor and acceptor for many different microbial reactions, and the interplay between the biota and iron is often environmentally significant. Of particular significance to the biota are the chelation reactions in which iron is stabilized via the formation of organic ligands; such organic chelates may be of great importance in making iron available to higher trophic levels which would be otherwise to scavenge iron from the aquatic environments in which they exist.

References

Aisen P. & Listkowsky I. (1980) Iron transport and storage proteins. *Ann. Rev. Biochem.* **49,** 357–93.

Aleem M.I.H., Lees H. & Nicholas D.J.D. (1963) Adenosinetriphosphate dependent reduction of nicotinamide adenine dinucleotide by ferrocytochrome c in chemoautotrophic bacteria. *Nature* **200,** 759–61.

Balashova V.V., Vedenina I.Y., Markosyan G.E. & Zavarzin G.A. (1974) *Leptospirillum ferroxidans* and peculiarities of its autotrophic growth. *Mikrobiologya* **43,** 581–5.

Balkwill D.L., Maratea D. & Blakemore R.P. (1980) Ultrastructure of a magnetotactic spirillum. *J. Bacteriol.* **141,** 720–9.

Barghoorn E.S. & Tyler S.A. (1965) Microorganisms from the gunflint chert. *Science* **147,** 563–7.

Beijerinck M.W. (1913) Oxydation des Mangancarbonates durch Bakterien und Schimmelpilze. *Folia Microbiologica* **2,** 123–34.

Blakemore R.P. (1975) Magnetotactic bacteria. *Science* **190,** 377–9.

Blakemore R.P., Maratea D. & Wolfe R.S. (1979) Isolation and pure culture of a fresh water magnetotactic spirillum in chemically defined medium. *J. Bacteriol.* **140,** 720–9.

Blaylock B.A. & Nason A. (1963) Electron transport systems of the chemoautotroph *Ferrobacillus ferrooxidans*. *J. Biol. Chem.* **238,** 3453–62.

Brierley C.L. & Brierley J.A. (1973) A chemoautotrophic and thermophilic microorganism isolated from an acid hot spring. *Can. J. Microbiol.* **19,** 183–8.

Brierley C.L. & Murr L.E. (1973) Leaching: use of a thermophilic and chemoautotrophic microbe. *Science* **179,** 488–90.

Brock T.D., Brock K.M., Belly R.T. & Weiss R.L. (1972) Sulfolobus—new genus of sulfur oxidizing bacteria living at low pH and high temperature. *Arch. Microbiol.* **84,** 54–6.

Broda E. & Peschek G.A. (1979) Did respiration or photosynthesis come first? *J. theor. Biol.* **81,** 201–12.

Bromfield S.M. (1954) Reduction of ferric compounds by soil bacteria. *J. Gen. Microbiol.* **11,** 1–16.

Byers B.R. & Arceneaux J.E.L. (1977) Transport and utilization of iron. In *Microorganisms and Minerals* (Ed. E.D.Weinberg), pp. 215–49. Marcel Dekker Inc, New York.

Cailleux A. (1969) Ein Beitrag zu Krumbein: Über den Einfluß der Mikroflora auf die exogene Dynamik (Verwitterung und Krustenbildung). *Geol. Rdsch.* **58,** 363–5.

Caldwell D.E. & Caldwell S.J. (1980) Fine structure of in situ microbial iron deposits. *Geomicrobiol. J.* **2,** 39–53.

Cloud P. (1973) Paleoecological significance of the banded iron-formation. *Economic Geol.* **68,** 1135–43.

Cloud P. (1974) Evolution of ecosystems. *Amer. Sci.* **62,** 54–66.

Cobley J.G. & Haddock B.A. (1975) The respiratory chain of *Thiobacillus ferrooxidans*: the reduction of cytochromes by Fe^{++} and the preliminary characterization of rusticyanin a novel blue copper protein. *FEBS Letters* **60,** 29–33.

Colliene R.H. (1983) Photoreduction of iron in the epilimnion of acidic lakes. *Limnol. Oceanogr.* **28,** 83–100.

Cox C.D. (1980) Iron reductases from *Pseudomonas aeruginosa*. *J. Bacteriol.* **141,** 199–204.

Crerar D.A., Knox G.W. & Means J.L. (1979) Biogeochemistry of bog iron in the New Jersey pine barrens. *Chem. Geol.* **24,** 111–35.

Crosa J.H. (1980) A plasmid associated with virulence in the marine fish pathogen *Vibrio anguillarum* specifies an iron-sequestering system. *Nature* **284,** 566–8.

DeCastro A.F. & Ehrlich H.L. (1970) Reduction of iron oxide minerals by a marine *Bacillus*. *Ant. Von. Leeuwenhoek* **36,** 317–27.

Deevey E.S. (1970) Mineral Cycles. *Scientific American* **223,** 148–58.

Din G.A., Suzuki I. & Lees H. (1967a) Ferrous iron oxidation by *Ferrobacillus ferrooxidans*: Purification and properties of Fe^{++}-cytochrome c reductase. *Can. J. Biochem.* **45,** 1523–46.

Din G.A., Suzuki I. & Lees H. (1967b) Carbon dioxide fixation and the phosphoenolpyruvate carboxylase in *Ferrobacillus ferrooxidans*. *Can. J. Microbiol.* **13,** 1413–20.

Drabkova V.G. & Stravinskaya E.A. (1969) Role of bacteria in the dynamics of iron in Lake Krasnoe. *Mikrobiologya* **38,** 300–9.

Dubinina G.A. (1969) Inclusion of *Metallogenium* among the Mycoplasmetales. *Dokl. Akad. Nauk. SSSR* **184,** 87–90.

Dubinina G. & Zhdanov A.V. (1975) Recognition of the iron bacteria '*Siderocapsa*' as *Arthrobacter* and description of *Arthrobacter* siderocapsulatus sp. nov. *J. Syst. Bacteriol.* **25,** 340–9.

Dugan P.R. & Lundgren D.G. (1965) Energy supply for the chemoautotroph *Ferrobacillus ferrooxidans*. *J. Bacteriol.* **89,** 825–34.

Eaton A. (1979) Removal of 'soluble' iron in the Potomac River Estuary. *Estuar. Coastal Mar. Sci.* **9,** 41–9.

Elwell L.P. & Shipley P.L. (1980) Plasmid-mediated factors associated with virulence of bacteria to animals. *Ann. Rev. Microbiol.* **34,** 465–96.

Frankel R.B., Blakemore R.P. & Wolfe R.S. (1979) Magnetite in freshwater magnetotactic bacteria. *Science* **203,** 1355–6.

Froelich P.N., Klinkhammer G.P., Bender M.L., Luedtke N.A., Heath G.R., Cullen D. & Dauphin P. (1979) Early oxidation of organic matter in pelagic sediments of the eastern equatorial Atlantic: suboxic diagenesis. *Geochim. Cosmochim. Acta* **43,** 1075–90.

Gale N.L. & Beck J.V. (1967) Evidence for Calvin cycle and hexose monophosphate pathway in *Thiobacillus ferrooxidans. J. Bacteriol.* **94,** 1052–60.

Garrels R.M. & Christ C.L. (1965) *Solutions, Minerals and Equilibria.* Harper & Row, New York.

Gebers R. & Hirsch P. (1978) Isolation and investigation of *Pedomicrobium* spp., heavy metal depositing bacteria from soil habitats. In *Environmental Biogeochemistry and Geomicrobiology*, Vol. 3 (Ed. W.E.Krumbein), pp. 911–22. Ann Arbor Science, Ann Arbor, Mich.

Ghiorse W.C. (1980) Electron microscopic analyses of metal-depositing microorganisms in surface layers of Baltic Sea ferromanganese concretions. In *Biogeochemistry of ancient and modern environments* (Eds P.A.Trudinger, M.R.Walter & O.J.Ralph), pp. 345–54. Springer, New York.

Ghiorse W.C. & Hirsch P. (1978) Iron and manganese deposition by budding bacteria. *Env. Biogeochem. and Geomicrobiol.* **3,** 897–909.

Ghiorse W.C. & Hirsch P. (1979) An ultrastructural study of iron and manganese deposition associated with extracellular polymers of *Pedomicrobium*-like budding bacteria. *Arch. Microbiol.* **123,** 213–26.

Ghiorse W.C., Wieczorek L. & Hirsch P. (1980) Ultrastructural analysis of *Leptothrix* in Fe-Mn containing surface films of two shallow ponds. *Proc. Ann. Meeting Amer. Soc. Microbiol.* **N-101,** 180.

Ghiorse W.C. & Hirsch P. (1982) Isolation and properties of ferromanganese-depositing budding bacteria from Baltic Sea ferromanganese concretions. *Appl. Env. Microbiol.* **43,** 1464–72.

Gregory E., Perry R.S. & Staley J.T. (1980) Characterization, distribution and significance of *Metallogenium* in Lake Washington. *Microbial Ecol.* **6,** 125–40.

Hallberg R. (1978) Metal–organic interaction at the redoxcline. In *Environmental Biogeochemistry and Geomicrobiology*, Vol. 3. (Ed. W.E.Krumbein), pp. 948–53. Ann Arbor Science, Ann Arbor.

Halvorson H.O. & Starkey R.L. (1927) Studies on the transformation of iron in nature II. Concerning the importance of microorganisms in the solution and reduction of iron. *Soil Sci.* **24,** 381–402.

Hanert H. (1968) Untersuchungen zur isolierung, stoffwechselphysiologie und morphologie von *Gallionella ferruginea. Arch. Mikrobiol.* **60,** 348–76.

Hantke K. & Braun V. (1975) Membrane receptor dependent iron transport in *Escherichia coli. FEBS Letters* **49,** 301–5.

Harder E.C. (1919) Iron depositing bacteria and their geologic relations. *U.S. Geol. Survey, Professional Paper No. 113.* Government Printing Office, Washington, D.C.

Hartmann A. & Braun V. (1980) Iron transport in *Escherichia coli*: uptake and modification of ferrichrome. *J. Bacteriol.* **143,** 246–55.

Hirsch P. (1974) Budding bacteria. *Ann. Rev. Microbiol.* **28,** 391–444.

Ingledew W.J., Cox J.C. & Halling P.J. (1977) A proposed mechanism for energy conservation during Fe^{++} oxidation by *Thiobacillus ferrooxidans*: Chemoosmotic coupling to net H^+ influx. *FEMS Micro. Lett.* **2,** 193–7.

James H.L. (1966) Chemistry of the iron rich sedimentary rocks. Data on geochemistry. *U.S. Geol. Survey. Prof. Paper* **440-W.**

Jones J.G., Gardner S. & Simon B.M. (1983) Bacterial reduction of ferric iron in a stratified entrophic lake. *J. Gen. Microbiol.* **129,** 131–9.

Kelly D.P., Norris P.R. & Brierley C.L. (1979) Microbiological methods for the extraction and recovery of metals. *Symp. Soc. Gen. Microbiol.* **29,** 263–308.

Krumbein W.E. (1969) Über den Einfluß der Mikroflora auf die exogene Dynamik (Verwitterung und Krustenbildung). *Geol. Rdsch.* **58,** 333–63.

Krumbein W.E. (1971) Manganese oxidizing fungi and bacteria in recent shelf sediments of the Bay of Biscay and the North Sea. *Naturwiss.* **58,** 56–7.

Krumbein W.E. (1972) Rôle des microorganisms dans la génèse, la diagénèse et la dégradation des roches en place. *Rev. Ecol. Biol. Sol* **9,** 283–319.

Krumbein W.E. (1983) Stromatolites—The challenge of a term in space and time. *Precambrian Research* **20,** 493–531.

Krumbein W.E. & Altmann H.J. (1973) A new method for the detection and enumeration of manganese oxidizing and reducing microorganisms. *Helgolander Wiss. Meeresunters.* **25,** 347–56.

Krumbein W.E. & Jens K. (1981) Biogenic Rock Varnish of the Negev Desert (Israel)—An ecological study on iron and manganese transformation by cyanobacteria and fungi. *Oecologia* **50,** 25–38.

Krumbein W.E., Gerdes G. & Holtkamp E. (1983) Stromatolites, stromatoloids and potential stromatolites and case studies of three microbial systems generating stromatolites. *Terra Cognita* **3,** 217–18.

Kucera S. & Wolfe R.S. (1957) A selective enrichment for *Gallionella ferruginea. J. Bacteriol.* **74,** 347–50.

Kuznetsov S.I. (1970) *The microflora of lakes and its geochemical activity*. University of Texas Press, Austin, Texas.

Lascelles, J. & Burke K.A. (1978) Reduction of ferric iron by L-lactate and D-L-glycerol-3-phosphate in membrane preparations from *Staphylococcus aureus* and interactions with the nitrate reductase system. *J. Bacteriol.* **134,** 585–9.

Lepp H. (1975) *Geochemistry of Iron*. John Wiley and Sons Inc., New York.

Lewis, G.L. & Goldberg E.D. (1954) Iron in marine waters. *J. Mar. Res.* **13,** 183–95.

Lundgren D.G. & Dean W. (1979) Biogeochemistry of iron. In *Biogeochemical Cycling of Mineral Forming Elements* (Eds P.A.Trudinger & D.J.Swaine) pp. 211–51. Elsevier, Amsterdam.

Lundgren D.G. & Silver M. (1980) Ore leaching by bacteria. *Ann. Rev. Microbiol.* **34,** 263–83.

Lundgren D.G., Vestal J.R. & Tabita F.R. (1974) The iron oxidizing bacteria. In *Microbial Iron Metabolism* (Ed. J.B.Neilands), pp. 457–72. Academic Press, New York.

Maciag J.W. & Lundgren D. (1964) Carbon dioxide fixation in the chemoautotroph *Ferrobacillus ferrooxidans*. *Can. J. Eng.* **48,** 669–76.

Moench T.T. & Konetzka W.A. (1978) A novel method for the isolation and study of a magnetotactic bacterium. *Arch. Microbiol.* **119,** 203–12.

Mulder E.G. (1972) Le cycle biologique tellurique et aquatique du fer et du manganese. *Rev. Ecol. Biol. Sol* **9,** 321–48.

Murray J.W. (1979) Iron oxides. In *Marine Minerals* (Ed. R.G.Burns), pp. 47–98. Mineral Soc. Amer., Washington, D.C.

Neilands J.B. (1973) Microbial iron transport compounds. In *Inorganic Biochemistry* (Ed. G.L.Eichorn) pp. 176–202. Elsevier, Amsterdam.

Neilands J.B. (Ed.) (1974) Iron and its role in microbial physiology. In *Microbial Iron Metabolism*, pp. 1–31. Academic Press, New York.

Neilands J.B. (1977) Siderophores: biochemical ecology and mechanisms of iron transport in *Enterobacteria*. In *Bioorganic Chemistry* (Ed. K.N.Raymond), pp. 3–22. Amer. Chem. Soc., Washington, D.C.

Obuekwe C.A., Westlake W.S. & Cook F.D. (1981) Effect of nitrate on reduction of ferric iron by a bacterium isolated from crude oil. *Can. K. Microbiol.* **27,** 692–7.

Ottow J.C.G. (1968) Evaluation of iron reducing bacteria in soil and the physiological mechanism of iron reduction in *Arthrobacter aerogenes*. *Zeitsch. für allg. Mikrobiol.* **8,** 441–3.

Ottow J.C.G. (1969) The distribution and differentiation of iron reducing bacteria in gley soils. *Z. Bakteriol. Parasitenk. Infektions. Hyg. Abt. II.* **123,** 600–15.

Ottow J.C.G. (1970) Selection characterization and iron reducing capacity of nitrate reductaseless (nit$^-$) mutants of iron reducing bacteria. *Zeitsch. für allg. Mikrobiol.* **10,** 55–62.

Perfil'ev B.V., Gabe D.R., Gal'perina A.M., Rabinovich V.A., Sapotniskii A.A., Sherman E.E. & Troshanov E.P. (1965) *Applied Capillary Microscopy*. (Ed. M.S.Gurevich) Consultants Bureau, New York (translated from Russian).

Pochon J., Chalvignac M.A. & Krumbein W.E. (1964) Récherches biologiques sur le mondmilch. *C.R. Acad. Sci. Paris* **258,** 5113–15.

Pringsheim E.G. (1949) Iron Bacteria. *Biol. Rev.* **24,** 200–50.

B.J.Ralph (Ed.) (1980) Leading conference. In *Biogeochemistry of Ancient and Modern Environments*. Springer, New York.

Raymond R.N. & Carrano C.J. (1979) Coordination chemistry and microbial iron transport. *Acc. Chem. Res.* **12,** 183–90.

Roberts J.L. (1947) Reduction of ferric hydroxide by strains of *Bacillus polymyxa*. *Soil Sci.* **63,** 135–40.

Rosenberg H. & Young J.G. (1974) Iron transport in the enteric bacteria. In *Microbial Iron Metabolism* (Ed. J.B.Neilands), pp. 67–82. Academic Press, New York.

Schweisfurth R., Eleftheriadis D., Gunlach H., Jacobs M. & Jung W. (1978) Microbiology of the precipitation of manganese. In *Env. Biogeochem. and Geomicrobiol.* Vol. 3 (Ed. W.E.Krumbein), pp. 923–8. Ann Arbor Science, Ann Arbor, Mich.

Silverman M.P. & Ehrlich H.L. (1964) Microbial formation and degradation of minerals. *Adv. Appl. Microbiol.* **6,** 153–206.

Silverman M.P. & Lundgren D.G. (1959) Studies on the chemoautotrophic iron bacterium *Ferrobacillus ferrooxidans* II. Manometric studies. *J. Bacteriol.* **78,** 326–331.

Singer P.C. & Stumm W. (1970) Acidic mine drainage: the rate determining step. *Science* **167,** 1121–3.

Sørensen J. (1982) Reduction of ferric iron in anaerobic, marine sediment and interaction with reduction of nitrate and sulfate. *Appl. Env. Microbiol.* **43,** 319–24.

Spiro T.G. & Saltman P. (1969) Polynuclear complexes of iron and their biological implications. *Struct. Bonding (Berlin)* **6,** 116–56.

Stuart S.J., Greenwood K.T. & Luke R.J.K. (1980) Hydroxamate-mediated transport of iron controlled by ColV plasmids. *J. Bacteriol.* **143,** 35–42.

Stumm W. & Lee G.F. (1960) Oxygenation of ferrous iron. *Ind. Eng. Chem.* **53,** 143–6.

Stumm W. & Morgan J.J. (1981) *Aquatic Chemistry.* Wiley Interscience, New York.

Sugimura Y., Suzuki Y. & Miyake Y. (1978) The dissolved organic iron in sea water. *Deep Sea Res.* **25,** 309–14.

Temple K.L. & Colmer A.R. (1951) The autotrophic oxidation of iron by a new bacterium: *Thiobacillus ferrooxidans. J. Bacteriol.* **62,** 605–11.

T'ieng Ha-Mung (1967) The biological nature of iron–manganese crusts of soil forming rocks in Sakhalin Mountain Soils. *Mikrobiologya (Transl.)* **37,** 749–53.

Troshanov E.P. (1965) Bacteria which reduce manganese and iron in bottom deposits. In *Applied Capillary Microscopy* (Ed. M.S.Gurevich), pp. 106–10. Consultants Bureau, New York (translated from Russian).

Troshanov E.P. (1968) Microorganisms reducing iron and manganese in ore containing lakes of the Karelian Isthmus. *Mikrobiologya* **37,** 934–40.

Troshanov E.P. (1969) The effect and conditions on the ability of bacteria to reduce iron and manganese in ore bearing lakes of the Karelian Isthmus. *Mikrobiologya* **38,** 634–43.

Trudinger P.A., Walter M.A. & Ralph B.J. (eds) (1980) *Biogeochemistry of modern and ancient environments.* Springer, New York.

Tuovinen O.H. & Kelly D.P. (1972) Biology of *Thiobacillus ferrooxidans* in the microbiological leaching of sulfide ores. *Z. allg. Mikrobiol.* **12,** 311–46.

Tyler P. & Marshall K. (1967a) Microbial oxidation of manganese in hydro-electric pipelines. *Ant. von Leeuwenhoek* **33,** 171–83.

Tyler P. & Marshall K. (1967b) Pleomorphy in stalked budding bacteria. *J. Bacteriol.* **93,** 1132–6.

vanVeen W.L., Mulder E.G. & Deinema M.H. (1978) The *Sphaerotilus–Leptothrix* group of bacteria. *Microbiol. Rev.* **42,** 329–56.

Walker J.C.G. (1980) Atmospheric constraints on the evolution of metabolism. *Origins of Life* **10,** 93–104.

Walsh F. & Mitchell R. (1972) An acid-tolerant iron-oxidizing *Metallogenium. J. Gen. Microbiol.* **72,** 369–76.

Wedepohl K.H. (1971) *Geochemistry.* Translated from German by E.Althaus. Holt, Rinehart and Winston, N.Y.

Weinberg E.D. (1978) Iron and infection. *Microbiol. Rev.* **42,** 45–66.

Wieczorek L., Ghiorse W.C. & Hirsch P. (1980) Occurrence and distribution of *Leptothrix* in a small forest pond. *Proc. Ann. Meeting Amer. Soc. Microbiol.* **N–102,** 180.

Williams P.H. & Warner P.J. (1980) ColV plasmid mediated colicin B independent iron uptake system of invasive strains of *E. coli*. *Infect. and Immun.* **29,** 411–16.

Wolfe R.S. (1963) Iron and manganese bacteria. In *Principles and Applications in Aquatic Microbiology* (Eds H.Heukelekian & N.C.Dondero), pp. 82–97. John Wiley & Sons, New York.

Woodrow G., Langman L., Young J.G. & Gibson F. (1978) Mutations affecting the citrate dependent iron uptake system in *Escherichia coli*. *J. Bacteriol.* **133,** 1524–6.

Yoch D.C. & Carithers R.P. (1979) Bacterial iron sulfur proteins. *Microbiol. Rev.* **43,** 422–42.

Zavarzin G.A. (1982) The genus *Metallogenium*. In *The Prokaryotes* (Eds M.P.Starr *et al.*), pp. 524–28. Springer, Berlin.

CHAPTER 7
THE MICROBIAL MANGANESE CYCLE

KENNETH H. NEALSON

7.1 INTRODUCTION

This chapter discusses the manganese cycle both on the global and local scale, including the influence of microorganisms on manganese transformations. Manganese is, next to iron, the most abundant heavy metal in the Earth's crust. It makes up about 0.4 to 1.6 per thousand of nearly all types of crustal rocks; ultrabasic and basaltic rocks are at the high end of this range. Manganese can vary in valence state from -2 to $+7$ although, in nature, only divalent (II) and tetravalent (IV) and, to a lesser extent, the trivalent (III) forms are found. The divalent form is soluble and the tetravalent form is insoluble. The most stable tetravalent state is usually represented by the dioxide MnO_2, and by the anionic $Mn^{IV}O_{1+x}$, in a variety of minerals. Manganese is soluble in environments with low pH and reducing conditions and rapidly precipitates under high pH and oxidizing conditions. Thus, its distribution is greatly affected by the conditions that govern its valence. Other factors also affect the distribution of manganese: it can be adsorbed on to both organic and inorganic substances to form a variety of complexes which may then either enhance or retard its oxidation. The hydrous oxides of manganese are themselves strong chelators of Mn^{2+} and other cations. Thus, adsorption-desorption processes play a major role in the geochemistry of manganese. In addition, organisms may play a large role in the manganese cycle interacting with manganese affecting its geochemistry in several ways.

Since organisms commonly accumulate manganese against environmental gradients, most or all organisms probably have manganese transport systems, several of which have been characterized in bacteria, yeast and animals (Hutner 1972, Bhattacharyya 1975, Jasper & Silver 1978). In fact, manganese is an essential trace element of virtually all living organisms, but the nature of the requirements are not well understood. Some specific involvements of manganese with enzyme systems are known, but in most cases, magnesium or other divalent cations can substitute for manganese (Hutner 1972, Mandelstam & McQuillen 1973, Gottschalk 1979). The enzymes enolase, superoxide dismutase, PEP carboxykinase, all require manganese, as do the enzymes in the early steps of fatty acid biosynthesis. Manganese is required at an early stage for the formation of spores by the genus *Bacillus*, and for the synthesis of cell wall pentapeptides in some bacteria. The reactions involved with RNA polymerization are greatly accelerated by manganese, and protein-bound manganese in photosystem II of plants participates, via valence changes, in the photoactivation reaction that results in oxygen release (Radmer & Kok 1975).

Manganese can be toxic in high concentrations; humans have been poisoned both by excessive inhalation of manganese dust or fumes and by ingestion of manganese-contaminated well water (National Academy of Sciences 1973). Manganese is both toxic and mutagenic for bacteria (Demerec & Hanson 1951), and is mutagenic for yeasts (Putrament *et al.* 1973). In general, the assimilation of trace quantities of Mn is required by all organisms, but large amounts are toxic.

In addition, many organisms oxidize or reduce a relatively large quantity of manganese without assimilating it. Such processes may profoundly effect the manganese cycle. The best understood of these transformations are those affected by bacteria, although fungi and algae have also been reported to catalyse manganese transformations (Beijerinck 1913, Krumbein 1971, Schweisfurth 1972). The various groups of organisms (Table 7.1) undoubtedly employ different mechanisms, although no manganese oxidase has yet been purified, and reports of manganese reductases in cell-free extracts are rare (Trimble & Ehrlich 1970, Ghiorse & Ehrlich 1974). In neither oxidation nor reduction reactions has the stoichiometry or even the reactants or products been firmly established.

The biological oxidation or reduction of manganese can be either direct or indirect. Many biological activities alter the Eh/pH of the environment and enhance valence changes indirectly (Fig. 7.1). Precipitates or reduced manganese formed indirectly are not necessarily cell associated, and may occur at considerable distances from the cells (Marshall 1979). Such indirect effects may be of great importance in nature, especially with regard to manganese reduction, which proceeds spontaneously under low Eh/pH conditions. The establishment of anaerobic zones via microbial oxygen consumption and the production of acids and sulphide can mobilize large quantities of manganese. Direct effects that are attributable to living organisms usually involve the production of proteins, carbohydrates or other materials that can bind, concentrate and thus enhance the oxidation of manganese. Even in laboratory studies, however, it is not always easy to distinguish indirect from direct effects. When manganese oxidation occurs much faster than would be predicted by thermodynamic considerations (concentrations of products and reactants, Eh and pH) and either in close association with cells or only when cells are present, a direct effect is suspected.

Table 7.1. Bacterial genera with members that oxidize and/or reduce manganese. (This table is not meant to be inclusive. It merely demonstrates the taxonomic and distributional versatility of manganese bacteria.)

Bacterial group (after Margulis 1974)	Genera	References: Soil	Fresh water	Marine
Oxidizers				
Pseudomonads	*Pseudomonas*	vanVeen 1973	Zavarzin 1962	Nealson 1978
Facultative Gram-negatives	*Aeromonas*			
	*Arthrobacter**	Bromfield 1974, van Veen 1973		Ehrlich 1966
Enterobacteria	*Flavobacterium*			Nealson 1978
	Oceanospirillum			Ehrlich 1978a
Sphaerotilus group	*Leptothrix*†	vanVeen 1973		
	Clonothrix		Perfil'ev & Gabe 1965	
Prosthecate bacteria	*Metallogenium*		Perfil'ev & Gabe 1965	
a	*Kuznetsovia*		Perfil'ev & Gabe 1965	
	Caulococcus		Perfil'ev & Gabe 1965	
	Pedomicrobium	Ghiorse & Hirsch 1979		
	Hyphomicrobium		Tyler & Marshal 1967a	
Actinobacteria	*Nocardia*	Schweisfurth 1968		
	Streptomyces	Bromfield 1980		
Gram-positive aerobes	*Bacillus*	vanVeen 1973		Nealson & Ford 1980
	Micrococcus			Ehrlich 1966
Reducers				
Many genera of both Gram-positive and Gram-negative bacteria are reducers			Troshanov 1965, Perfil'ev & Gabe 1965	

* Also called *Siderocapsa* (Perfil'ev & Gabe 1965).
† *Leptothrix* and *Sphaerotilus* are used synonymously (van Veen *et al.* 1978).

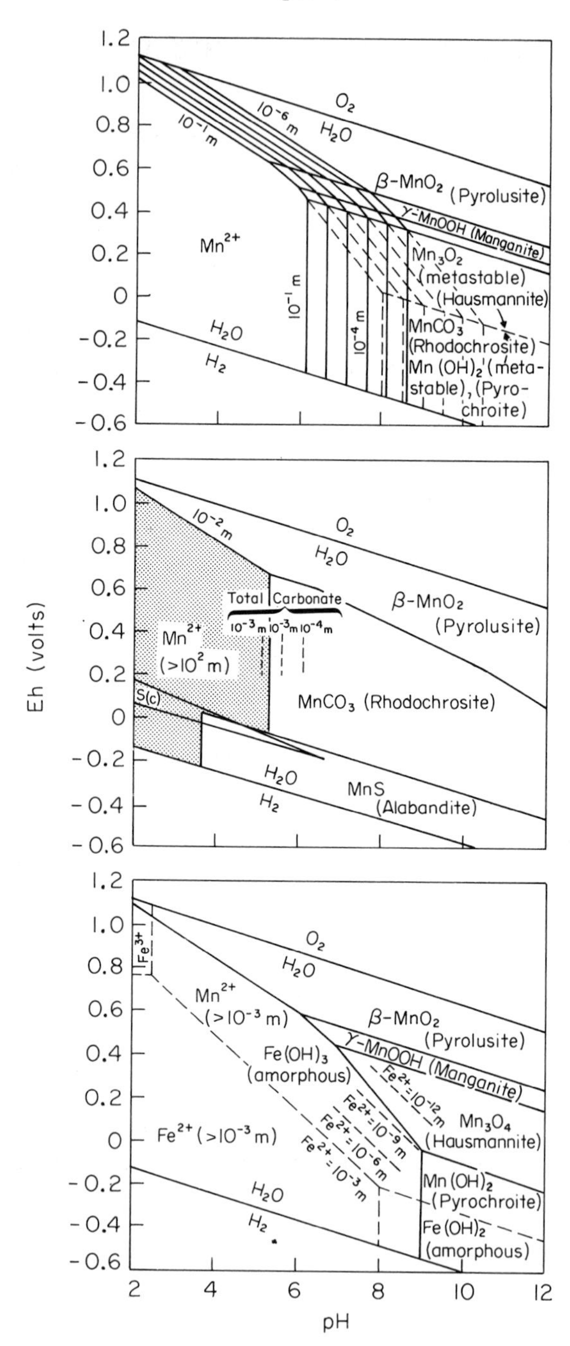

Fig. 7.1. Stability diagrams for the phases and minerals of manganese as a function of Eh, pH and other chemical species present.

7.2 CHEMISTRY, ABUNDANCE AND DISTRIBUTION OF MANGANESE

Manganese comprises about 0.1% of the total mass of the Earth. It is the twelfth most abundant element; only aluminum, iron, magnesium and titanium are more abundant metals. Although it is not magnetic, manganese, which is adjacent to iron in the periodic table, shares several chemical properties with iron. The distribution and biological activity of manganese is, at least in part, a function of its valence; many salts of Mn(II) are stable and highly soluble, and as the cation, Mn(II) is available for the trace requirements of living organisms. Mn(IV) on the other hand, is usually not biologically available; it forms a variety of crystalline or amorphous precipitates, usually represented as MnO_2, although in nature it occurs in various degrees of oxidation ranging from $MnO_{1.3}$ to $MnO_{1.9}$. At least 100 minerals, some of which are shown in Table 7.2, contain manganese as an essential element.

Table 7.2. Some common manganese-containing minerals. (Based on data from Burns & Burns 1979, Marshall 1979, Dr. R.Potter, personal communication.)

Mineral	Formula
Oxides and hydroxides	
Birnessite (delta MnO_2)	$(Na,K,Ca)(Mg,Mn^{2+})Mn_6O_{14}\cdot 5H_2O$
Buserite	Na-Mn oxide hydrate
Hausmannite	Mn_3O_4
Hollandite	$(Ba,K)_{12}Mn_8O_{16}\cdot xH_2O$
Manganite	MnOOH
Manganosite	MnO
Psilomelane	$(Ba,K,Mn^{2+},Co)_2Mn_5O_{10}\cdot xH_2O$
Pyrolusite (Rhamsdellite)	MnO_2
Pyrochroite	$Mn(OH)_2$
Todorokite	$(Na,K,Ca)(Mg,Mn^{2+})Mn_5O_{12}\cdot xH_2O$
Iron and iron-silicates	
Jacobsite	$MnFeO_4$
Pyromanganite	$(Mn,Fe)SiO_3$
Rhodonite	$(Mn,Fe,Ca)SiO_3$
Carbonate	
Rhodochrosite	$MnCO_3$
Sulphide	
Albandite	MnS

The chemical conversions between the various forms of manganese can be described by equilibrium chemistry; such descriptions, based on thermodynamic grounds are straightforward if solutions containing only manganese are considered (Morgan & Stumm 1965, Stumm & Morgan 1981). However, as more inorganic chemicals are added to such systems, the complexity of the oxidation-reduction reactions, and the predictions, become much more difficult; the amounts of energy available, the predicted phases and the mineral products can differ (Fig. 7.1). Complexing with inorganic ligands may play important roles in stabilizing the Mn(II) by decreasing and/or changing its rate of oxidation (Hem 1963, Morgan 1967). Thus, with either bicarbonates or sulphates, manganese can form complexes that are more resistant to oxidation than is free Mn(II). Although similar complexes with chloride ions are less stable, they can still be formed with resultant stabilization of Mn(II). Manganese also forms ligands with a variety of organic compounds, including amines, organic acids and amino acids (Nakhshina 1975). Thus, although Eh and pH will ultimately determine the valence and stability of manganese, the chemical behaviour of the element in nature may vary widely from that predicted from equilibrium chemistry because of the presence of many other interacting chemicals.

Furthermore, thermodynamic considerations give no information as to the rates at which manganese oxidation and reduction occurs; for example, even though the conversion of Mn(II) to Mn(IV) is thermodynamically favourable under many natural conditions, due to the high energy of activation of this reaction, the reaction will proceed slowly unless catalysed (Crerar *et al.* 1972, 1979). In fact, it is the catalysis, presumably via changes in activation energies, that highlight the effects that certain inorganic and organic chemicals and the biota can have on manganese chemistry.

Equilibrium predictions are further complicated because in its oxidized states, manganese forms hydrates and oxides that interact with many organic and inorganic molecules changing its overall reactivity. The hydrous oxides of manganese, the properties of which have been recently reviewed (Murray 1974), are so active in their ability to bind cations that they have been called 'the chemical scavengers of the sea' (Goldberg 1954).

Despite these complications, given the Eh, pH and metal concentrations in the environment, some manganese equilibria can be predicted. Manganese tends toward the solid-phase oxidized forms of manganate minerals represented as MnO_2, in high Eh and pH environments, while in low pH-Eh environments, it tends toward soluble Mn(II), the distribution of which is controlled by diffusive and mixing processes. This is illustrated below:

high pH,Eh,O_2

Mn(II) (soluble) ⟶ Mn(II) (absorbed) ⟶ Mn(IV)O_x (precipitated)

⟶ manganese oxidizers ⟶

⟵ manganese reducers ⟵

None of the abundant manganese compounds are volatile; in fact, there is no major atmospheric form of manganese, either gaseous or particulate.

The small quantity of particulate atmospheric manganese that exists is most likely either anthropogenic (from blast furnace release and other pollutants) or volcanic. Particulates are usually rapidly removed from the atmosphere within a short distance of the sources of production by precipitation (rain), or particulate fall-out. Thus, atmospheric circulation of manganese is not a major component of the manganese cycle. In this sense, the manganese cycle differs from geochemical cycles of many other essential elements, lacking a rapidly transported volatile phase. On the global scale it cycles slowly with rates that are primarily determined by geological processes. The major tectonic activities, oceanic sedimentation, crustal plate uplifting and subduction and chemical weathering apparently constitute the major modes of manganese redistribution between the continents and the oceans.

Nevertheless, manganese is not an immobile element. Under anaerobic conditions it is mobilized and its distribution, which is a reflection of its solubility under different conditions, is far from uniform. It is generally abundant in soils and sediments but occurs only in trace amounts in aerated fresh and marine waters (Fig. 7.2). Most of the transfer between the sedimentary and aqueous environments occurs through changes in valence (and solubility) as described above.

In the oceans, an apparent discrepancy in the manganese budget has been noted for many years; the amount of precipitated manganese in deep-sea sediments is much greater than can be accounted for by river input to surface waters (Bender *et al.* 1977). A major source of manganese to surface waters is postulated to be atmospheric particulates (Klinkhammer & Bender 1980). The manganese in surface waters is in low concentration, but has a short residence time (5–25 years). Deep-sea thermal vents, recently described by Edmond *et al.* (1979), may also be supplying manganese to the oceans via manganese-rich efflux. If the estimates of these authors are correct, these vents more than account for the manganese that is precipitated throughout the world's oceans.

On the global scale, manganese appears to cycle only over geological time but does it cycle rapidly in local environments? There is good evidence that local environments such as soil systems (Mann & Quastel 1946, vanVeen 1973, Bromfield & David 1978), estuaries (Duinker *et al.* 1979) and lakes (Delfino & Lee 1968, Kuznetsov 1970, 1975, Shterenberg *et al* 1975, Dean and Ghosh 1978) exhibit substantial manganese cycling. The details of the manganese cycles in these systems differ between the various environments, depending on physical, chemical and biological parameters. A hypothetical manganese cycle (Fig. 7.3) modified from the description of Kuznetzov (1970) illustrates some of the activities in different microenvironments and the interactions that might be expected to be found in such a system which result in the manganese cycle within a lake. Table 7.3 shows the other redox

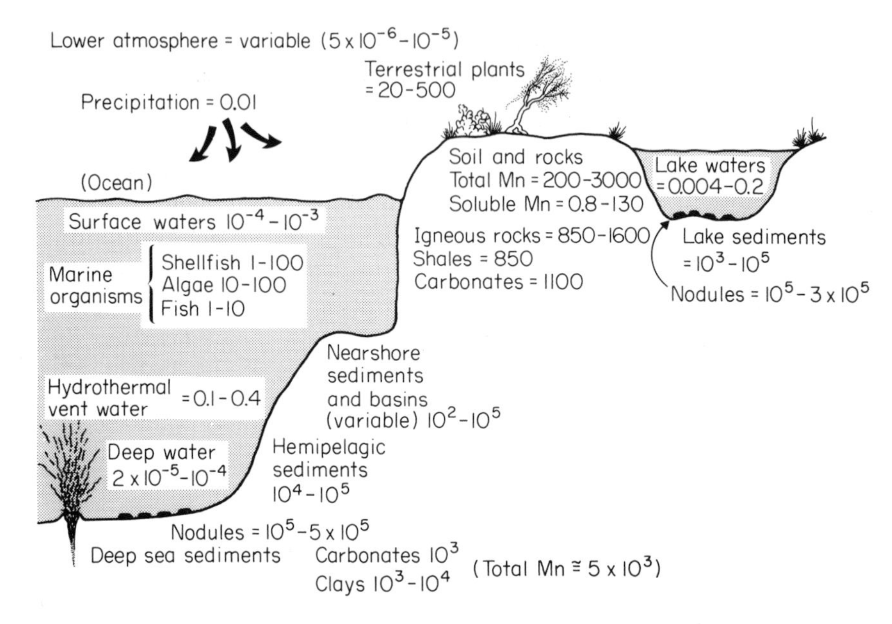

Fig. 7.2. Distribution and abundance of manganese. All values, which have been obtained from many different references cited in this chapter, are presented in parts per million (ppm). River (10^2–10^3 ppm) and ground water (1–10 ppm) are not shown on the figure.

reactions that result from the oxidation of organic carbon by microorganisms. The quality of organic carbon plays a major role in determining the manganese oxidation state. If carbon input is low, the environment does not become anaerobic enough for the reduction of manganese to occur, while if it is high, manganese reduction may be essentially complete either in the water column or in the shallow layers of the sediment. Manganese thus

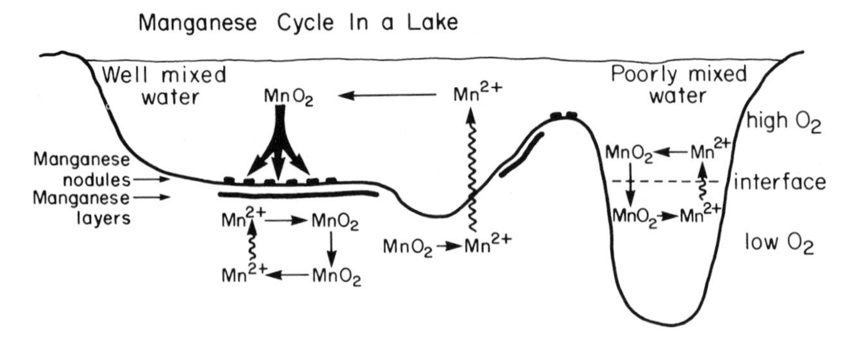

Fig. 7.3. Hypothetical manganese cycle in a lake environment (after Kuznetsov 1970).

Table 7.3. Bacterial redox reactions.

Process	Electron acceptors	Reaction products	Eh range
Aerobic respiration	O_2	CO_2, H_2O, biomass	+500 to +800
Nitrate reduction	NO_3^-	NO_2^-	+300 to −500
Metal reduction	$\left\{\begin{array}{l} MnO_2 \\ Fe_2O_3 \end{array}\right.$	$\left.\begin{array}{l} Mn^{2+} \\ Fe^{2+} \end{array}\right\}$	+100 to +400
Sulphate reduction	SO_4^{2-}	SH^-, acetate	−100 to −400
Methanogenesis	CO_2	CH_4	−800
Fermentation	Organic carbon	H2, CO_2 Organic acids	Variable

cycles between the water column, and the upper and deeper sediments, with rates that depend on the quantities of the organic carbon, of oxygen, and with the extent of mixing and other variables. Reactions in the sediments may lead to the formation of manganese-rich subsurface zones (Gabe *et al.* 1965), or to complete reduction and mobilization back into the water column. Missing from the hypothetical manganese cycle in the lacustrine environment of Fig. 7.3 are other inorganic ions, organic compounds and microorganisms that might modulate the inorganic processes shown.

7.3 LABORATORY STUDIES OF MANGANESE BACTERIA

Although the role of the biota in the manganese cycle can only be understood through studies of the environment, the laboratory investigations of many workers form the basis on which predictions can be made, experiments designed and hypotheses tested. These studies indicate that manganese is assimilated by all organisms as an essential mineral, while it is oxidized and reduced in a dissimilatory fashion by far fewer organisms, mainly microbes. In the study of biogeochemistry, all types of organisms must be considered; fungi, algae and bacteria have all been shown to catalyse manganese transformations. However, for the purpose of this review, only the bacteria which are the most abundant, widely distributed and best studied will be discussed.

7.3.1 Indirect mechanisms of manganese transformation

Since virtually all microbes alter Eh/pH of their environment during growth, it is not surprising that manganese transformations occur as an indirect consequence of microbial growth and metabolism. In stratified environments, where transportive elements are diffusion limited, a series of redox fronts may be expected with the Mn^{+2}/MnO_2 couple included. As the Eh and pH decrease in response to heterotrophic oxygen utilization and acid excretion, manganese, after oxygen and nitrate, and before iron and sulphate, will be reduced and mobilized (Table 7.3). Both oxygenic photosynthesis and ammonia excretion should favour manganese oxidation. Indirect oxidation and reduction are often visualized in the laboratory as reactions that occur some distance from the bacterial colonies growing on a solid medium; haloes of manganese oxide or reduced zones appear around the colonies, while direct effects are usually cell associated. However, the criterion of cell association may not always be definitive. For the case of one *Arthrobacter* species, the same bacteria had cell-associated and/or unassociated manganese oxides; these varied with growth medium and conditions (Bromfield 1974). Unassociated precipitates may occur at considerable distances from the living cells and still be the result of diffusable cellular products; for instance, a high molecular weight (non-dialysable) extracellular substance from a *Streptomyces* sp. that catalyses manganese oxidation even at a pH's below 5.0 has recently been reported (Bromfield 1979).

7.3.2 Direct catalysis of manganese oxidation

The studies of the mechanisms of direct microbial Mn oxidation reveal two classes of direct effects: enzymatic catalysis and specific binding by a variety of cell-associated materials that enhance auto-oxidation.

Some manganese oxidizing bacteria synthesize proteins, carbohydrates or other materials that bind, concentrate, and thus enhance auto-oxidation of manganese in, or on cells. An extracellular protein from *Pseudomonas manganoxydans* that binds manganese and enhances its oxidation has been isolated (Jung & Schweisfurth 1979). Proteins are probably involved in the oxidation of manganese by bacillus spores (Rosson & Nealson 1982), and on the sheaths of *Leptothrix discophorous* (vanVeen *et al.* 1978). Whether or not coupling to cellular metabolic processes occurs in many of the above examples is not known. *Pedomicrobium* species produce extracellular acidic polysaccharides upon which manganese accumulates and presumably oxidizes (Ghiorse & Hirsch 1978, 1979). Since, in many cases, dead bacteria and cell extracts catalyse the manganese binding and oxidation, these reactions may operate simply by binding manganese and thus lowering the energy of activation of oxidation or raising the concentrations of reactants such that the reaction can proceed more readily (Silverman & Ehrlich 1974).

Other bacteria directly oxidize manganese enzymatically. *Arthrobacter* extracts catalyse manganese oxidation if solid MnO_2 is provided in the reaction mixture; this cell-free activity has been partially purified (Ehrlich 1968). *Leptothrix*, *Arthrobacter* and *Metallogenium* are all thought to directly oxidize manganese via a catalase reaction (Dubinina 1978), but since neither the catalases nor the manganese oxidases have been purified, such catalytic activities remain unproven. Manganese oxidase activity with a Km in the molecular range and maximum rates of manganese oxidation of 1–2 μmol/ml per hr were reported for extracts of a soil bacterium (Douka 1977, 1980).

The distinction between manganese binding and the catalysed manganese oxidation cannot always easily be made. An enzyme, even when separated from its cofactors, may still bind its substrate (manganese) and may thus catalyse this transformation if auto-oxidation is enhanced.

7.3.3 Direct catalysis of manganese reduction

The only reported examples of direct reduction of manganese involve a few strains of marine bacteria. These bacteria possess manganese reductases which are inducible by MnO_2, and their activities have been demonstrated in cell-free extracts (Trimble & Ehrlich 1968, 1970, Ghiorse & Ehrlich 1974).

7.3.4 Physiology of manganese oxidation

The effects of pH, temperature, oxygen concentration and other variables upon the ability of bacteria to oxidize manganese have been investigated. However, studies were done in different laboratories, using a variety of organisms and conditions, making generalizations and extrapolations to the natural environment difficult. Some of the organisms that oxidize manganese are shown in Table 7.1.

Nutrients

In general, for soil bacteria (Bromfield 1956, 1974), *Leptothrix* (vanVeen *et al.* 1978), marine bacteria (Nealson 1978) and fungi (Schweisfurth 1972), as the concentrations of nutrients are increased in the medium, growth is favoured and manganese oxidation is inhibited. The mechanism of nutrient inhibition is unknown; the effects may be either at the level of enzyme synthesis, metabolic inhibition, or elsewhere. Experiments with glucose inhibition of manganese oxidation and the reversal of this inhibition by the addition of cyclic-AMP suggest that different mechanisms operate in different bacterial species (Nealson 1978). In some cases, oxidation is enhanced by nutrient addition (Nealson 1978).

Oxygen, manganese and pH effects

Although there are several reactions that can be written for manganese oxidation, almost all involve molecular oxygen as the oxidant; for example:

$$Mn^{+2} + 1/2\ O_2 + 2OH^- \rightleftharpoons MnO_2 + H_2O.$$

Although the reaction predicts that raising manganese, oxygen or hydroxyl ion concentrations will drive the reaction to the right, in the case of bacteria this does not always occur. Low oxygen can favour manganese oxidation, and many manganese oxidizers are apparently microaerophilic (Uren & Leeper 1978, Marshall 1979). Furthermore, high concentrations of manganese can inhibit bacterial manganese oxidation (Bromfield 1956, Ali & Stokes 1972, vanVeen *et al.* 1978), as can pH's greater than 7.5. Biological oxidation generally occurs between pH's of 6.5 and 7.5 (vanVeen 1972, Bromfield & David 1976, Uren & Leeper 1978).

Interfaces and surface effects

Attachment of bacteria to solid surfaces can dramatically accelerate manganese oxidation. A marine *Bacillus* oxidizes Mn much more rapidly when it is attached to one of a variety of surfaces including glass, sand grains and calcite crystals (Nealson & Ford 1980, Kepkay & Nealson 1982). The enrichment methods used by Tyler and Marshall (1967a) indicate that Mn(II) oxidation catalysed by freshwater bacteria is also accelerated by bacterial attachment to surfaces. The encrustation of water pipes by *Hyphomicrobium*-like forms that precipitate manganese is well known (Schweisfurth & Mertes 1962, Tyler & Marshall, 1967a,b), and these bacteria are apparently enhanced in their oxidizing activity through surface attachment. The nature of the physiological changes that occur when bacteria attach to surfaces are not well understood, and in relation to manganese removal, physiological responses such as the deposition of polysaccharides that might chelate manganese will be important to understand.

An apparently different kind of surface effect has been described for marine bacteria: the requirement for MnO_2 in the solid phase before manganese oxidation can occur. This is hypothesized to be a requirement of all marine bacteria, and related to the removal of Mn(II) from solution by adsorption on to MnO_2 before bacterial activity can occur (Ehrlich 1963, 1968). Two reactions are thus required, only one of which is catalysed by bacteria (reaction II below).

$$\text{I.}\quad Mn(II) + H_2MnO_3 \longrightarrow MnMnO_3 + 2H^+.$$

$$\text{II.}\quad MnMnO_3 + 1/2O_2 + 2H_2O \longrightarrow 2\ H_2MnO_3.$$

The hypothesis that all marine bacteria use this mechanism and that none can remove Mn(II) from solution without the presence of MnO_2, is not correct in its entirety, as Rosson and Nealson (1982) have reported a marine *Bacillus* that oxidizes free manganous iron. As more strains are studied it may well be that surface effects will be important in both freshwater and marine environments and in several ways.

Autotrophy, lithotrophy and mixotrophy

One of the most controversial areas of manganese bacterial physiology involves the question of whether or not bacteria can obtain energy for growth by the oxidation of manganese. Do manganese lithotrophs or mixotrophs exist? A lithotroph is an organism that can gain energy from the oxidation of inorganic compounds. If its carbon comes from CO_2, it is called an autotroph. A true manganese chemoautotroph should thus be able to grow on manganese as the sole source of energy, and obtain all of its cell carbon from CO_2. The cell yields should be consistent with the amount of manganese added as an energy source. In no case have these criteria been met.

The existence of manganese mixotrophs, by definition bacteria that derive some of their energy from the lithotrophic oxidation of manganese and some carbon and energy from organic carbon, is much harder to establish. The demonstration of growth stimulation by the addition of manganese, and growth in proportion to the amount of manganese oxidized should suffice, but such experiments in batch culture are often quite difficult, and chemostat studies, which would be definitive (Gottschal & Kuenen 1980), have not been reported.

7.3.5 Physiology of manganese reduction

Although the study of manganese reduction began in 1913 (Beijerinck 1913), very little is known about the relationship between manganese reduction and cell metabolism. Manganese dioxide has been suggested to be an alternate electron acceptor analogous to nitrate, which under anaerobic conditions allows growth of organisms at the expense of organic carbon (Trimble & Ehrlich 1968, 1970). This intriguing suggestion although consistent with the thermodynamics of manganese (Fig. 7.1) is not supported by the data. The cells in which manganese reductase activities have been detected catalyse reduction equally well in the presence or absence of molecular oxygen (Trimble & Ehrlich 1968). Since manganese reduction by bacteria is widespread among many different bacteria, only a few of which have been studied, significant differences in physiology may exist. In those studied, manganese reduction occurs at more rapid rates, and goes further towards completion, as the concentration of organic substrates is increased. Thus, whether manganese dioxide serves directly as an electron acceptor, or indirectly as an environmental electron sink, the effect is the same: as organic carbon is heterotrophically metabolized, manganese dioxide is reduced and released as Mn(II) in solution. Recent studies by Buridge (1983) have shown that sulphate reducers may play a major role in the Mu reduction in anoxic marine environments, and that the Mu reducing activity of enrichment cultures is strongly inhibited by oxygen and azide, suggesting that manganates can indeed serve as alternate electron acceptors.

However, laboratory studies cannot always be directly extrapolated to nature; the concentration of manganese used in the laboratory studies is usually far higher than that found in nature. Furthermore, some bacteria growing on media containing high Mn(II) can catalyse both oxidation and reduction in the laboratory depending on conditions (Bromfield & David 1976). Given the low concentration of manganese in nature, these bacteria may be predominantly manganese reducers, whereas they are detected as manganese oxidizers in the laboratory.

7.3.6 Structural studies of manganese bacteria and their precipitates

Manganese-reducing bacteria isolated and grown as colonies free of manganese precipitates, present only standard rod, coccoid and spirillum morphologies. Manganese oxidizers on the other hand, are characteristically pleomorphic, and hence their morphologies change when they are grown under different laboratory conditions. In fact, the changes are sometimes as extreme as to lend some doubt as to the taxonomic status of the cells (which are established on morphological grounds). One culture, *Hyphomicrobium*, strain 737, can qualify either as a *Hyphomicrobium* or *Pedomicrobium* depending on growth conditions (Bauld & Tyler 1971). With methanol as the carbon source, the *Hyphomicrobium*-facies predominates, while with methanol-methylamine, or when manganese is being oxidized, the *Pedomicrobium*-facies is seen. Further studies may reveal that all of these organisms are variants of a single *Hyphomicrobium*-facies as proposed by Tyler and Marshall (1967c). When manganese oxidation occurs for these organisms, the situation is even more complex.

Extreme pleomorphism of single strains of *Hyphomicrobium*-like organisms isolated from manganese encrustations of water pipes were seen by Tyler and Marshall (1967a,c). Morphological variations of *Pedomicrobium*-like bacteria during manganese oxidation in the laboratory were also documented (Gebers & Hirsch 1978, Ghiorse & Hirsch 1978, 1979). As cultures of manganese-oxidizing bacteria age, they become encrusted with precipitate displaying gradually indecipherable, often varying, surface morphologies (Nealson & Tebo 1980; Fig. 7.4). There is no evidence that these precipitates are genetically determined, rather their morphology is sensitive to environmental conditions such as age of colony, concentration of manganese, presence and nature of solid substrate, bacterial growth rate, and so forth. Unless the precipitates are dissolved, recognition of the bacteria by light microscopy and scanning electron microscopy is limited.

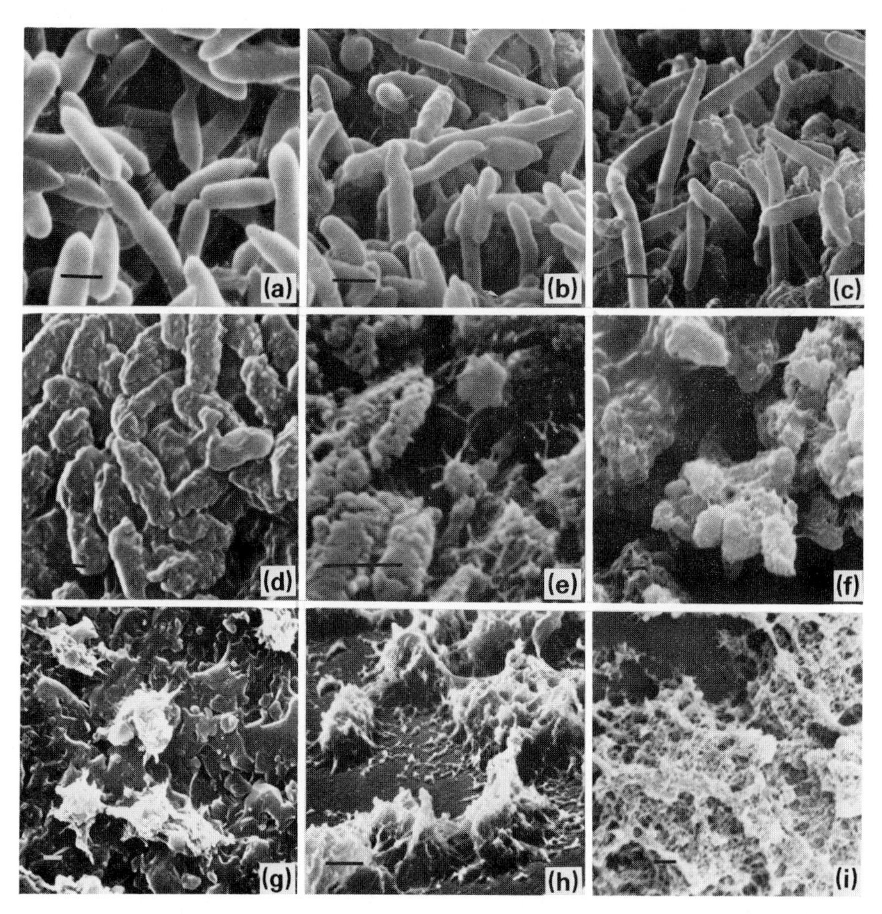

Fig. 7.4. Scanning electron micrographs of a manganese-oxidizing *Bacillus* (SG-1) under different conditions (bars = 1.0 μm). (a) Agar medium without Mn, after one week. (b) Agar medium with Mn, after one day of growth. (c) Agar medium with Mn after one week. (d) Top of colony grown on solid medium with Mn, after one month. (e) Bottom of colony as in (d). (f) Clump of bacteria from sea-water culture after several months of growth. (g) Clumps of MnO_2-coated bacteria on sand grains in sea-water after one week of growth. (h) Clumps of MnO_2-coated bacteria on glass slides after one week of growth. (i) Clumps of MnO_2-coated bacteria on glass slides after three months of growth (after Nealson & Tebo 1980).

On the other hand, thin section analysis by transmission electron microscopy may offer a powerful tool for the study of manganese bacteria. The structural relationship of microbes with biologically formed laboratory precipitate displaying gradually indecipherable, often varying, surface Gram-negative bacteria (Nealson & Tebo 1980, Fig. 7.5). Both a marine bacillus (SG-1; Nealson & Ford 1980, Nealson & Tebo 1980, Tebo 1983) and *Pedomicrobium* (Gebers & Hirsch 1978, Ghiorse & Hirsch 1978) produced precipitates that looked more and more like inorganic minerals after some time. Generally, the manganese must be removed in order to reveal the microbial remains by SEM and light microscopy; an example of such removal is shown in Fig. 7.6. Several different reagents can be used including oxalic acid (Schweisfurth & Mertes 1962), hydrochloric acid (Tyler & Marshall 1967a) and leukoberbelin blue reagent (Krumbein & Altmann 1973, Ghiorse & Hirsch 1978). Transmission electron microscopic (TEM) analyses may show unequivocally the association of bacteria with manganese precipitates (Ghiorse & Hirsch 1978, 1979, Ghiorse 1980, Nealson & Tebo 1980).

7.4 CRITERIA OF BIOGEOCHEMICAL ACTIVITY

On what basis can it be suggested that organisms play important roles in the geochemical behaviour of manganese? Although no definitive answers are available, several criteria should be met.

(1) Manganese-oxidizing organisms should be associated with naturally occurring precipitates that are chemically and structurally recognizable. Ideally, the same precipitates should be produced in the laboratory. If precipitates are dissolving in nature, manganese reducers should be present.

(2) The conditions of pH, Eh and manganese concentration in the natural environment should correspond to those under which the organisms in the laboratory actively transform manganese.

(3) To determine that a given manganese transformation in nature was biologically catalysed, the activities of the organisms must be measured directly.

Each of these criteria alone has its limitations but taken together they form the basis for indicating the biological involvement in geochemical processes. In no single comprehensive study have all these criteria been met. The field data which relate to the above criteria are discussed in the following sections.

7.5 ABUNDANCE AND DISTRIBUTION OF MANGANESE-ACTIVE BACTERIA

Manganese-oxidizing or -reducing bacteria are often identified in high numbers from naturally occurring manganese precipitates (Table 7.4). This

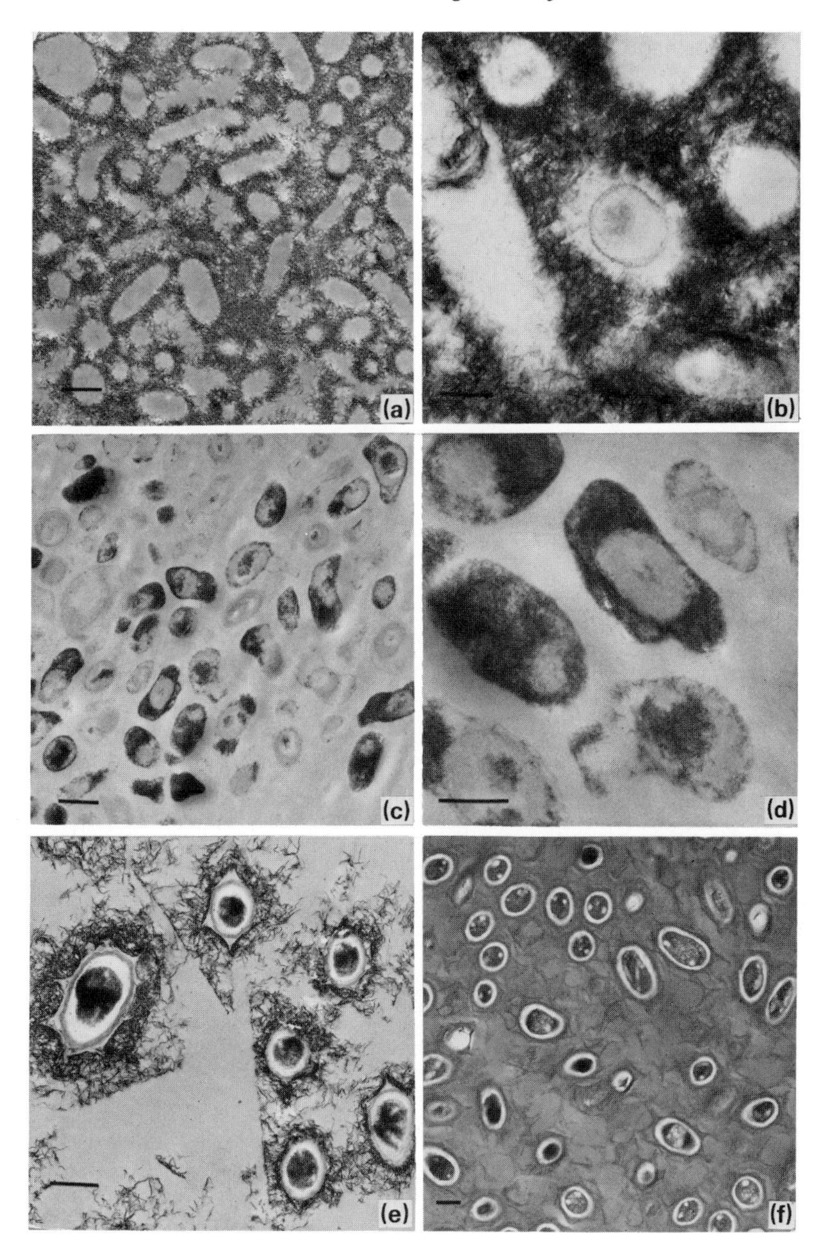

Fig. 7.5. Transmission electron micrographs of manganese precipitation by bacteria. (a) A Gram-negative rod (56A) forms an extracellular precipitate of MnO_2 (bar = 0.5 μm. (b) Same as (a) but bar = 0.25 μm. (c) A second Gram-negative rod (45B) that accumulates Mn inside the cells (bar = 0.5 μm) (d) Same as (c) but bar = 0.25 μm. (e) Spores of a marine *Bacillus* that accumulate Mn around them (bar = 0.5 μm). (f) Same as (e), but the spores have been treated with Leukoberbelin blue to remove MnO_2 (bar = 0.5 μm) (after Nealson & Tebo 1980).

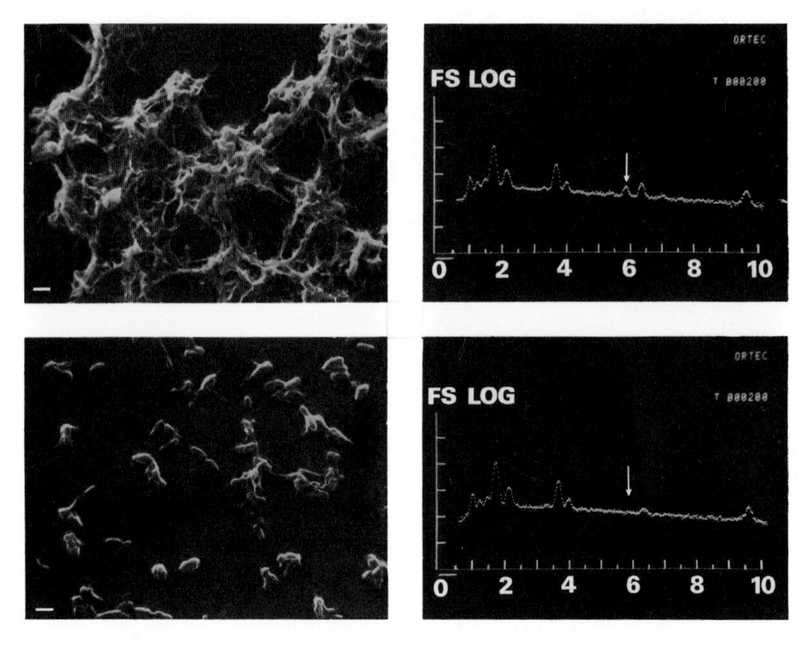

Fig. 7.6. Strain SG-1 attached to glass, before (top) and after (bottom) treatment with HCl to remove MnO_2. EDS analyses (right side) show that manganese (arrow) is present before but not after the treatment (bars = 1 μm; Nealson & Ford 1980).

identification can be either by obtaining viable manganese oxidizers or reducers from the precipitates, or by visual identification. However, an organism oxidizing or reducing manganese in culture does not necessarily indicate that it does so in nature; the association of manganese bacteria with natural precipitates may be a result rather than a cause of the precipitation. Nevertheless, the presence of high numbers of active bacteria associated with natural precipitates is suggestive of a biogeochemical role, and studies in which temporal variations and abundances are correlated with activity patterns would be quite convincing.

One difficulty in assessing biogeochemical activity via cultural methods is the fact that some manganese oxidizers, such as the stellate forms of microcolonies, *Metallogenium*, are difficult to isolate in pure culture. Several species of this genus have been reported only on the basis of microscopic analyses of natural precipitates (Kuznetsov 1970). Stellate forms isolated from widely separated geographical locations have been placed in the genus *Metallogenium* solely on the basis of morphology of what is essentially the manganese precipitates. A life cycle of *Metallogenium* is shown in Fig. 7.7 (Perfil'ev & Gabe 1965). *M. symbioticum* (Zavarzin 1964) grows easily as

Table 7.4. Distribution and abundance of manganese precipitating bacteria.

Environment	Methods*	Population estimates	References†
Soil			
Rhizosphere	C	10^7–10^8 g^{-1}	Timonin (1950)
Non-rhizosphere	C	10^7 g^{-1}	Timonin (1950)
Garden soil	C	10^4 g^{-1}	Schutt & Ottow (1978)
Sandy soil	C	10^6 g^{-1}	Schutt & Ottow (1978)
Fresh waters			
Lake surface waters	V	0–10^2 ml^{-1}	Kuznetsov (1970)
	C	0–10^2 ml^{-1}	Chapnick *et al.* (1982)
Lake bottom waters	V	10^3–10^5 ml^{-1}	Kuznetsov (1970)
	C	10^4–10^6 ml^{-1}	Chapnick *et al.* (1982)
Lake sediments	C	10^5–10^6 ml^{-1}	Chapnick *et al.* (1982)
Lake nodules	C	10^5–10^7 ml^{-1}	Chapnick *et al.* (1982)
Pipeline deposits	V, C	10^5 g^{-1}	Tyler & Marshall (1967a,b)
Marine			
Surface waters	C	None detected	Sorokin (1971) Ehrlich *et al.* (1972) Nealson (1978)
Deep waters, fresh‡	C	None detected	Nealson (unpublished data)
stored	C	10^5–10^6 ml^{-1}	Schutt & Ottow (1978)
Sediments, fresh‡	C	None detected	Nealson (unpublished data)
stored	C	10^3	Ehrlich *et al.* (1972)
stored	C	10^3–10^7 g^{-1}	Schutt & Ottow (1978)
Nodules, fresh‡	V, C	None detected	Sorokin (1971)
fresh	C	None detected	Nealson (unpublished data)
stored	C	10^4	Ehrlich *et al.* (1972)
stored	C	10^4–10^6 g^{-1}	Schutt & Ottow (1978)
stored	C	10^5–10^7 g^{-1}	Nealson (unpublished data)

* The methods of population estimation were either cultural (C) or visual (V).

† The estimates given are not meant to be complete, but to demonstrate the range of variability encountered in this kind of work. In some instances, the numbers listed are calculated from percentages given in the original work, and may therefore be subject to some error.

‡ Fresh samples are those examined immediately upon retrieval, while stored samples are those that have been maintained at 4 °C for weeks or months until examination for viable manganese bacteria.

characteristic microcolonies in co-culture with a fungus using acetate as the sole carbon source. On the basis of penicillin resistance, electron microscopic analysis showing no cell walls, and growth only on a medium containing serum, *Metallogenium* was placed in the group Mycoplasmatales (Dubinina 1969). However, the validity of *Metallogenium* as a bacterial genus has been disputed (Schweisfurth 1978, Schweisfurth *et al.* 1978). Until the bacteria are grown in pure culture and the characteristic stellate forms are shown to be the result of bacteria metabolism, either direct or indirect, the identification of these microcolonies will remain obscure. This is an especially important issue because Barghoorn and Tyler (1965) described similar stellate forms in the Precambrian rocks of the gunflint iron formation of Western Ontario. These stellate forms, distributed in thin sections of 1.9 billion-year-old rocks in a pattern that strikingly resembles *Metallogenium* grown on agar plates, were called *Eoastrion* by Barghoorn and Tyler (1965). Since then, very similar putative microfossils have been observed in another Precambrian iron formation in the Nabberu Basin in Western Australia, dated about 1.6 billion years old. *Eoastrion* has now become synonymous with *Metallogenium* (Barghoorn 1977), on the basis of morphological similarities. So far, all *Eoastrion* types are reported in iron facies and *Metallogenium* types are involved primarily with manganese precipitates, although some natural *Metallogenium* precipitates contain up to 30% iron (Shapiro 1965). If *Metallogenium* is indeed a modern analogue of *Eoastrion*, it should precipitate iron as well as manganese. Iron precipitation by *Metallogenium* in laboratory culture has not been reported and the interpretation of the Precambrian samples may be greatly aided by an understanding of the interaction of *Metallogenium* with oxygen, iron and manganese.

The *Metallogenium* controversy discussed above serves to illustrate the problems inherent in enumeration by cultural methods. Complications arise because the media and conditions used for culturing the bacteria may select only a few of the species that are actually present. In fact, they may select against the forms that predominate in nature. Some of the discrepancies in abundances of manganese bacteria reported in the literature must be due to the differences in cultivability.

Such considerations are particularly important when bacteria from extreme environments, such as the deep ocean, are considered. When manganese nodules were retrieved from the deep ocean, the total numbers of bacteria on the nodules were reported to be low, and then increased with time after retrieval. As the manganese nodules sat at 1 atmosphere of pressure at 4°C, the number of manganese oxidizers increased as a function of the time after sampling (Ehrlich *et al.* 1972, Rosson & Nealson 1982a) and the qualitative nature of the populations on the nodules changed with time. Since high pressures and other conditions characteristic of the deep sea were not imposed, it is unlikely that the indigenous population of the bacteria on the nodules were those studied.

Furthermore, the question of whether or not manganese bacteria were enriched on the manganese nodules in relation to the sediments surrounding

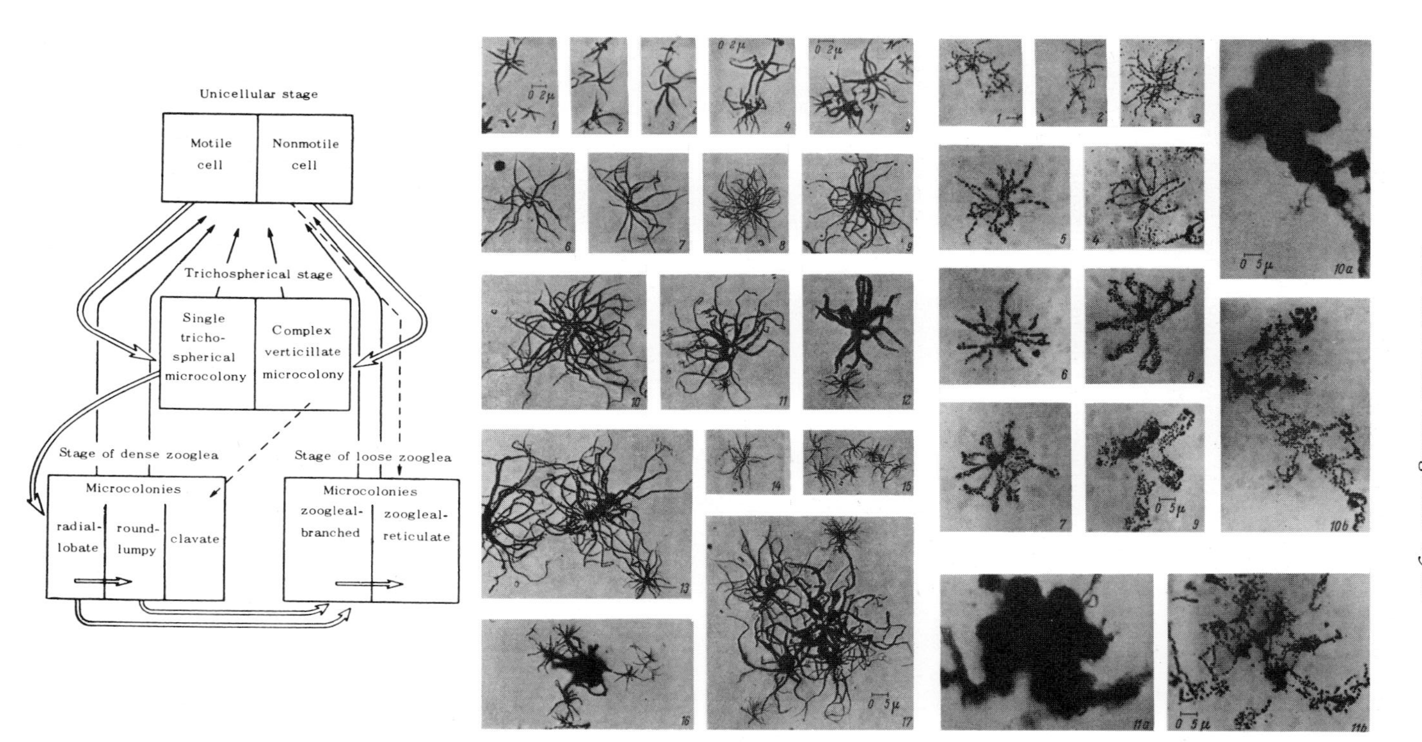

Fig. 7.7. Life cycle of *Metallogenium* (from Perfil'ev & Gabe 1965) showing light micrographs of some of the various stages in the diagram at the left.

the nodules is also not resolved. Some reports indicate that manganese oxidizers and reducers are associated with deep-sea manganese nodules, and are enriched on nodules in relation to the sediments around them (Ehrlich *et al.* 1972), while other workers (Schutt & Ottow 1977, 1978) reached the conclusion that the manganese bacteria in the deep sea formed a continuum; they were enriched neither in the sediment nor on the manganese nodules themselves. Similar lack of enrichment was observed for the manganese nodules and sediments in Oneida Lake, New York (Chapnick *et al.* 1982). However, all of these studies on the abundance and distribution of culturable manganese bacteria are subject to the limitations discussed above, and while suggestive of a role for bacteria in manganese biochemistry, should be viewed with some caution.

Despite all of these problems, the temptation to attribute to bacteria causative biogeochemical roles on the basis of numbers alone is great and several such discussions of the roles of organisms in precipitate and ore formation have appeared (Gabe *et al.* 1965, Dubinina *et al.* 1974, Ehrlich 1974, Kuznetsov 1970, 1975, Shterenberg *et al.* 1975, Klaveness 1977).

Another method of assessment of abundance is the microscopic detection of manganese oxidizers. However, the direct correlation between manganese-oxidizing bacteria and manganese oxides in nature is far more difficult to establish than one might think. Manganese-oxidizing bacteria tend to attach to or become embedded in manganese precipitates, preventing their recognition as bacteria. Removal of these precipitates to reveal the bacteria that form them often destroys the bacteria. Even when it does not, in the absence of the manganese precipitates, the manganese oxidizers are indistinguishable from other bacteria (Table 7.1).

Hyphomicrobium-like bacteria were seen to be abundant in manganese encrustations of water pipes (Tyler & Marshall 1967a, b). They are highly pleomorphic forms, exhibiting several stages of development and manganese accretion. The bacteria were visualized with the aid of treatments that removed the precipitated manganese, revealing the intimate association of the bacteria with their precipitates, and suggesting the involvement of the organisms in the formation of the precipitates.

Combining thin section TEM analysis, histological and geological staining techniques of natural manganiferous deposits with precipitate removal showed convincing evidence of the involvement of *Pedomicrobium*-like bacteria in the formation of natural manganese precipitates (Ghiorse & Hirsch 1979).

Bacteria-like particles are abundant on the surfaces of marine manganese nodules from the Blake plateau as determined by scanning electron microscopy (SEM; LaRock & Ehrlich 1975). Manganese-oxidizing bacteria detected by cultural methods on the samples were considerably fewer than those observed with SEM; perhaps the bacteria on the nodules are dead or non culturable on the media chosen for their isolation. In fact, the bacteria-like particles on the manganese nodules in nature may not even be bacteria.

TEM analyses of naturally occurring manganese oxides in Saanich Inlet,

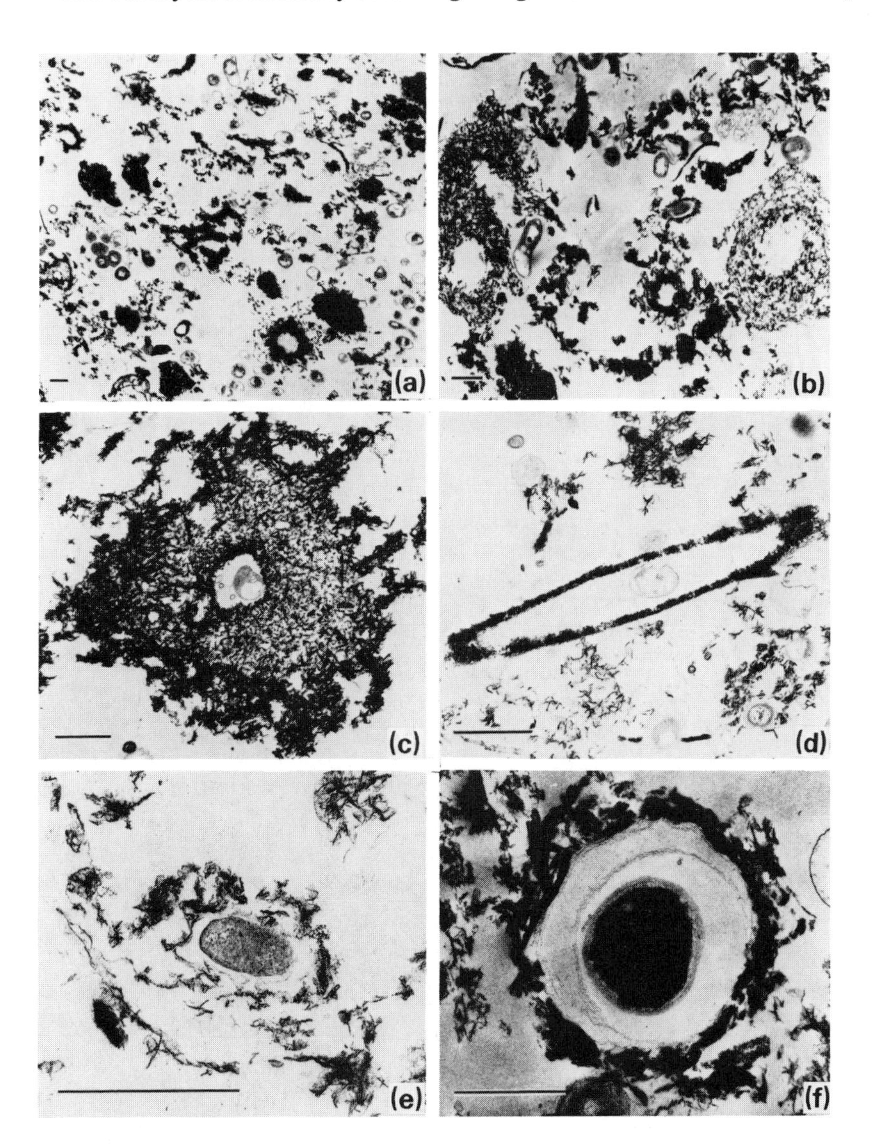

Fig. 7.8. Transmission electron micrographs of natural marine manganese precipitates from Saanich Inlet, British Columbia (bars = 1 μm). (Photo kindly supplied by B.Tebo.)

British Columbia, revealed the presence of a variety of bacteria-like forms associated with the manganese-rich particulate layer (Fig. 7.8; Emerson *et al.* 1982). Analyses of the surface deposits of manganese concentrations and nodules from the Baltic Sea showed similar encrustation of bacterial remains (Ghiorse 1980). Manganese-oxidizing bacteria, similar to those studied in the laboratory may well be responsible for the formation of such natural precipitates. As catalogues of structural types of bacterial manganese precipitates are assembled, it may be possible to identify groups of organisms from field samples based on structural properties of the bacteria and the precipitates they accumulate.

7.6 ACTIVITY MEASUREMENTS

The most convincing criterion that geochemical transformations of manganese are biologically mediated is the direct measurement of those activities, and the demonstration that such activities do not occur to a significant extent in the absence of organisms. The methods that are used for the assessment of manganese oxidative activity are based largely upon laboratory experiments. The usual methods include monitoring the disappearance or uptake of manganese from solution. This can be done by either adding a radioactive tracer (^{54}Mn), or by following total manganese in solution by atomic absorption spectrophotometry. Care must be taken because it is often difficult to distinguish between manganese binding and removal from manganese oxidation. Other methods include the direct monitoring of MnO_2 formation, or the uptake of molecular oxygen, which is presumably used for the formation of MnO_2. This latter method suffers from the fact that all known manganese oxidizers are also heterotrophs, using oxygen for respiratory processes as well as manganese oxidation. Several methods have been used to distinguish between absorption and oxidation of manganese by washing the manganese-containing precipitates with a variety of solutions (Bromfield & David 1976).

Poisons such as sodium azide and iodoacetate blocked manganese oxidation in soil perfusion experiments over a seven-day period (Mann & Quastel 1946), indicating that metabolically active organisms were responsible for the observed manganese oxidation. If poisons were added to thoroughly mixed lake sediment samples which originally contained microzones of oxidized manganese, no zones of oxidized manganese were reformed in the laboratory (Gabe & Gal'perina 1965, Troshanov 1965). In the unpoisoned controls, zones of oxidized manganese were reformed within one year, demonstrating that manganese oxide zone stratification is, therefore, produced or at least accelerated by microbial activities.

Manganese-oxidizing activity of estuarine samples was also inhibited by poisons, suggesting to Wollast *et al.* (1979) that the biota is important in the estuarine manganese cycle. Similarly, poisons inhibited virtually all of the manganese oxidation that can be measured *in situ* in both freshwater (Chapnick *et al.* 1982) and marine (Emerson *et al.* 1982) samples. Since many antibiotics interact with manganese and can alter the rates of manganese oxidation or reduction even in the absence of living organisms, careful testing of all components must be done, and 'poison control' experiments must be cautiously interpreted (Tebo 1983).

The rates of manganese oxidation in nature which lead to the distribution of manganese phases that are out of equilibrium are often too fast to be accounted for purely by chemical mechanisms. That biological activity is important in accelerating manganese oxidation in both marine (Emerson *et al.* 1979) and freshwater estuarine (Boulegue, personal communication) environments is, therefore, inferred. Environments with high rates of manganese precipitation should be examined by both geochemical and microbiological *in situ* activity measurements to substantiate such inferences. Such collaborative work has recently been done in Saanich Inlet, British Columbia. Both geochemical and microbiological observations from this stratified fjord, which has an active anaerobic–aerobic interface, indicate that microorganisms catalyse rapid manganese oxidation (Emerson *et al.* 1982). An abundance of manganese oxide-coated bacteria at the interface (Fig. 7.8) and many manganese oxidizers present as judged by cultural methods, support this interpretation of the *in situ* activity measurements.

Manganese reduction has rarely been studied in the field (*in situ*), although the laboratory experiments of Gabe *et al.* (1965), Troshanov (1965) and more recently Burdige (1983) suggest that in nature the quantity of manganese reduction due to bacterial catalysis could be substantial. To fully understand both oxidative and reductive parts of the manganese cycle, field measurements must be correlated with the presence and physiology of the responsible biota.

7.7 SUMMARY

While on the global scale manganese moves slowly between the land and the sea due to lack of a rapidly cycling phase it is, nevertheless, a dynamic element. As the soluble Mn(II) phase, it is quite mobile and, as such, can move to new environments and become highly concentrated. In the laboratory, microbes have been shown to transform both soluble and solid manganese; thus, they potentially have great effects on local manganese cycles. Physiological, biochemical and structural studies of manganese oxidizers and reducers in the laboratory form the basis on which models of the participation of microbes in the cycling of manganese have been proposed. Field analyses of the distribution of manganese oxidizers and reducers, structural properties of manganese precipitates, and *in situ* activity measurements support the hypothesis that microbes play an integral role in the cycling of manganese in some environments.

Acknowledgements

I wish to thank the many colleagues who supplied me with references and information in response to my general enquiry. Special thanks are due to those who supplied me with unpublished information: S. M. Bromfield, S. Chapnick, D. A. Crerar, W. Ghiorse, K. Marshall, W. Moore, R. Rosson, R. Schweisfurth, B. Tebo and P. Tyler. Helpful comments on the manuscript were made by B. Tebo, R. Rosson, J. Gieskes, and L. Margulis. Financial support for part of the work described here was obtained from the National Science Foundation, IDOE, *MANOP* (manganese nodule program) Grant number NSF OCE 77-11530.

References

Ali S.H. & Stokes J.L. (1972) Stimulation of heterotrophic and autotrophic growth of *Sphaerotilus discophorus* by manganous ions. *Antonie van Leeuwenhoek* **37,** 519–28.

Barghoorn E.S. (1977) Eoastrion and the *Metallogenium* problem. In *Chemical Evolution of the Precambrian.* Academic Press, New York.

Barghoorn E.S. & Tyler S.A. (1965) Microorganisms from the Gunflint Chert. *Science* **147,** 563–77.

Bauld J. & Tyler P.A. (1971) Taxonomic implications of reproductive mechanisms of *Hyphomicrobium*-facies and *Pedomicrobium*-facies of a pleomorphic budding bacterium. *Antonie van Leeuwenhoek* **37,** 417–24.

Beijerinck M.W. (1913) Oxydation des Mangancarbonates durch Bakterien und Schimmelpilze. *Folio Microbiol.* (Delft) **2,** 123–234.

Bender M.L., Klinkhammer G.P. & Spencer D.W. (1977) Manganese in seawater and the marine manganese balance. *Deep-Sea Res.* **24,** 799–812.

Bhattacharyya P. (1975) Active transport of manganese in isolated membrane vesicles of *Bacillus subtilus. J. Bacteriol.* **123,** 123–7.

Bromfield S.M. (1956) Oxidation of manganese by soil microorganisms. *Austr. J. Biol. Sci.* **9,** 238–52.

Bromfield S.M. (1974) Bacterial oxidation of manganous ions as affected by organic substrate concentration and composition. *Soil Biol. Biochem.* **6,** 383–92.

Bromfield S.M. (1979) Manganous ion oxidation at pH values below 5.0 by cell free substances from *Streptomyces* sp. cultures. *Soil Biol. Biochem.* **11,** 115–8.

Bromfield S.M. & David D.J. (1976) Sorption and oxidation of manganous ions and reduction of manganese oxide by cell suspensions of a manganese-oxidizing bacterium. *Soil Biol. Biochem.* **8,** 37–43.

Bromfield S.M. & David D.J. (1978) Properties of biologically formed manganese oxide in relation to soil manganese. *Austr. J. Soil Res.* **16,** 79–89.

Burdige D.J. (1983) The Biogeochemistry of Manganese Redox Reactions: Rates and Mechanisms. PhD Thesis, University of California, San Diego, 245 pp.

Burns R.G. & Burns V.M. (1979) Manganese oxides. In *Marine Minerals* (Ed. R.G.Burns), pp. 1–40. Mineralogical Society of America, Washington, D.C.

Chapnick S.D., Moore W.S. & Nealson K.H. (1982) Microbially mediated manganese oxidation in a freshwater lake. *Limnol. Oceanogr.* **27,** 1004–14.

Crerar D.A., Cormick R.K. & Barnes H.L. (1972) Organic controls on the sedimentary geochemistry of manganese. *Acta Minerologica Petrographica, Szeged.* **22,** 217–26.

Crerar D.A., Cormick R.K. & Barnes H.L. (1979) Geochemistry of manganese: An overview. In *Geology and Geochemistry of Manganese* (Ed. I.M.Varentsov). Hungarian Academy of Sciences, Budapest.

Dean W.E. & Ghosh S.K. (1978) Factors contributing to the formation of ferromanganese modules in Oneida Lake, N.Y. *J. Res. U.S. Geol. Survey* **6,** 231–240.

Delfino J.J. & Lee G.F. (1968) Chemistry of Mn in Lake Mendota, Wisconsin. *Environ. Sci. Technol.* **3,** 1094–100.

Demerec M. & Hanson J. (1951) Mutagenic action of manganous chloride. *Cold Spring Harbor Symposium Quantitative Biology* No. 16, pp. 215–28.

Douka C. (1977) Study of bacteria from manganese concretions. *Soil Biol. Bioch.* **9,** 89–97.

Douka C. (1980) Kinetics of manganese oxidation by cell-free extracts of bacteria isolated from manganese concretions from soil. *Appl. Env. Microbiol.* **39,** 74–80.

Dubinina G.A. (1969) Inclusion of *Metallogenium* among the Mycoplasmatales. *Dokl. Akad. Nauk. SSSR* **184,** 87–90. (English translation.)

Dubinina G.A. (1978) Mechanism of the oxidation of divalent iron and manganese by iron bacteria growing at neutral pH of medium. *Microbiology* **47,** 471–8.

Dubinina G.A., Golenko V.M. & Suleimanov Y.I. (1974) A study of microorganisms involved in the circulation of manganese, iron, and sulfur in meromictic Lake Gek-Gel. *Microbiology* **42,** 817–22.

Duinker J.C., Wollast R. & Billen G. (1979) Behavior of Mn in the Rhine and Scheldt estuaries. II. Geochemical cycling. *Estuar. Coastal Mar. Sci.* **9,** 727–738.

Edmond J.M., Measures C., Mangum B., Grant B., Sclater F.R., Collier R., Hudson A., Gordon L.I. & Corliss J.B. (1979) On the formation of metal rich deposits at ridge crests. *Earth Plan. Sci. Letters* **46,** 19–30.

Ehrlich H.L. (1963) Bacteriology of manganese nodules. 1. Bacterial action on manganese in nodule formation enrichment. *Appl. Microbiol.* **11,** 15–19.

Ehrlich H.L. (1966) Reaction with manganese by bacteria from marine ferromanganese nodules. *Dev. Ind. Microbiol.* **7,** 279–86.

Ehrlich H.L. (1968) Bacteriology of manganese nodules. II. Manganese oxidation by cell-free extracts from a manganese nodule bacterium. *Appl. Microbiol.* **16,** 197–202.

Ehrlich H.L. (1974) The formation of ores in the sedimentary environment of the deep sea with microbial participation: The case for ferromanganese concretions. *Soil Sci.* **119,** 36–41.

Ehrlich H.L. (1976) Manganese as an energy source for bacteria. In *Environmental Biochemistry* vol. 2 (Ed. J.O.Nriagu), pp. 633–44. Ann Arbor Press, Ann Arbor, Michigan.

Ehrlich H.L. (1978a) Conditions for bacterial participation in the initiation of manganese deposition around sediment particles. *Env. Biogeochem. Geomicrobiol.* **3,** 839–45.

Ehrlich H.L. (1978b) Inorganic energy sources for chemolithotrophic and mixotrophic bacteria. *Geomicrobiol. J.* **1,** 65–83.

Ehrlich H.L., Ghiorse W.C. & Johnson G.L. (1972) Distribution of microbes in manganese nodules from the Atlantic and Pacific oceans. *Dev. Ind. Microbiol.* **13,** 57–65.

Emerson S., Cranston R.E. & Liss P.S. (1979) Redox species in a reducing fjord: equilibrium and kinetic considerations. *Deep-Sea Res.* **26,** 859–78.

Emerson *et al.* (1982) Environmental oxidation rate of manganese (II): bacterial catalysis. *Geochim. Cosmochim. Acta* **46,** 1073–9.

Gabe D.R. & Gal'perina A.M. (1965) The development of the microzonal and profile in the absence of microflora. In *Applied Capillary Microscopy* (Ed. M.S.Gurevich), pp. 110–20. Consultants Bureau, New York. (transl. from Russian)

Gabe D.R., Troshanov E.P. & Sherman E.E. (1965) The formation of manganese-iron layers in mud as a biogenic process. In *Applied Capillary Microscopy* (Ed. M.S.Gurevich), pp. 88–105. Consultants Bureau, New York. (transl. from Russian)

Gebers R. & Hirsch P. (1978) Isolation and investigation of *Pedomicrobium* spp., heavy metal-depositing bacteria from soil habitats. In *Environmental Biogeochemistry and Geomicrobiology*, vol. 3 (Ed. W.E.Krumbein), pp. 911–22. Ann Arbor Science, Ann Arbor, Michigan.

Ghiorse W.C. (1980) Electron microscopic analysis of metal-depositing microorganisms in surface layers of Baltic Sea ferromanganese concretions. In *Biogeochemistry of Ancient and Modern Environments* (Eds. P.Trudinger, M.Walter & B.Ralph), pp. 345–54. Australian Acad. of Sciences.

Ghiorse W.C. & Ehrlich H.L. (1974) Effects of seawater cations and temperature on manganese dioxide-reductase activity in a marine *Bacillus. Appl. Microbiol.* **28,** 785–92.

Ghiorse W.C. & Hirsch P. (1978) Iron and manganese deposition by budding bacteria. In *Environmental Biogeochemistry and Geomicrobiology*, vol. 3 (Ed. W.E.Krumbein), pp. 879–909. Ann Arbor Science, Ann Arbor, Michigan.

Ghiorse W.C. & Hirsch P. (1979) An ultrastructural study of iron and manganese deposition associated with extracellular polymers of *Pedomicrobium*-like budding bacteria. *Arch. Microbiol.* **123,** 213–26.

Goldberg E.D. (1954) Marine geochemistry I. Chemical scavengers of the sea. *J. Geol.* **62,** 249–65.

Gorlenko V.M., Dubinina G.A. & Kuznetsov S.N. (1977) *Ecology of Aquatic Microorganisms*. Nauka Press, Leningrad (in Russian).

Gottschal & Kuenen (1980) Selective enrichment of facultatively chemolithotrophic thiobacilli and related organisms in continuous culture. *FEMS Microbiol. Letters* **7,** 241–7.

Gottschalk A. (1979) *Bacterial Metabolism*. Springer-Verlag, New York.

Hajj H. & Makemson J. (1976) Determination of the growth of *Sphaerotilus discophorous* in the presence of manganese. *Appl Environ. Microbiol.* **32,** 699–702.

Hem J.D. (1963) Chemical equilibria and rates of manganese oxidation. *U.S. Geol. Survey Water Supply Pap. 1667-A.*

Hem J.D. (1964) Deposition and solution of manganese oxides. *U.S. Geol. Survey Water Supply Paper 1667-B.*

Hutner S.H. (1972) Inorganic nutrition. *Ann. Rev. Microbiol.* **26,** 313–46.

Jasper P. & Silver S. (1978) Devalent cation transport systems of *Rhodopseudomonas capsulata. J. Bacteriol.* **133,** 1323–8.

Jung W.K. & Schweisfurth R. (1979) Manganese oxidation by an intracellular protein of a *Pseudomonas* species. *Zeit. für Allg. Mikrobiol.* **19,** 107–15.

Kepkay P. & Nealson K. (1982) Surface enhancement of sporulation and manganese oxidation by a marine *Bacillus. J. Bacteriol.* **151,** 1022–36.

Klaveness D. (1977) Morphology, distribution and significance of manganese-accumulating microorganism metallogenium in lakes. *Hydrobiologia* **56,** 25–33.

Klinkhammer G.P. & Bender M.L. (1980) The distribution of manganese in the Pacific Ocean. *Earth Plan. Sci. Lett.* **46,** 361–84.

Krumbein W.E. (1971) Manganese oxidizing fungi and bacteria in recent shelf sediments of the Bay of Biscay and the North Sea. *Die Naturwissenschaften* **58,** 56–7.

Krumbein W. & Altmann H.J. (1973) A new method for the detection and enumeration of manganese oxidizing and reducing microorganisms. *Helgo. Wiss. Meeresunters.* **25,** 347–56.

Kuznetsov S.I. (1970) *Lake microflora and its geochemical activity.* Nauka Press, Leningrad (in Russian).

Kuznetsov S.I. (1975) The role of microorganisms in the formation of lake bottom deposits and their diagenesis. *Soil Sci.* **119,** 81–8.

LaRock P.A. & Ehrlich H.L. (1975) Observations of bacterial microcolonies on the surface of ferromanganese nodules from Blake Plateau by scanning electron microscopy. *Microbiol. Ecol.* **2,** 84–96.

Mandelstam J. & McQuillen K. (1973) *Biochemistry of Bacterial Growth,* 2e. Blackwell Scientific Publications, Oxford.

Mann P.J.G. & Quastel J.H. (1946) Manganese metabolism in soils. *Nature* **158,** 154–6.

Margulis L. (1974) The classification and evolution of prokaryotes and eukaryotes. *Handbook of Genetics* **1,** 1–41.

Marshall K.C. (1976) *Interfaces in Microbial Ecology.* Harvard University Press, Cambridge, Massachusetts.

Marshall K.C. (1979) Biogeochemistry of manganese minerals. In *Biogeochemical Cycling of Mineral-Forming Elements* (Eds P.A.Trudinger & D.J.Swaine). Elsevier, Amsterdam.

Morgan J.J. (1967) Chemical equilibria and kinetic properties of Mn in natural water systems. In *Equilibrium Concepts in Natural Water Systems* (Ed. R.F.Gould). American Chemical Society, Washington, D.C.

Morgan J.J. & Stumm W. (1965) Analytical chemistry of aqueous manganese. *J. Am. Water Works Assoc.* **57,** 107–19.

Mulder E.G. & vanVeen W.L. (1968) Effect of microorganisms on the transformation of mineral fractions in soil. *Trans. 9th Int. Cong. Soil Sci., Adelaide* **4,** 651–60.

Murray J.W. (1974) The surface chemistry of hydrous manganese dioxide. *J. Colloid Interface Sci.* **46,** 357–71.

Nakhshina Y.P. (1975) Manganese in freshwater. *Hydrobiological Journal* **11,** 77–90.

National Academy of Sciences (1973) *Report of the Committee on Medical and Biological Effects of Environmental Pollutants. Manganese.* National Academy of Sciences, Washington, D.C.

Nealson K.H. (1978) The isolation and characterization of marine bacteria which catalyze manganese oxidation. In *Environmental Biogeochemistry and Geomicrobiology* vol. 3 (Ed. W.E.Krumbein), pp. 847–58. Ann Arbor Science, Ann Arbor, Michigan.

Nealson K.H. & Ford J. (1980) Surface enhancement of bacterial manganese oxidation: Implications for aquatic environments. *Geomicrobiol. J.* **2,** 21–37.

Nealson K.H. & Tebo B.M. (1980) Structural features of manganese precipitating bacteria. *Origin of Life* **10,** 117–26.

Okorokov L.A., Kadometseva V.M. & Tiyouskii B.I. (1979) Transport of manganese into *Saccharonyces cerevisiae. Folia Microbiol.* **24,** 240–6.

Perfil'ev B.V. & Gabe D.R. (1965) The use of the microbial landscape method to investigate bacteria which concentrate manganese and iron in bottom deposits. In *Applied Capillary Microscopy*, pp. 9–52. Consultants Bureau, New York. (transl. from Russian.)

Putrament A., Baranowska H. & Prazmo W. (1973) Induction by manganese of antibiotic resistance mutations in yeast. *Mol. Gen. Genetics* **126,** 357–66.

Radmer R. & Kok B. (1975) Energy capture in photosynthesis: Photosystem II. *Ann. Rev. Biochem.* **44,** 409–33.

Robbins J.A. & Callender E. (1975) Diagenesis of manganese in Lake Michigan sediments. *Am. J. Sci.* **275,** 512–33.

Rosson R. & Nealson K.H. (1982) Manganese binding and oxidation by spores of a marine *Bacillus. J. Bacteriol.* **151,** 1037–44.

Rosson R.A. & Nealson K.H. (1982a) Manganese bacteria and the marine manganese cycle. In *The Environment of the Deep Sea* (Eds J.G.Morin & W.G.Ernst), pp. 206–16. Pentice-Hall Inc., New Jersey.

Schutt C. & Ottow J.C.G. (1977) Mesophilic and psychrophilic manganese-precipitating bacteria in manganese nodules of the Pacific Ocean. *Zeit. für Allgemeine Mikrobiologie* **17,** 611–16.

Schutt C. & Ottow J.C.G. (1978) Distribution and identification of manganese-precipitating bacteria from noncontaminated ferromanganese nodules. *Env. Biogeochem. Geomicrobiol.* **3,** 869–78.

Schweisfurth R. (1968) Untersuchungen über manganoxydirende und -reduzierende Mikroorganismen. *Mitt. Int. Ver. Theor. Angew. Limnol.* **14,** 179–86.

Schweisfurth R. (1972) Manganoxydirende Pilze. II. Untersuchungen an Laboratoriumskulturen. *Zeit. für Allgemeine Mikrobiol.* **12,** 667–71.

Schweisfurth R. (1978) Microbial manganese oxidation. *Gesellschaft für Ökologie Verhandlungen, Keil,* **1977,** 281–3.

Schweisfurth R., Eleftheriddis D., Gundlach H., Jacobs M. & Jung W. (1978) Microbiology of the precipitation of manganese. *Env. Biogeochem. Geomicrobiol.* **3,** 923–8.

Schweisfurth R. & Mertes R. (1962) Mikrobiologische und chemische Untersuchungen über Bildung und Bekämpfung von Manganschlamm-Ablagerungen in einer Druckleitung für Talsperrenwasser. *Arch. Hyg. Bakteriol.* **146,** 401–17.

Shapiro N.I. (1965) The chemical composition of deposits formed by *Metallogenium* and *Siderococcus*. In *Applied Capillary Microscopy*, pp. 82–8. Consultants Bureau, New York. (transl. from Russian)

Shterenberg L.E., Dubinina G.A. & Stepanova K.A. (1975) The formation of flat Fe-Mn nodules. In *Problems of Lithology and Geochemistry in the Precipitation of Rocks and Ores* (Ed. A.V.Peive). Nauka Press, Moscow.

Silverman M.P. & Ehrlich H.L. (1964) Microbial formation and degradation of minerals. *Adv. Appl Microbiol.* **6,** 153–206.

Sorokin (1971) Microflora of iron manganese concretions from the Ocean floor. *Microbiol.* **40,** 493–5.

Stumm W. & Morgan J.J. (1981) *Aquatic Chemistry*. Wiley Interscience, New York.

Tebo B.B. (1983) The Ecology and Ultrastructure of Marine Manganese-Oxidizing Bacteria. PhD Thesis, University of California, San Diego, 220 pp.

Timonin M.I. (1950) Soil microflora and manganese deficiency. *Trans. 4th Int. Congr. Soil Sci., Amsterdam* **3,** 97–9.

Trimble R.B. & Ehrlich H.L. (1968) Bacteriology of manganese nodules. III. Reduction of MnO_2 by strains of nodule bacteria. *Appl. Microbiol.* **16,** 695–702.

Trimble R.B. & Ehrlich H.L. (1970) Bacteriology of manganese nodules. IV. Induction of an MnO_2-reductase system in a marine bacillus. *Appl. Microbiol.* **19,** 966–72.

Troshanov E.P. (1965) Bacteria which reduce manganese and iron in bottom deposits. In *Applied Capillary Microscopy* (Ed. S.M.Gurevich), pp. 106–10. Consultants Bureau, New York. (transl. from Russian)

Tyler P.A. & Marshall K. (1967a) Form and function in manganese oxidizing bacteria. *Arch. Microbiol.* **56,** 344–53.

Tyler P.A. & Marshall K. (1967b) Hyphomicrobia—a significant factor in manganese problems. *J. Am. Water Works Assoc.* **59,** 1043–8.

Tyler P.A. & Marshall K. (1967c) Pleomorphy in stalked budding bacteria. *J. Bacteriol.* **93,** 1132–6.

Uren N.C. & Leeper G.W. (1978) Microbial oxidation of divalent manganese. *Soil Biol. Biochem.* **10,** 85–7.

vanVeen W.L. (1972) Factors affecting the oxidation of manganese by *Sphaerotilus discophorus*. *Antonie van Leeuwenhoek* **38,** 623–6.

vanVeen W.L. (1973) Biological oxidation of manganese in soils. *Antonie van Leeuwenhoek* **39,** 657–62.

vanVeen W.L., Mulder E.G. & Deinema M.H. (1978) The *Sphaerotilus-Leptothrix* group of bacteria. *Microbiol. Rev.* **42,** 329–56.

Wollast R., Billen G. & Duinker J.C. (1979) Behavior of manganese in the Rhine and Scheldt estuaries. I. Physico-chemical aspects. *Estuar. Coastal Mar. Sci.* **9,** 161–9.

Zavarzin G.A. (1962) Symbiotic oxidation of manganese by two species of *Pseudomonas*. *Microbiology* **31,** 481–2.

Zavarzin G.A. (1964) *Metallogenium symbioticum*. *Z. Allg. Microbiol.* **4,** 390–5.

CHAPTER 8
MICROBIAL WEATHERING PROCESSES

JACQUES BERTHELIN

8.1 INTRODUCTION

Weathering is considered here under the aspect well defined by a geochemist (Pedro 1971) as the chemical and mineralogical modifications of the different types of rocks and minerals within the biosphere and involve essentially the solubilization processes of major mineral elements (Si, Al, Fe, Mn, Mg, Ca, K, Na, Ti) from silicates, oxides, phosphates, carbonates, sulphides.

In the biosphere, in order to generate energy by oxidation reactions, microorganisms break down nutrient materials and form new chemical compounds (new cell structures, fermentation end-products, excess biosynthetic products, etc.). Autotrophic microorganisms can oxidize large amounts of inorganic compounds (see Chapters 2, 3, 4, 5) and are directly (enzymatic processes) or indirectly (non enzymatic processes involving products of their metabolism) involved in mineral transformations. However, heterotrophic microorganisms, which represent the larger part of the microflora, are most active in the biosphere by transforming the majority of the energy available as organic compounds (essentially plant materials). This very important heterotrophic microbial activity can act directly or indirectly on mineral transformations. Thus microbial activity appears as one of the most important factors occurring in weathering of rocks and minerals in the biosphere.

Although many geologists and soil scientists, discussing the mechanisms of minerals weathering, have repeatedly suggested the important influence of biological processes, most of the reviews or books concerning weathering of rocks and minerals consider only the chemical and physical processes. However, some reviews or books such as those of Kuznetsov *et al.* (1963), Silverman and Ehrlich (1964), Zajic (1969), Dommergues and Mangenot (1970), Doetsch and Cook (1973), Alexander (1977), Brierley (1978), Iverson and Brinckman (1978), Summers and Silver (1978), Brock (1979), Lundgren and Silver (1980), Murr (1980) have discussed the role of microorganisms in the transformation of mineral elements in nature, but with the main emphasis on:

(1) solubilization or concentration of elements of economic importance (in particular the sulphur cycle from the viewpoint of sulphide leaching or sulphide deposits, or fossil fuel formation);

(2) soil fertility or soil formation;
(3) metabolism of inorganic compounds;
(4) toxicity and pollution;
(5) removal of toxic metals or trace elements from industrial and municipal wastes.

Only a very few number of reviews (e.g. Krumbein 1972, Bertrand 1972, Berthelin 1977, Berthelin & Toutain 1979, Murr 1980) give data concerning mineralytic effects of microorganisms.

This chapter is not intended as an exhaustive survey of the literature but considers those papers which provide a basis for our present knowledge of the mineralytic effects of microorganisms.

8.2 METHODS USED TO INVESTIGATE AND TO MEASURE MICROBIAL WEATHERING OF MINERALS

As microbial activity is not always correlated with the presence of microorganisms, even if the populations are important (Domsch 1968, Dommergues *et al.* 1969, Berthelin & Cheikhzadeh-Mossadegh 1977), studies of microbial populations by counting and plating enrichment techniques are not enough and need complementary studies. Scanning electron microscopy (SEM) eventually associated with an electron microprobe, light microscopy, analysis of nutrients and metabolic compounds and other methods derived from soil microbiology and general microbial ecology are very helpful. However, direct measurements of microbial weathering activity and quantitative tests using laboratory or field experiments are of great interest.

As in these investigations, it is important to separate direct or indirect microbiological processes from strictly chemical or physical processes, experiments must include sterile controls (without microorganisms).

8.2.1 Batch cultures

To date, microbial weathering of minerals (essentially solubilization of mineral elements from different types of rocks but also insolubilization processes and transformations of minerals) has been studied more intensively under static conditions (e.g. Silverman & Munoz 1970, Berthelin 1971, Daragan 1971, Krumbein 1971, Moureaux 1972). In such studies, flasks containing samples of rocks or minerals receive nutrient media and are inoculated with one strain, or a mixed population of microorganisms. In some experiments, to prevent contact between minerals and microorganisms and to separate direct (enzymatic) from indirect (role of metabolic compounds) processes, or to distinguish the influence of large molecular weight compounds from low molecular weight compounds, mineral samples were placed in dialysis bags (Weed *et al.* 1969, Berthelin 1976). Such experimental devices are used as simplified or complex experimental models (depending on the complexity of mineral samples, microbial populations and nutrient media). They allow study of different aspects of microbial activity on minerals (solubilization or insolubilization of mineral elements; degradation of minerals; influence of metabolic compounds, of the source of energy, of the size of mineral particles, of the nature of the microorganisms, of the nature of minerals, of the composition of the medium). Such devices (including 'fermenters'), however, are closed and confined whereas most of the rock weathering systems in nature are open. Therefore, for some studies, continuous or semi-continuous flow methods in percolation devices appear to be most appropriate and presumably closest to natural conditions.

8.2.2 Laboratory and field perfusion or percolation devices

Most of the percolation tests concerning biogeochemistry were performed under conditions where the effluents were recycled (Lees & Quastel 1944, Audus 1946, Bryner *et al.* 1954, Gundersen 1960, Sperber & Sykes 1964, Fisher 1966, Kaufman 1966, Wagner & Schwartz 1967, Wildung *et al.* 1969, Wright & Clark 1969, Moureaux 1970). However, the continuous flow method (Macura & Malek 1958, Macura 1961) appeared to be most appropriate because this device (which is an open system) was presumably closest to natural conditions. Berthelin and Kogblevi (1972) and Berthelin *et al.* (1974) have presented experiments based on a method differing from Macura's only in that perfusion was semi-continuous instead of continuous. Distilled water or nutrient solutions were perfused at regular intervals through columns containing soils, rocks or minerals. In some experiments plant materials (litters) placed at the top of the column are the sole source of nutrients (Fig. 8.1). Sterile controls were obtained by autoclaving or by adding anti-microbial compounds to the perfusing solution.

Experiments in the field were based on a method relative to the perfusion one. Lysimetric columns containing granite sand were placed in the field and every year received plant materials (litters or roots) or glucose as nutrients. Controls without litter received only rainfalls and abiological controls were obtained by regular addition of a solution of antiseptic (sodium merthiolate; Fig. 8.2; Berthelin 1976, Berthelin *et al.* 1981).

8.2.3 Models of investigations of microbial weathering in the rhizosphere

To study microbial weathering in the rhizosphere, plants are grown in

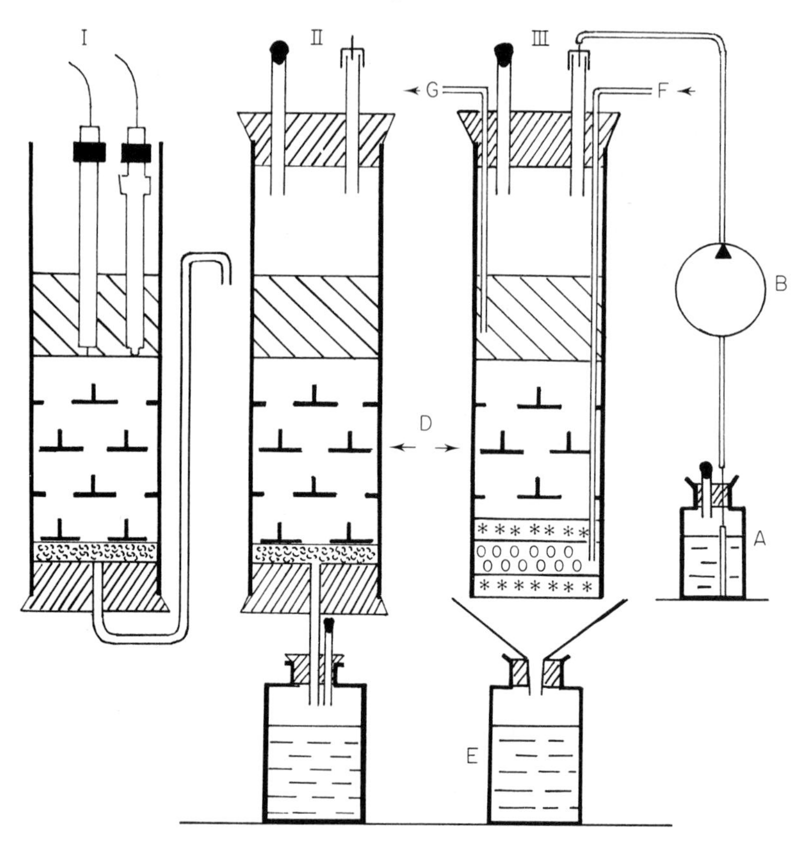

Fig. 8.1. Semi-continuous flow device: I waterlogged with electrodes (pH, Eh); II semi-aerated; III aerated. A = liquid nutrient medium reservoir; B = multi-channel peristaltic pump with automatic time switch; D = soil or rock column; E = effluent vessel; F = air inlet; G = air outlet.

axenic conditions (without microorganisms). In some treatments the rhizosphere is inoculated with a symbiotic or non-symbiotic microflora. Only few studies (Azcon *et al.* 1976, Ralston & McBride 1976, Mojallali & Weed 1978, Powell & Daniel 1978) concern microbial weathering (solubilization of phosphates and biotite) in the rhizosphere because the rhizospheric microflora and the rhizospheric sterility are quite difficult to control. It seems easier to determine only the effect of the symbiotic microflora than to distinguish those of symbiotic from non-symbiotic or from both microflora. Devices similar to the one presented in Fig. 8.3 were used to study solubilization of phosphorus from insoluble phosphates and of potassium from biotite by symbiotic and non-symbiotic rhizospheric microflora of pine and maize (Berthelin & Leyval 1980, Leyval *et al.* 1981).

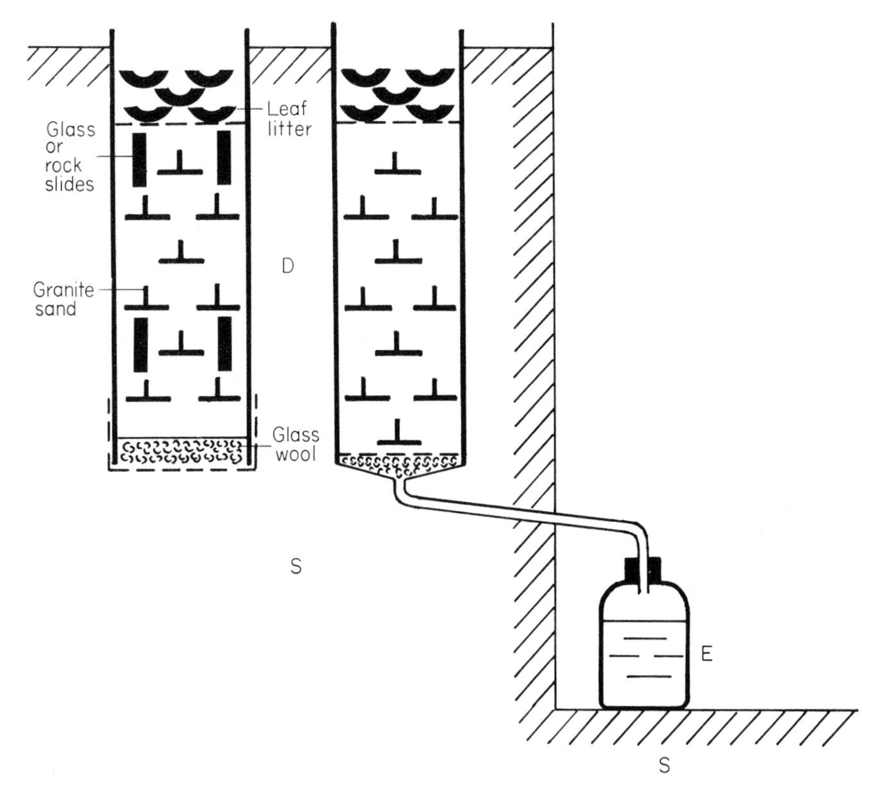

Fig. 8.2. Experimental devices in the fields: D = rock column; E = effluent vessel; S = soil.

8.2.4 Enrichment, counting and plating techniques

Microbial populations involved in the solubilization of phosphates, carbonates and silicates, in the oxidation and the reduction of iron, manganese and sulphur minerals and in the biodegradation of aluminium or iron organic complexes were studied in solid- or liquid-specific media.

Plating techniques concerning phosphate solubilization were described by Louw and Webley (1958), Sperber (1958), Kobus (1962), Tardieux-Roche (1966); those concerning carbonates by Smyk and Drzal (1964) and those concerning silicates by Aleksandrov and Zak (1950), Tesic and Todorovic (1958), Webley *et al.* (1960), Aleksandrov *et al.* (1963), Jackson and Voigt (1971), Wood and MacRae (1972). The method of Savostin (1976) without organic compounds as source of energy must be carefully considered because most microorganisms solubilizing silicates are heterotrophic.

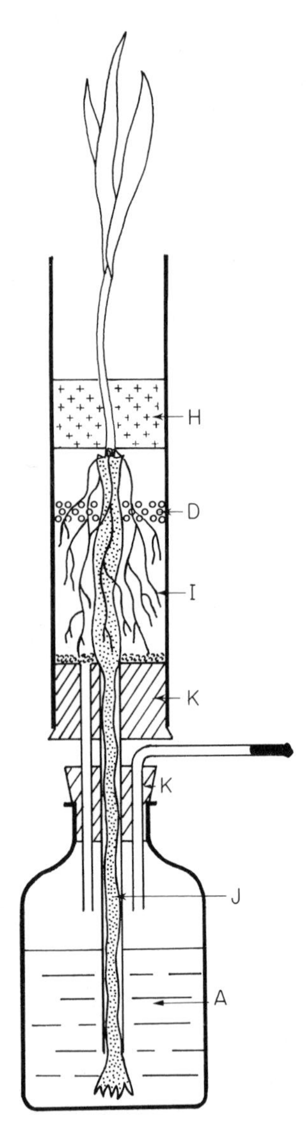

Fig. 8.3. Model of investigation of microbial weathering in the rhizosphere: A = liquid nutrient medium reservoir; D = mineral; H = sand coated with hydrofugeant; I = sand coated with agar; J = glass fibre sleeving; K = rubber plug.

Methods used to study biodegradation of iron and aluminium organic complexes were described by Aristovskaya (1964), Dommergues and Duchaufour (1965), Clark *et al.* (1967), Bruckert and Dommergues (1968), Fisher and Ottow (1972), Gordienko *et al.* (1972), MacRae *et al.* (1973), Berthelin and Cheikhzadeh-Mossadegh (1977), Aristovskaya and Zykina (1978), Cheikhzadeh-Mossadegh *et al.* (1981).

Oxidation and reduction of iron, manganese and sulphur are reported in Chapters 4, 6 and 7, but we can mention Ehrlich (1966), Krumbein and Altmann (1973), Bolotina and Mirchink (1975), Agate and Deshpande (1977), Bromfield (1978) for oxidation and reduction of manganese; Silverman and Lundgren (1959), Hirsch (1968) for iron oxidation; Bromfield (1954), Ottow (1968) for iron reduction; Baldensperger (1976), Mouraret and Baldensperger (1977) for sulphur oxidation; Abd-el-Malek and Rizk (1958) and Starkey (1966) for sulphur reduction.

8.2.5 Measurements of microbial weathering of minerals

Measurements of microbial weathering are performed on a liquid and a solid phase. The liquid phase contains the solubilized products. The solid phase is formed by two main types of products: the residual minerals which remain *in situ* after destruction of the initial minerals, and the insolubilized elements that will be deposited *in situ* or after migration. The residual and the insolubilized elements are able to recombine in structures of neogenesis. Therefore, these measurements concern solubilization and insolubilization of mineral elements in the media or in the effluents; immobilization on or in the microbial cells; changes in mineral micromorphology, crystal structure, chemical composition; plant growth and plant mineral nutrition.

8.3 PRINCIPAL MECHANISMS OF MICROBIAL WEATHERING OF MINERALS AND THEIR MINERALYTIC EFFECTS

Separately or simultaneously, microorganisms can develop different solubilization or insolubilization mechanisms regarding mineral elements. Solubilization includes processes defined by geochemists as acidolysis, complexolysis, alkalinolysis, corresponding to the formation of acidic, complexing, and alkaline metabolic compounds solubilizing rocks. Other solubilization processes are direct (enzymatic) or indirect (non enzymatic) reduction of manganese and iron, oxidation of sulphur compounds, and exchange uptake of ions (adsorption to cell surfaces or absorption by microorganisms).

Insolubilization processes are promoted by the oxidation of iron and manganese, reduction of sulphur compounds, biodegradation of metal organic complexes or metal organic chelates, adsorption or absorption of mineral elements on or in microbial cells, formation of carbonates (calcite, aragonite), biosynthesis of insoluble, high molecular weight compounds, e.g. fulvic and humic acids. These form coatings of organo-metallic compounds around mineral or soil particles (Berthelin & Dommergues 1976).

Microbial insolubilization processes are discussed further in Chapters 2, 4, 5, 6, 7 and 9 and promote mainly the formation of deposits. So these will be mentioned here only when they occur in mineralysis or in important pedogenetic (soil formation) processes.

All these microbial weathering processes depend, of course, on environmental conditions (e.g. composition of organic and mineral nutrients; composition and size of minerals; influence of antimicrobial compounds; conditions of hydromorphy and aeration).

8.3.1 Acidolysis

As summarized by the following general reaction,

$$\text{(Mineral)}^{-}\ M^{+} + H^{+}\ R^{-} \rightarrow H^{+}\ \text{(Mineral)}^{-} + M^{+}\ R^{-}$$
$$(R^{-} = NO_3^{-},\ R_1COO^{-},\ HCO_3^{-},\ SO_4^{2-})$$

acidolysis of minerals is defined as a process involving proton activity and solubilizing mineral elements from rocks in ionized form. Acidification may result either from the formation of an acidic metabolite or from a preferential utilization of alkaline substrates. Microbial oxidation of inorganic (sulphide, sulphur, ammonia) and organic compounds may produce non-complexing or weakly complexing acids (carbonic, nitric, sulphuric, formic, acetic, butyric, lactic, succinic, gluconic, etc.). Another method of acidification is, for instance, the assimilation of ammonium ions leaving acidic anions (sulphate, chloride) in solution.

Mineralytic effect of mineral acidolysis

During laboratory incubations of acid soils, increase of nitrification was accompanied by an increase of exchangeable aluminium (Juste 1965). During biodegradation of litters of *Festuca silvatica* (a graminea characteristic of brown acid forest soils) in field experiments, nitrate formation is correlated ($\alpha = 0.99$) with calcium and magnesium solubilization from a granite sand containing anorthite and biotite (Fig. 8.4; Berthelin *et al.* 1981).

For Kauffmann (1960) nitrification undoubtedly participates in weathering and decay of limestone of buildings and monuments because he always

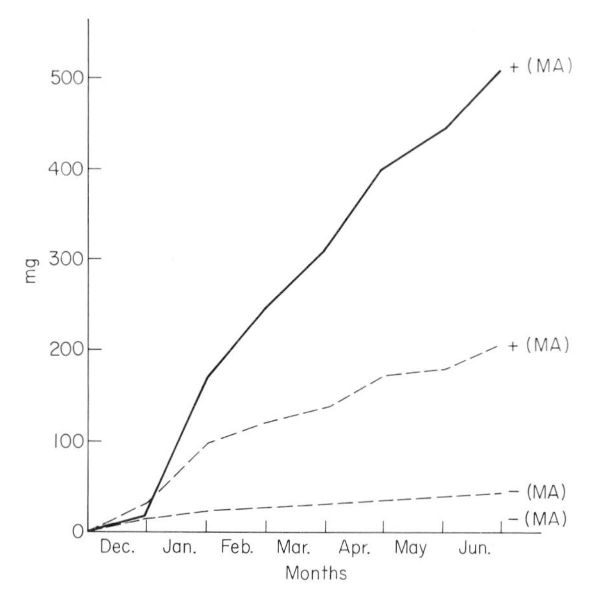

Fig. 8.4. Influence of nitrification on calcium solubilization (cumulative curves) from a granite sand during litter decomposition in the fields (nitrification in control = 0). (– – –) = Ca^{2+}; (——) = NO_3^-; (MA) = microbial activity.

found nitrate in all lesions he observed while sulphate seldom occurred. Experimental data confirmed the possibility of destruction of limestone by nitrifying bacteria using gaseous or liquid ammonia as energy. Nitrifying bacteria also occur in weathering of serpentinized ultrabasic rocks (Lebedeva *et al.* 1978).

For Pochon *et al.* (1960), Pochon and Jaton (1967), *Thiobacilli* were responsible for definite types of lesions on limestone or on sandstone containing limestone by transformation of calcium carbonate (calcite and aragonite) in calcium sulphate (gypsum). However, Jaton (1972) by chemical and microbiological analysis in a church concluded that both microbiological production of nitrate and sulphate were responsible for weathering. The same effects were observed by Krumbein and Pochon (1964) and Krumbein (1969).

Thiobacilli can also be involved in soil degradation via acid production; for instance treatments of vineyards with copper sulphate and sulphur may cause a degradation of certain types of soils. The principal cause of this is not the copper but the sulphur which brings about an increase in *Thiobacilli* population (Pochon & Chalvignac 1964). Other degradation processes occurred after drainage of hydromorphic soils containing reduced sulphur compounds. Aeration of the soil promotes *Thiobacilli* activity and oxidation of sulphur compounds with sulphate formation. However, *Thiobacilli* can also promote mineral nutrition of plants by solubilizing phosphates (Lipman *et al.* 1916) or by promoting reduction of manganese (Quastel 1965).

The examination of *Thiobacillus* by scanning electron microscopy during oxidation of colloidal sulphur suggested physical attachment of *Th. denitrificans* to colloidal sulphur (Baldensperger *et al.* 1974). Brock (1979) has observed attachment of the sulphur-oxidizing *Sulfolobus acidocaldarius* to a crystal of elemental sulphur. Murr and Berry (1976) have also observed a selective attachment of a similar bacteria on low-grade sulphide ores suggesting a direct contact mechanism of bacterial oxidation as mentioned by Ehrlich (1978).

A copper ore after leaching with *Th. ferrooxidans* was partially coated with various amorphous iron precipitates (but some natrojarosite was present) which block rock surfaces (Le Roux & Mehta 1978).

Basic ferric sulphates are formed through the action of sulphuric acid on minerals and rocks (Holler 1967), but they are also formed (jarosite, natrojarosite, ammoniojarosite) by the iron oxidizing bacterium *Thiobacillus ferrooxidans* (Ivarson 1973, Torma 1976, Tuovinen & Carlson 1979) when K, Na or NH_4 are supplied in the nutrient medium. When K is supplied by glauconite, illite or microcline and Na by albite instead of salts (and in the absence of NH_4 which blocks the release of K from these minerals), acidification of the medium, by formation of sulphuric acid, released K and Na from these minerals (Ivarson *et al.* 1978). Jarosite and natrojarosite were formed. Simultaneously glauconite, illite and microcline were weathered. Glauconite weathers at a faster rate than illite or microcline, and albite weathers more easily than microcline. X-ray analyses show a slight expansion in illite but a noticeable one in glauconite. K is less supplied from microcline because it is more strongly bonded in the framework structure.

Mineralytic effect of organic acidolysis

Mainly *Clostridia* and *Bacillus* produce volatile and semi-volatile organic acids (formic, acetic, butyric, lactic) that solubilize significantly Ca and K from a granite sand, but are less active in Al and Fe solubilization than complexing agents such as oxalic acid (Table 8.1; Berthelin & Dommergues 1972, Berthelin 1976). During chemical weathering experiments of micas, Razzaghe-Karimi and Robert (1975) have shown that volatile acids and also malonic, hydroxybenzoïc, fumaric, malic, succinic, vanillic, aspartic acids are less active in aluminium solubilization from phlogopite than citric, oxalic, tartaric, salicylic acids. This difference depends on the chelating ability of the compounds involved. They have also observed that non-

Table 8.1. Effect of metabolic products on the solubilization of mineral elements from a granite sand: positive correlation (P) or absence of correlation (A) (test of Spearman for $P = 0.05$) between microbial solubilization and metabolic compounds simultaneously formed.

	Si	Al	Fe	Mn	Mg	Ca	K
Total microbial acidity	P	P	P	P	P	P	P
Total volatile microbial acidity	A	A	A	P	A	P	P
Oxalic microbial acidity	P	P	P	P	P	A	A

chelating or complexing acids promote an evolution of micas with expansion of the layers and formation of a hydroxyaluminous vermiculite (Fig. 8.5; Robert & Razzaghe-Karimi 1975, Razzaghe-Karimi 1976). Water containing biogenic CO_2 is also able to transform biotite and phlogopite in hydroxyaluminous vermiculite (Razzaghe-Karimi 1976). However, in nature, processes are more complex and simultaneous with microbial production of formic, acetic, butyric and lactic acids by *Bacillus* (facultative anaerobic bacteria) and *Clostridia* (anaerobic bacteria) in waterlogged soils, the microbial reduction of iron and manganese promotes the destruction of primary chlorite, vermiculite and biotite and thus the transformation of biotite into vermiculite and of illite-vermiculite into montmorillonite as a degradation product (Berthelin & Boymond 1978).

All organic acids mentioned here were found in different soils (Stevenson 1967, Bruckert 1971, Flaig 1971), but they are more abundant in

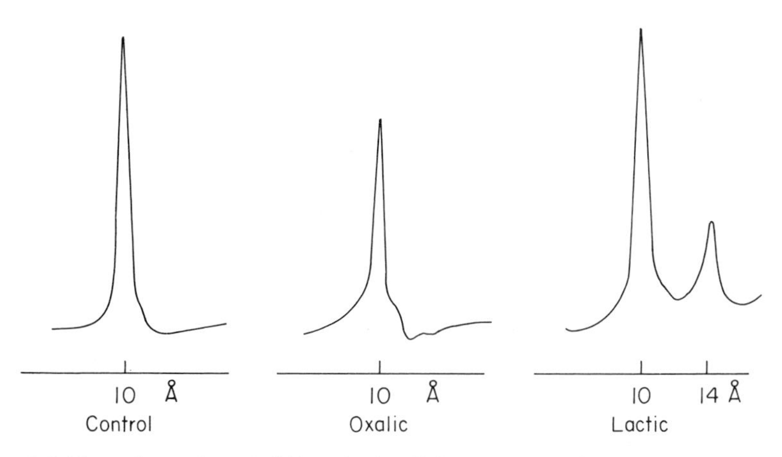

Fig. 8.5. Transformation of phlogopite by different acids (after Robert & Razzaghe-Karimi 1975, Razzaghe-Karimi 1976). (X-ray diagrams: 10Å peak = phlogopite; 14Å peak = vermiculite.)

hydromorphic soils (Flaig 1971, Yoshida 1975). Juste (1965) found that more than 50% of the organic acids in litters of hydromorphic soils were organic volatile acids. Wagner and Schwartz (1967) have identified lactic acid of microbial origin as the weathering agent of granite, orthoclase and biotite. Agbim and Doxtader (1975) have found lactic and acetic acid during solubilization of zinc silicates and Kiel (1977) obtained a successful leaching of silicate and carbonate copper ores with lactic acid in her experiments on ore enrichment by microbial rather than chemical treatment of low-grade ores.

8.3.2 Complexolysis

Complexolysis is a process corresponding to microbial formation of complexing or chelating agents that solubilize mineral elements (iron, aluminium, copper, zinc, nickel, manganese, calcium, magnesium). Thus metal-organic complexes or metal-organic chelates are formed. Chelation may be defined as the equilibrium reaction between a metal ion and a complexing agent characterized by the formation of more than one bond between the metal and the molecule of the complexing agent and resulting in the formation of a ring structure incorporating the metal ion. Metal ions are bound strongly to chelating agents and many of their properties are modified so that the stability of the solution will cover a large range of pH, Eh. This increasing stability of metal ions (depending on the number of rings formed by one molecule of chelating agent with the metal ion, the size of the rings and the nature of the donor atoms) promotes the possibilities of migration of a metal ion in different natural systems (Lehman 1963, Mortensen 1963, Schnitzer & Skinner 1964, Schatz *et al.* 1964, Perlman 1965).

Among the chelating compounds formed by microorganisms are simple organic acids (citric, oxalic, 2 keto-gluconic, tartaric) and phenols (salicylic 2-3 dihydroxybenzoic acids). All these compounds were found in soils (Stevenson 1967, Bruckert 1971, Flaig 1971, Védy & Bruckert 1979) or in weathered sites of monuments (Eckhardt 1978). In soils we have also found, as did Powell *et al.* (1980), strong chelating agents such as trihydroxamic compounds (Berthelin & Emery, unpublished data). The major chelating agents in soil, however, are certainly humic substances either of plant or microbial origin (Schnitzer & Khan 1972, Schnitzer 1978). Complexing agents were found to be responsible for microbial weathering in different experiments. 2 ketogluconic acid was identified during microbial weathering of silicates (Webley *et al.* 1963) or during solubilization of calcium carbonate, montmorillonite and hydroxyapatite (Duff & Webley 1959). Oxalic and citric acids were involved in microbial weathering of biotite

(Boyle *et al.* 1967), of basalt and granodiorite (Silverman & Munoz 1970), of different silicates (Henderson & Duff 1963), of granite sand and of phyllosilicates from granites and granite sand (Berthelin 1971, Berthelin *et al.* 1974, Berthelin 1976, Berthelin & Belgy 1979), and of granite, orthoclase and biotite (Wagner & Schwartz 1967).

Tsyurupa (1964) has observed the formation of non identified Si, Al, Fe, Mg and Ca organomineral complexes by 'silico-bacteria' during albite weathering. Fe (Arrieta & Grez 1971, Daragan 1971) and Fe and Al (Berthelin & Belgy 1979) from different minerals (silicates, sulphides, oxides) were also found to be solubilized by fungi and bacteria as organomineral complexes.

Wood and McRae (1972) have studied the deterioration of sandstone in railway tunnels and did not find acid-producing *Thiobacilli* but have shown that silicate-dissolving microorganisms can contribute to the corrosion of sandstone by production of organic acids such as 2 ketogluconic. Weathering of sandstone monuments was probably affected by fungi which enhanced the deterioration of both cementing material and aluminium silicates by producing organic acids (citric and oxalic; Eckhardt 1978).

Berthelin and Dommergues (1972) have observed correlations between production of oxalic acid and solubilization of Si, Al, Fe, Mn, Mg from a granite sand (Table 8.1). During chemical weathering experiments, Al from phlogopite is faster solubilized by complexing agents than by non-complexing products (citric $>$ oxalic $>$ tartaric $>$ salicylic $>$ lactic $\gg$ butyric; Razzaghe-Karimi & Robert 1975). However, for Boyle *et al.* (1974) oxalic acid extracted much more polyvalent cations from biotite than citric, malonic, malic, lactic and propionic acids. Results of such investigations suggest that acids affect the weathering of mica minerals by the following processes. First, H^+ ions of the acids replace the interlayer K^+ ions in a manner similar to the replacing actions of other cations and render layers expansible. Second, H^+ ions migrate into the octahedral and tetrahedral positions extracting the multivalent cations therein and rendering the weathered edge fragile. Extraction of multivalent cations is further facilitated by the chelating properties of organic acids. Robert and Razzaghe-Karimi (1975) have observed that this formation of chelates destroys the micas (Fig. 8.5). Boyle *et al.* (1967) did not observe interlayer expansion during weathering by oxalic and citric acid of synthetic and biological origin (*Aspergillus niger*) but found removal of octahedral ions leaving a fragile matrix of amorphous material.

In a more complex system (e.g. simulated soil in perfusion devices), in the presence of a microflora obtained by partial sterilization of soil and forming essentially complexing agents, Berthelin *et al.* (1974) and Berthelin and Belgy (1979) have observed destruction of primary chlorite, vermiculite

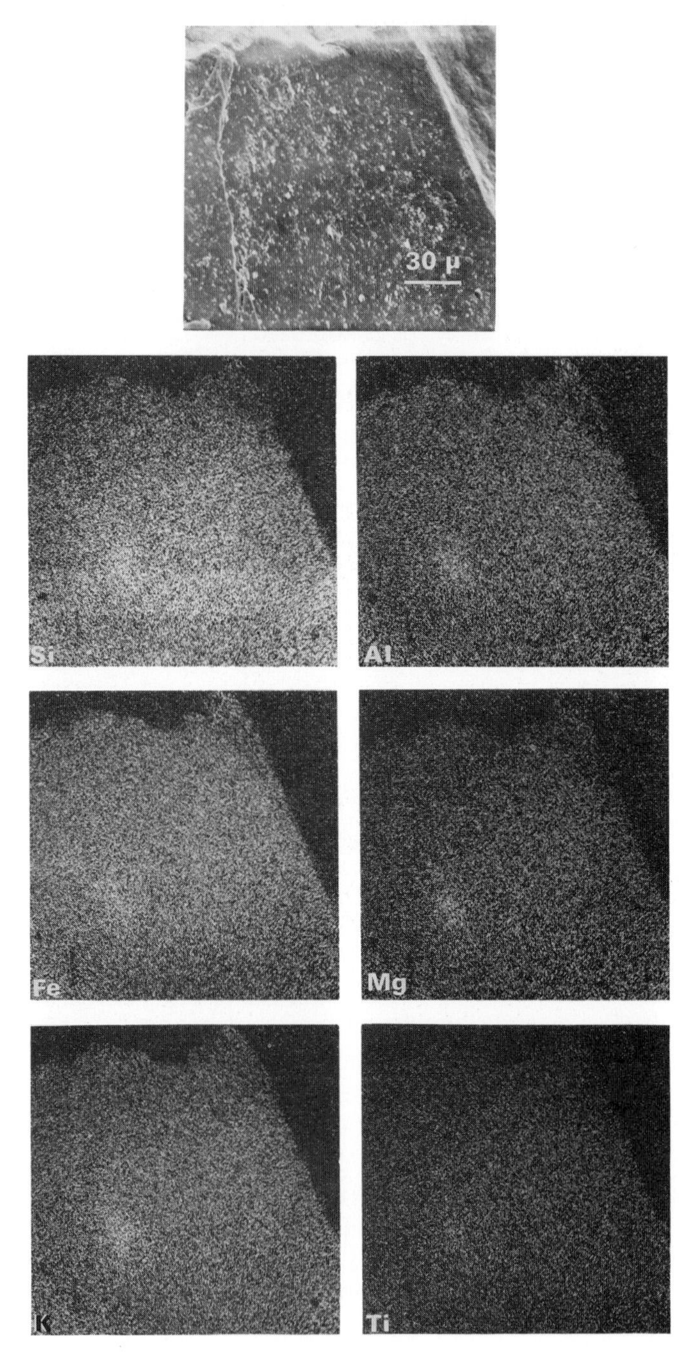

Fig. 8.6. Observation of microbial weathering of biotite by scanning electron microscopy connected with an electron microprobe; aspect of the grain and distribution of elements Si, Al, Fe, Mg, K, Ti. (a) Unweathered biotite.

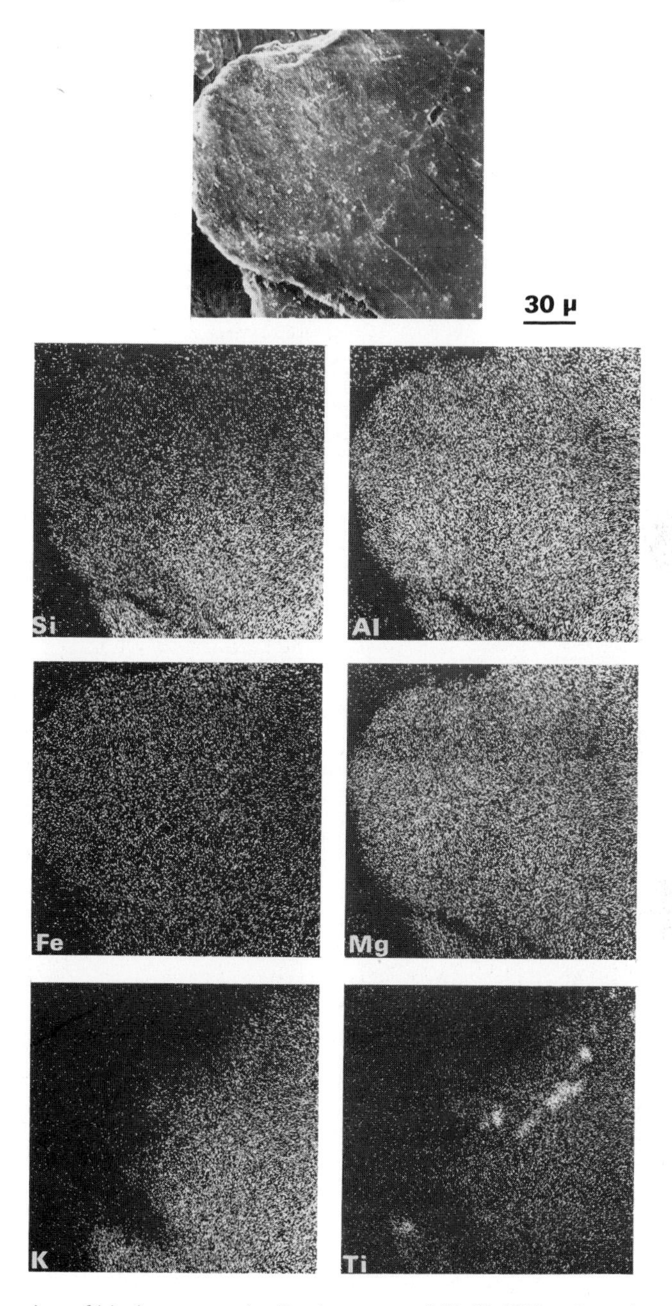

(b) Conversion of biotite to vermiculite (contents of Si, K, Ti has been lowered).

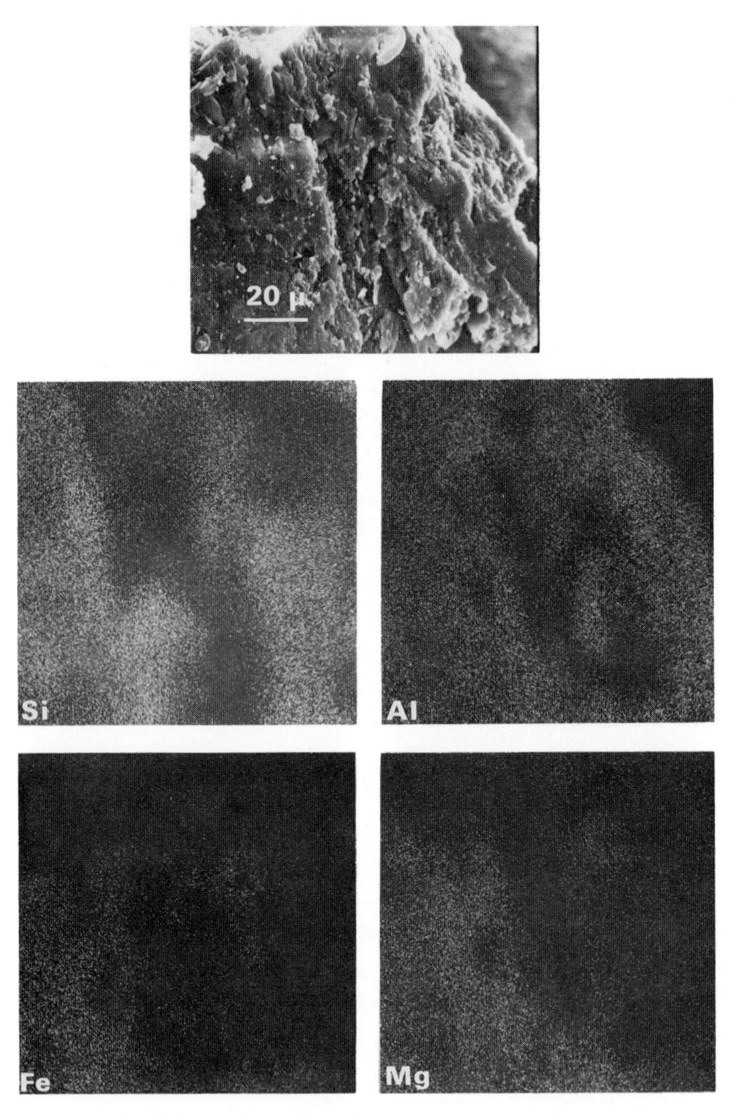

(c) Destruction of biotite, white, brittle Si–Al residue from biotite.

and biotite, and transformation of biotite into vermiculite and into a white and brittle residue of Si and Al (Fig. 8.6). Also transformation of interstratified mineral (illite-vermiculite) into vermiculite was observed.

8.3.3 Comparison of the effects of different acids on mineralysis

The influence of different types of acids on phlogopite weathering can be summarized by Table 8.2 (Robert & Razzaghe-Karimi 1975).

Another type of phyllosilicate transformation (i.e. biotite into smectite by galacturonic acid) has been shown by Vicente Hernandez and Robert (1975) who observed the following biotite transformation:

Biotite {
- strong complexing acid (oxalic)→destruction of biotite
- galacturonic acid→vermiculite Al^{3+} H^{+}→smectites
- weak complexing acid (lactic)→hydroxy-aluminous vermiculite (Al–OH).

Table 8.2. Influence of different types of acids on trioctahedric micas (phlogopite; after Robert & Razzaghe-Karimi 1975). (Acid concentration N/1000, except for ⋆ (N/200).)

Oxalic, citric, tartaric, salicylic	destruction of micas important chemical solubilization by chelation
Quinic, lactic, formic, sulphuric, hydrochloric, malic, fumaric, acetic, hydroxybenzoïc⋆, vanillic⋆, gluconic, malic	expansion of mica interlayers middle chemical solubilization
Aspartic, alanine, butyric	no detectable transformation of micas weak solubilization

During the solubilization of plagioclases, the efficiency of organic acids was established as follows by Huang and Kiang (1972): acetic < aspartic < salicylic < citric.

8.3.4 Alkalinolysis

Heterotrophic microorganisms may form organic bases (ammonia during biodegradation of nitrogen compounds, sodium carbonate during biodegradation of sodium salts of organic acids) that raise the pH and thus promote the solubilization of silica. Silica from nepheline, plagioclase, quartz and phytoliths is extracted by ammonia formed intensively by *Sarcina urea* growing in meat peptone broth (Aristovskaya & Kutuzova 1968). Alkalis have a weaker action than acids on nepheline and plagioclase but a stronger action on quartz and phytoliths. Laboratory experiments have shown that the culture medium became strongly alkaline during the decomposition of sodium acetate and citrate by mycobacteria. This, in turn, led to the release of silica from the silicates of the glass flasks wherein the experiment was performed and from nepheline (Kutuzova 1973).

8.3.5 Reduction of iron and manganese

Large-scale microbial iron and manganese solubilization by indirect (metabolic products) or direct (enzymatic processes) reduction (Ehrlich 1981, Ottow & Munch 1981) can be done by heterotrophic bacteria. Under semi-anaerobic or anaerobic conditions, different *Bacilli* and *Clostridia* can use ferric ion from ferric hydroxide or from ferric oxides and also Mn from MnO_2 as an electron acceptor during anaerobic respiration. Thus, they form large amounts of soluble ferrous ions in anaerobic environments.

Some of these iron-reducing bacteria also reduce nitrate and results from Munch and Ottow (1977) suggest that perhaps the same enzyme system is involved in both reductions. According to these authors an unknown enzyme system 'ferri-reductase' occurred in iron-reducing bacteria without nitrate reductase. Non-crystalline oxides of iron are reduced in preference to the crystalline forms (Munch *et al.* 1978). Boquel and Suavin (1974) have shown that bacteria (*Bacillus* sp.) are able to solubilize iron from ferric oxide and ferrous sulphide in medium containing glucose and litter extracts. Such strains probably act by enzymatic reduction of ferric oxide but also by metabolic products that solubilize ferrous iron from sulphide.

8.3.6 Exchange uptake of ions

Microorganisms can absorb mineral elements for their mineral nutrition (Hutner 1972, Heinen 1974), or eventually in excess of their mineral nutrition (Tardieux-Roche 1966, MacRae & Edwards 1972, Ennever *et al.* 1974). Ions may be also incorporated into microbial cell walls (Beveridge & Murray 1976). Such processes correspond to ion immobilization or insolubilization in microbial cells and provide an ion sink that may increase the release of ions from minerals following the sequence: minerals→ions in solution→microbial sink.

As the mechanism of K release from mica is basically a cation exchange phenomenon, Weed *et al.* (1969) have shown that fungi, grown in nutrient solution buffered at pH 6.0, used three octahedral micas and one dioctahed-

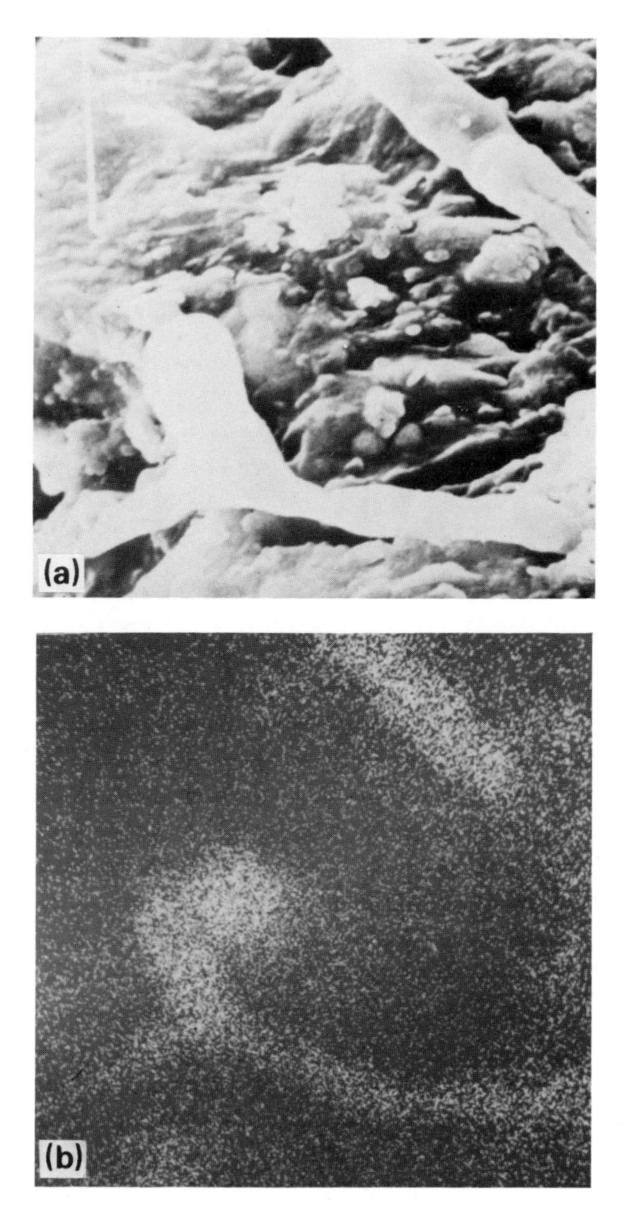

Fig. 8.7. Immobilization of uranium by fungi growing on mining residue (SEM connected with a microprobe; by permission of Capus & Munier). (a) View of the fungi. (b) Uranium content. (Diameter of the mycelium is approximately 1.5 to 2.5 μm.)

ral mica as K^+ sources and that vermiculite was formed because the Na^+ of the solution was exchanged for mica K^+ with the fungi functioning as K^+ sinks. This mineral transformation is similar to those observed during acidolysis. Fig. 8.7 shows immobilization of uranium by fungi growing on mining residue (Capus & Munier 1978).

8.4 FACTORS INFLUENCING MICROBIAL WEATHERING

8.4.1 Microorganism species

Among the heterotrophic microorganisms fungi are considered as the most active ones solubilizing minerals (Henderson & Duff 1963, Müller & Förster 1963, 1964, Webley *et al.* 1963, Boyle *et al.* 1967, Aristovskaya *et al.* 1969, Silverman & Munoz 1970, 1971, Arrieta & Grez 1971, Goni *et al.* 1973, Agbim & Doxtader 1975, Mehta *et al.* 1978, Rossi 1978, Eckhardt 1979). However, we have found (Berthelin 1976) that some *Pseudomonas* are able to solubilize Al, Fe and Mg from silicates to a large extent, and Le Roux *et al.* (1978) have isolated an unidentified bacteria able to leach nickel and copper efficiently from a copper/nickel concentrate.

It is known that autotrophic microorganisms such as nitrifying bacteria and *Thiobacilli* contribute to the weathering of minerals (see section 8.3.1.) but the algal flora (algae *sensu stricto* and cyanobacteria) and lichens may also be involved in such processes. Algae contribute to the mobilization of elements and reduce the stability of cement and rock particles of buildings (Krumbein & Lange 1978). They seem to contribute significantly to the weathering of rocks in desert ecosystems (cold or warm deserts; Krumbein 1969, Friedmann 1971, Friedmann & Ocampo 1976). The weathering action of lichens attributed to the lichen acids which they contain was reported by Schatz (1963). Recent observations of Robert *et al.* (1979) by scanning electron microscopy and electron microprobe on mountain rocks show that lichens seem to be active agents of weathering (Fig. 8.8). Jones and Wilson (1980) have shown the weathering of a basalt by oxalic acid generated by a lichen mycobiont, but we have only few data on the weathering capability of algae and lichens. It is important to mention that mutualism can stimulate microbial mineral solubilization (Tsuchiya *et al.* 1974).

8.4.2 Possible stimulating effect of nutrient deficiencies

Microbial weathering depends on the mineral and organic composition of the nutrient medium, but one very important factor is the deficiency in

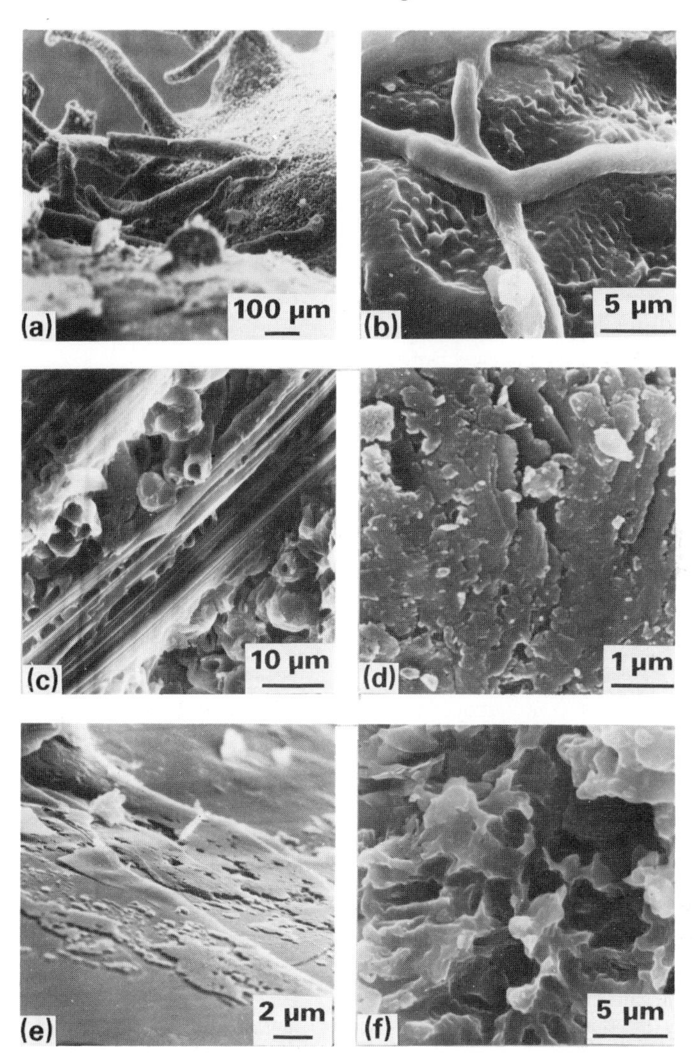

Fig. 8.8. Possible occurrence of lichens in mountain rock weathering (after Robert *et al.* 1979). (a) *Umbilicaria cylindrica* on a mica. (b) Hypha of *Rhizocarpon* on a feldspath. (c) Penetration of *Rhizocarpon geographicum* into micas. (d) Weathered mica under lichens. (e) Solubilization and reprecipitation on a mica. (f) Feldspath solubilization (Na–Ca).

soluble mineral elements. Some bacteria and fungi are able to produce larger amounts of chelating agents in deficient media (Peters & Warren 1970, Emery 1974). Aristovskaya (1956) has observed that deficiencies in Ca, K and Mg promote microbial acid formation. A soil yeast *Lipomyces starkeyi* in deficient media (glucose alone) solubilizes more mineral elements from clay

Table 8.3. Influence of deficiency on the microbial solubilization of elements from a granite sand. (n.d. = non determined; acidity in milliequivalents by culture flask.)

		Non-deficient medium		Deficient medium	
		Days of incubation			
		21	69	21	69
Microbial solubilization	SiO_2		6703	2124	9133
μg/culture flask	Al_2O_3	1493	3383	3511	7329
	Fe_2O_3	189	580	5606	8037
	MnO	904	1239	1750	1955
	MgO	952	1764	1255	1101
	CaO	1707	3134	4742	5819
	K_2O	39	135	95	71
pH (non sterile)		3.4	3.0	3.2	2.9
pH (sterile controls)		4.7	4.4	4.7	4.5
Total microbial acidity		0.797	1.277	2.029	2.660
Total volatile microbial acid		n.d.	0.165	0.757	0.955
Fumaric microbial acidity		n.d.	0	0.213	0
Lactic and succinic microbial acidity		n.d.	0.094	0.292	0.282
Oxalic microbial acidity		n.d.	0.054	0.146	0.275
Citric microbial acidity		n.d.	0	0.018	0

minerals than in non-deficient media (glucose + mineral salts; Berthelin 1976) because larger amounts of glucuronic and volatile acids seem to be produced. Deficient media also promote the development of microbial populations (bacteria and fungi from an acid brown forest soil) that produce larger amounts of organic acids and solubilize many more mineral elements from a granite sand (Berthelin 1976; table 8.3). Torma and Tabi (1973) also reported that the amount of nitrogen $(NH_4)_2$ SO_4 and phosphorus (K_2H PO_4) influence considerably the activity of *Thiobacillus ferrooxidans* during solubilization of sulphides.

8.4.3 Nature and particle size of minerals

Solubilization of mineral elements by autotrophic or heterotrophic microorganisms is a surface phenomenon. Müller and Förster (1963) reported the highest release of K from fine material and the lowest from coarse material during weathering of feldspars by fungi. We have also observed (Berthelin 1971, 1976) that the solubilization of elements from a granite sand by a heterotrophic microflora depends on the particle size.

Torma and Tabi (1973), and Torma and Guay (1976) have demonstrated that the effect of specific surface area, total surface area and particle size are in good agreement with a Monod type equation applicable to describe quantitative biodegradation of sphalerite and zinc extraction by *Thiobacillus*. Solubilization depends also on the solubility product (Torma & Sakaguchi 1978). During clay weathering, chemical and microbial solubilization follow the same decreasing order (Berthelin 1976):

$$\begin{pmatrix}\text{corrensite}\\ +\text{illite}\end{pmatrix} > (\text{montmorillonite}) > \begin{pmatrix}\text{illite}\\ +\text{`illite-chlorite'}\\ +\text{`illite-vermiculite'}\end{pmatrix} > (\text{illite}) > (\text{kaolinite}).$$

During alteration of three mineral fractions (magnetite, pyroxene-amphibole and plagioclases) by *Aspergillus niger* isolated from a diabase rock, magnetite was the most weathered mineral and plagioclase remained practically unaltered (Ribeiro *et al.* 1976). Leleu *et al.* (1973) reported that orthoclase and albite were differently weathered by *Aspergillus niger*. Solubilization of albite was slow and congruent but orthoclase solubilization was fast and incongruent with a superficial silica enrichment.

8.4.4 Effect of waterlogging

During anaerobic microbial decomposition of plant materials in flooded soils, iron and manganese from soils, metal oxides and granite sand are more readily dissolved (Betrémieux 1951, Kee & Blomfield 1962) than under aerobic conditions. Waterlogged conditions decrease the aerobic microbial population and favour anaerobic (*Clostridia*) and facultatively anaerobic bacteria (*Bacillus*) that promote the formation of organic acids (formic, acetic, propionic, butyric, lactic; Berthelin & Kogblevi 1974). Iron solubilization under the form of non-complexed ferrous ions increases (Fig. 8.9). Since Fe reduction did not occur in sterile controls, even when pH and Eh were low, heterotrophic microorganisms are thought to play a major role in the reduction of Fe (Ottow 1969, Berthelin & Kogblevi 1974, Berthelin & Boymond 1978, Munch *et al.* 1978,). Such anaerobic or facultative anaerobic microflora can destroy primary chlorite, vermiculite and transform interstratified minerals (illite-vermiculite) into montmorillonite (Fig. 8.10) and biotite into vermiculite (Berthelin & Boymond 1978).

8.4.5 Influence of antimicrobial compounds

Partial sterilization of soils by some natural antimicrobial compounds did not significantly affect the total numbers of microorganisms but reduced the

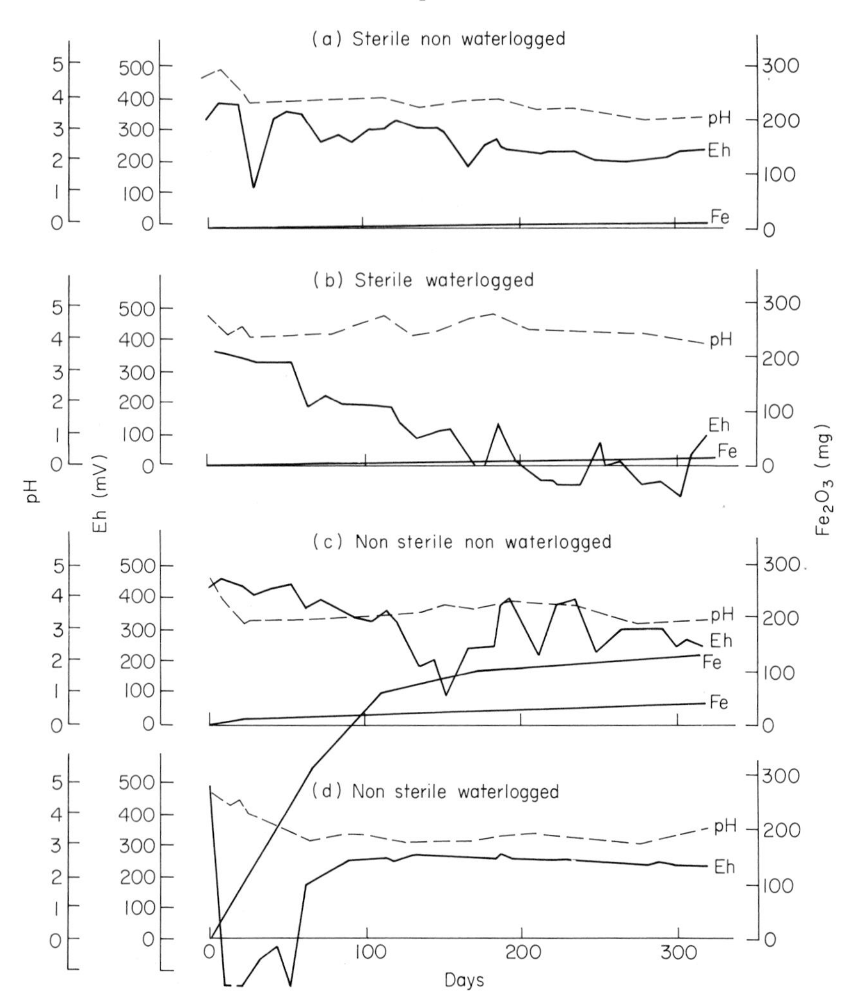

Fig. 8.9. Influence of waterlogging on microbial iron solubilization in soil.

number of species. The remaining species synthesized larger amounts of complexing agents (e.g. oxalic acid) that promote the solubilization of mineral elements (Si, Al, Fe, Mn, Mg, Ca, K) from granite rocks. These acidic compounds solubilized appreciable amounts of ferromagnesian minerals (biotite) and destroyed primary chlorite and some vermiculite. Illite-vermiculite was transformed into vermiculite and biotite into vermiculite or into a white and brittle residue of Si and Al (Berthelin *et al.* 1974, Berthelin & Belgy 1979; see Fig 8.6).

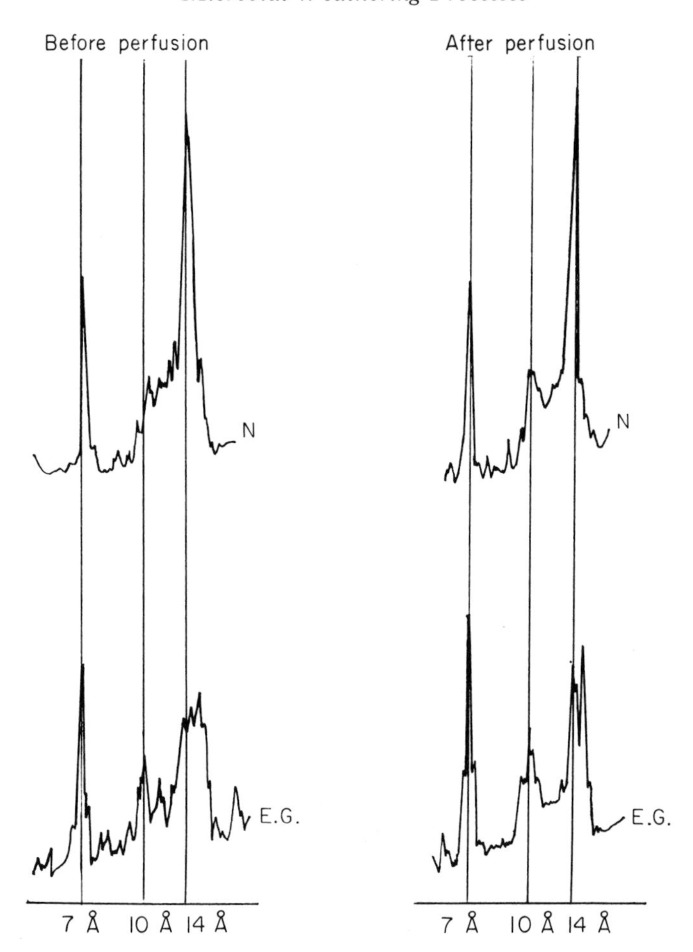

Fig. 8.10. Microbial transformation of (illite-vermiculite) into montmorillonite in waterlogged soils. X-ray diagrams of powder: N = normal; E.G. = ethylene glycol. (X-ray diagrams of sterile control were similar to those before perfusion.)

8.5 SOME IMPLICATIONS OF THE MICROBIAL WEATHERING OF MINERALS

Implications of microbial weathering of minerals are undoubtedly of great importance under environmental and economical points of view. As mentioned previously the mineralytic effect of autotrophic (nitrifying bacteria, *Thiobacilli*, algal microflora) and heterotrophic microflora, bacteria and fungi affect limestone and sandstone monuments. Other materials and objects of art such as mineral paints are also affected (see sections 8.3.1 and 8.3.2). However, implications are of greater importance in soil formation

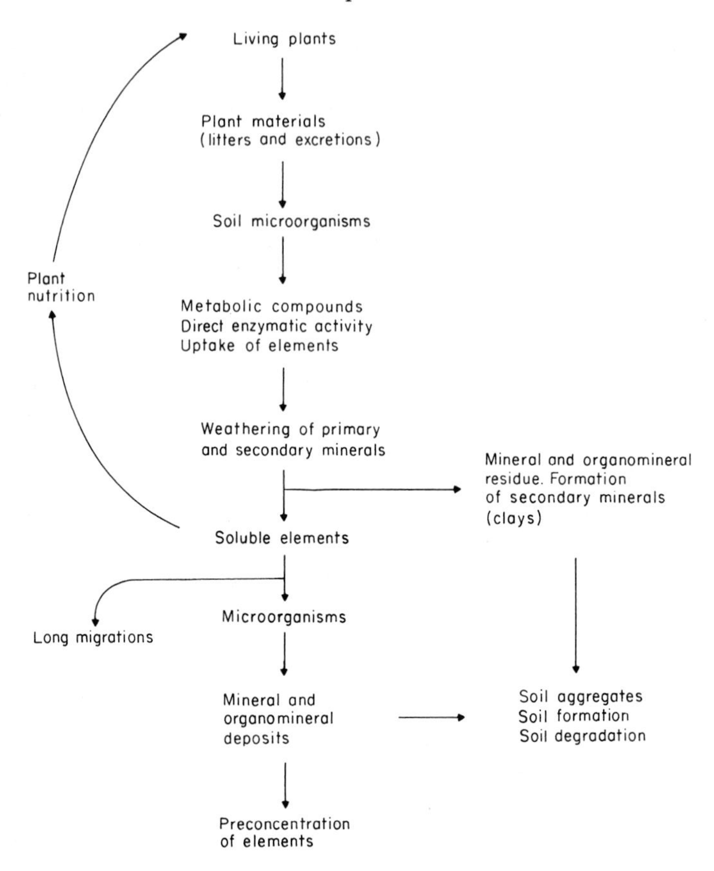

Fig. 8.11. General aspect of the implications of microbial weathering processes in soil ecosystems.

and degradation, soil fertility, plant growth and plant nutrition, metal extraction, metal concentration and formation of deposits. Figure 8.11 summarizes some implications of microbial weathering in soil ecosystems.

8.5.1 Soil formation and degradation

The podzolization processes can be dependent on two microbial mechanisms.

(a) The decrease of biodegradation of organometallic complexing compounds in the soils where podzolization occurs, promotes mineral weathering by complexing agents and migration of elements as organometallic complexes that will be deposited in the accumulation horizons 'Bh' and 'Bs'.

(b) The deficiency in soluble mineral elements and the partial sterilization by antimicrobial natural compounds (pine resin) can promote development of microorganisms forming larger amounts of aggressive compounds or can modify the microbial metabolism by stimulating the biosynthesis of chelating agents (these processes occur in other deficient soils or microhabitats).

Such processes promote mineral degradations similar to those observed during podzolization in the field (Berthelin & Belgy 1979).

Waterlogging of soil profiles promotes the development of an anaerobic microflora that solubilizes mineral elements by 'acidolysis' and by direct and indirect reduction of Mn and Fe. The microbial activity promotes soil gleyification, and weathering of phyllosilicates (Berthelin & Boymond 1978). Such processes should be important in anaerobic horizons of degraded paddy soils, in reduced niches containing organic matter and possibly in reducing marine sediments, and could participate in the degradation of hydromorphic soils by the acid organic-reducing water table.

In acid sulphate soils, *Thiobacillus ferrooxidans*, through the formation of basic ferric sulphates, is able to weather 2:1 phyllosilicates that supply K^+ for jarosite formation. Kaolinite is formed or is the remaining clay mineral (Ivarson *et al.* 1978). *Thiobacilli* can also be involved in soil degradation if important acidification occurs through sulphur and sulphide oxidation.

The formation of alkaline compounds (ammonia, soluble carbonates, sodium or calcium hydroxide and bicarbonates) through the biodegradation of organic compounds or through sulphide formation (sulphate reduction) significantly increases the pH and is a possible way of silica solubilization in alkaline soils.

8.5.2 Rhizospheric microbial weathering and plant growth

The ability of root systems to weather minerals was reported by Mortland *et al.* (1956), Spyridakis *et al.* (1967), Conyers and McLean (1968), Juang and Uehara (1968), Kabata-Pendias (1971), Boyle and Voigt (1973). These, however, did not distinguish the role of the rhizospheric microflora from that of the plants themselves.

Since the experiments of Gerretsen (1948) who observed a better growth of plants inoculated with phosphate-solubilizing bacteria and which were growing in sand plus rock phosphate (this may be due to an improved phosphate nutrition by microbial solubilization of phosphate or to a hormonal effect), we have few data concerning microbial weathering in the rhizosphere.

Most of the studies in the rhizosphere concern essentially the effect of microorganisms on the uptake of soluble mineral elements by plants (Barber 1978). Other studies concern counting of microbial populations solubilizing phosphates (Sperber 1958, Swaby & Sperber 1958, Louw & Webley 1959, Raghu & MacRae 1966, El Gibaly *et al.* 1977) or silicates (Jackson & Voigt 1971). The organisms solubilizing insoluble phosphates and silicates were consistently present in higher proportions in rhizosphere isolates than in those from nearby soil. Microorganisms involved were aerobic or anaerobic. Some experiments such as those of Alexsandrov and Zak (1950) and Holobrady *et al.* (1970) gave different types of results and did not show a similar effect of inoculation of bacteria in the rhizosphere.

Bromfield (1958) considering the rhizosphere of oat, thought essentially root excretions but not the rhizospheric microflora were responsible for the increase of availability of manganese to the plants.

Azcon *et al.* (1976) and Powell and Daniel (1978) have shown that vesicular arbuscular mycorrhizal fungi (*Glomus*) stimulate the uptake of soluble and insoluble phosphate. A bacteria able to solubilize rock phosphate *in vitro*, associated to mycorrhizal fungi enhanced mycorrhization and phosphate uptake (Azcon *et al.* 1976). Inoculation of calcium-phosphate-dissolving bacteria on to seedling cultures of *Pinus resinosa* in a soil deficient in soluble phosphate, but enriched with insoluble calcium phosphate, enhanced seedling growth as well or better than soluble phosphate fertilizers (Ralston & McBride 1976).

Mojallali and Weed (1978) reported that the inoculation of soybean plants by *Glomus* endomycorrhizae increased the uptake of K from biotite, but considering their results it appears that the difference of mycorrhizal versus non-mycorrhizal K uptake is most important in treatment with soil alone (soil without mica). This result shows that K of biotite is easily solubilizable (probably exchange uptake of ions). If the endomycorrhizal fungi (*Glomus mosseae*, *Glomus epigaeus*) are able to stimulate maize growth and K uptake from biotite, a non-symbiotic microflora is able to enhance biotite solubilization and K uptake (Table 8.4; Berthelin & Leyval 1980, Leyval 1981). Recent results (Leyval *et al.* 1981) showed that ectomycorrhizae (*Pisolithus tinctorius*) and bacteria-solubilized insoluble phosphates

Table 8.4. Influence of symbiotic (*Glomus mosseae*) and non-symbiotic microflora of the rhizosphere of corn on plant growth and K uptake from biotite.

Weight and mineral uptake (mg/plant)	Corn	Corn +mycorrhizae	Corn +non-symbiotic microflora	Corn +mycorrhizae +non-symbiotic microflora
Top weight	456	1100	979	1170
K uptake	1.6	1.9	2.6	2.8

Table 8.5. Influence of ectomycorrhizae (*Pisolithus tinctorius*) and bacteria-solubilizing phosphates (*Bacillus* sp.) on growth and mineral nutrition of *Pinus caribea* in the presence of insoluble phosphate ($Ca_3(PO_4)_2$).

Weight and mineral uptake (mg/plant)	Pinus	Pinus + mycorrhizae	Pinus + bacteria	Pinus + mycorrhizae + bacteria
Top weight	400	1200	1000	1900
P uptake	0.05	0.26	0.24	0.28
K uptake	2.2	5.5	4.5	7.5
Mg uptake	0.5	1.2	1.0	1.6
N uptake	1.5	7.5	6.0	15.0

(mineral and organic phosphates) in the rhizosphere of *Pinus caribea* and promoted plant growth and plant mineral nutrition (Table 8.5).

Such microbial rhizospheric weathering processes possibly involve solubilization of mineral elements by microbial compounds, but also production of hormonal-like substances that primarily promote plant growth and/or plant mineral nutrition. If the processes of solubilization of mineral elements by heterotrophic microorganisms (symbiotic or non-symbiotic) is able to promote the growth and mineral nutrition of plants by releasing assimilable mineral elements, it could also be unfavourable to this nutrition by transforming an assimilable into an unassimilable element (e.g. manganese deficiency). However, these processes including microbial ecology (population equilibria) are not well understood and further studies are needed to find agronomical (i.e. fertilization) applications. One might think that greenhouses and nursery plantations, reforestry of degraded soils and mining residues will be certainly the first possible applications.

8.5.3 Geological and leaching implications

Leaching with *Thiobacilli* is actually used to extract copper and uranium from low-grade ores. Recent reviews give data concerning the organisms, the minerals, the methods and the mechanisms involved (Lundgren & Silver 1980, Murr 1980, Laboureur 1981). Improvements seem possible under the aspects of ecology, engineering and genetic manipulation. Leaching is undoubtedly accompanied by weathering of all the minerals present in the dumps. The knowledge of these processes, which lead to the formation of amorphous precipitates and clays that block rock surfaces and adsorb metals, could be of use to improve metal extraction.

From an economic point of view, leaching is only possible with *Thiobacilli* but will actually be too expensive with heterotrophic microorganisms. However, it is foreseeable that other specific microbial processes (for instance metal uptake) or specific microbial compounds (specific chelating agents) will be used for some mineral purifications or some mineral concentrations, particularly if the supplying of certain metals becomes more difficult.

We consider now that some deposits are essentially of microbial origin (sulphides, oxides, carbonates; Ehrlich 1981) and results, such as those from Magne *et al.* (1974), Berthelin *et al.* (1977), Munier-Lamy (1980), suggest that microbial weathering by heterotrophic microorganisms is involved in the superficial processes of mobilization and concentration of metallic elements and eventually in reworking deposits.

References

Abd-el-Malek Y. & Rizk S.C. (1958) Counting of sulphate reducing bacteria in mixed bacterial populations. *Nature, London* **182,** 538–9.

Agate A.D. & Deshpande H.A. (1977) Leaching of manganese ores using arthrobacter species. *Conference Bacterial Leaching* (Ed. W.Schwartz), pp. 243–50. Verlag Chemie Publishers, New York.

Agbim N.N. & Doxtader K.G. (1975) Microbial degradation of zinc silicates. *Soil Biol. Biochem.* **7,** 275–80.

Aleksandrov V.G., Ternovskaja M.J. & Blagodyr R.N. (1963) Transformation du potassium en une forme assimilable lors de la décomposition des alumino-silicates par les bactéries. *Vest Sil'Skokholz Nauk. SSSR* **11,** 95–7.

Aleksandrov V.G. & Zak G.A. (1950) Les bactéries destructrices des aluminosilicates. *Mikrobiologya* **19,** 97–104.

Alexander M. (1977) *Introduction to soil microbiology*, 2e. John Wiley & Sons Inc., New York.

Aristovskaya T.V. (1956) *The role of microorganisms in podzol forming process.* Sixième Congrès de la Science du Sol, Paris, Vol. C, Commission III, 43, pp. 263–9.

Aristovskaya T.V. (1964) Decomposition of organic mineral compounds in podzolic soils. *Soviet Soil Science* **1,** 20–9.

Aristovskaya T.V., Daragan A. Yu., Zykina L.V. & Kutuzova R.S. (1969) Microbial factors in the movement of some mineral elements in the soil. *Soviet Soil Science* **5,** 538–46.

Aristovskaya T.V. & Kutuzova R.S. (1968) Microbial factors in the extraction of silicon from slightly-soluble natural compounds. *Soviet Soil Science* **12,** 1653–9.

Aristovskaya T.V. & Zykina L.V. (1978) Biological factors of aluminium migration and accumulation in soils and weathering crusts. *Problems of Soil Science.* Soviet pedologists to the XI International Congress of Soil Science. Publishing Office Nauka, pp. 175–82.

Arrieta L. & Grez R. (1971) Solubilization of iron-containing minerals by soil microorganisms. *Appl. Microbiol.* **22,** 487–90.

Audus L.J. (1946) A new soil perfusion apparatus. *Nature, London* **158,** 419.

Azcon R., Barea J.M. & Hayman D.S. (1976) Utilisation of rock phosphate in alkaline soils by plants inoculated with mycorrhizal fungi and phosphate solubilizing bacteria. *Soil Biol. Biochem.* **8,** 135–8.

Baldensperger J. (1976) Use of respirometry to evaluate sulphur oxidation in soils. *Soil Biol. Biochem.* **8,** 423–7.

Baldensperger J., Guarraia, L.J. & Humphreys W.J. (1974) Scanning electron microscopy of *Thiobacilli* grown on colloidal sulfur. *Arch. Microbiol.* **99,** 323–9.

Barber D.A. (1978) Nutrient uptake. In *Interactions between non pathogenic soil microrganisms and plants* (Eds Y.Dommergues & E.S.Krupa), Ch. IV, pp. 131–62. Elsevier, Amsterdam.

Berthelin J. (1971) Altération microbienne d'une arène granitique. Note préliminaire. *Science du Sol* **1,** 11–29.

Berthelin J. (1976) Etudes expérimentales de mécanismes d'altération des minéraux par des microorganismes hétérotrophes. Doctoral thesis, University of Nancy I.

Berthelin J. (1977) Quelques aspects des mécanismes de transformation des minéraux des sols par les microorganismes hétérotrophes. *Science du Sol* **1,** 13–24.

Berthelin J. & Belgy G. (1979) Microbial degradation of phyllosilicates during simulated podzolisation. *Geoderma* **21,** 297–310.

Berthelin J., Belgy G. & Magne R. (1977) Some aspects of the mechanisms of solubilization and insolubilization of uranium from granites by heterotrophic microorganisms. *Conference on bacterial leaching* (Ed. W.Schwartz), pp. 251–60. Verlag Chemie Publishers, New York.

Berthelin J., Belgy G. & Wedraoggo F.X. (1981) Nitrification au cours de la biodégradation de litières forestières: incidence sur l'altération d'une arène granitique. *Colloque Humus et Azote*, Reims, France, pp. 112–17, Commissions I et II. Station Science du Sol INRA, 51000 Chalons-sur-Marne. Service Science du Sol, Ensaia-Inpl, 54000 Nancy.

Berthelin J. & Boymond D. (1978) Some aspects of the role of heterotrophic microorganisms in the degradation of minerals in waterlogged acid soil. In *Environmental Biogeochemistry and Geomicrobiology*, vol. 2, *The Terrestrial Environment* (Ed. W.E.Krumbein), pp. 659–73. Ann Arbor Science, Ann Arbor.

Berthelin J. & Cheikhzadeh-Mossadegh D. (1977) Biodégradation de complexes organoferriques dans un sol brun acide et dans un sol podzolique sur granite. *Soil organic matter studies*, vol. I, pp. 413–24. I.A.E.A., Vienna.

Berthelin J. & Dommergues Y. (1972) Rôle des produits du métabolisme microbien dans la solubilisation des minéraux d'une arène granitique. *Rev. Ecol. Biol. Sol* **9,** 397–406.

Berthelin J. & Dommergues Y. (1976) The role of heterotrophic microorganisms in the deposition of iron and carbon in soil profiles. In *Environmental Biogeochemistry* (Ed. J.O.Nriagu) pp. 609–31. Ann Arbor Science, Ann Arbor.

Berthelin J. & Kogblevi A. (1972) Influence de la stérilisation partielle sur la solubilisation microbienne des minéraux dans les sols. *Rev. Ecol. Biol. Sol* **9,** 407–19.

Berthelin J. & Kogblevi A. (1974) Influence de l'engorgement sur l'altération microbienne des minéraux dans les sols. *Rev. Ecol. Biol. Sol* **11,** 499–509.

Berthelin J., Kogblevi A. & Dommergues Y. (1974) Microbial weathering of a brown forest soil. Influence of partial sterilization. *Soil Biol. Biochem.* **6,** 393–9.

Berthelin J. & Leyval C. (1982) Ability of symbiotic and non symbiotic rhizospheric microflora of maize (*Zea mays*) to weather micas and to promote plant growth and plant nutrition. *Plant and Soil*, **68**, 369–77.

Berthelin J. & Toutain F. (1979) Biologie des Sols. In *Pédologie*, vol. 2, *Constituants et propriétés du sol* (Eds M.Bonneau & B.Souchier), pp. 123–60. Masson, Paris.

Bertrand D. (1972) Interactions entre éléments minéraux et microorganismes du sol. *Rev. Ecol. Biol. Sol* **9**, 349–96.

Betrémieux R. (1951) Etude expérimentale de l'évolution du fer et du manganèse dans les sols. *Ann. Agron.* **2**, 193–5.

Beveridge T.J. & Murray R.G.E. (1976) Uptake and retention of metals by cell walls of Bacillus subtilis. *J. Bacteriol.* **127**, 1502–18.

Bolotina I.N. & Mirchink T.G. (1975) Distribution of manganese-oxidizing microorganisms in soils. *Soviet Soil Sci.* **7**, 325–38.

Boquel G. & Suavin L. (1974) Solubilisation du fer par deux souches bactériennes en presence de litière de teck. *Rev. Ecol. Biol. Sol* **11**, 187–95.

Boyle J.R. & Voigt G.K. (1973) Biological weathering of silicate minerals. Implications for tree nutrition on soil genesis. *Plant and Soil* **38**, 191–201.

Boyle J.R., Voigt G.K. & Sawhney B.L. (1967) Biotite flasks: alteration by chemical and biological treatment. *Science* **155**, 193–5.

Boyle J.R., Voigt G.K. & Sawhney B.L. (1974) Chemical weathering of biotite by organic acids. *Soil Science* **117**, 42–5.

Brierley C.L. (1978) Bacterial leaching. *CRC Crit. Rev. Microbiol.* **6**, 207–62.

Brock T.D. (1979) *Biology of microorganisms*. Prentice-Hall Inc., Englewood Cliffs, New Jersey.

Bromfield S.M. (1954) The reduction of iron oxide by bacteria. *J. Soil Sci.* **5**, 129–39.

Bromfield S.M. (1958) The properties of a biologically formed manganese oxide, its availability to oats and its solubilization by root washing. *Plant and Soil* **9**, 325–37.

Bromfield S.M. (1978) The oxidation of manganous ions under acid conditions by an acidophilous actinomycetes from acid soil. *Aust. J. Soil Res.* **16**, 91–100.

Bruckert S. (1971) Influence des composés organiques solubles sur la pédogenèse en milieu acide. I.—Etudes de terrain. II.—Expériences de laboratoire. Modalités d'action des agents complexants. *An. Agron.* **21**, 421–52, 725–57.

Bruckert S. & Dommergues Y. (1968) Importance relative de l'immobilisation physico-chimique et de l'immobilisation biologique du fer dans les sols. *Science du Sol* **1**, 19–27.

Bryner L., Beck J., Davis D. & Wilson D. (1954) Microorganisms in leaching sulfide minerals. *Ind. and Eng. Chem.* **46**, 2587.

Capus G. & Munier C. (1978) Mise en évidence de fortes concentrations d'uranium dans les corps microbiens de microorganismes actuels. *C.R. Acad. Sci., Paris*, **287**, D, 191–4.

Cheikhzadeh-Mossadegh D., Chone T. & Berthelin J. (1981) Influence de l'activité microbienne sur la migration et le devenir du fer et du carbone d'un complexe organo-ferrique dans des sols acides et un sol carbonaté. Colloque International C.N.R.S., n° 303 *Migrations organo-minérales dans les sols tempérés*, Nancy, 199–207.

Clark F.M., Scott R.M. & Bone E. (1967) Heterotrophic iron precipitating bacteria. *J. Amer. Water Works Ass.* **59,** 1036–42.

Conyers E.S. & McLean E.O. (1968) Effect of plant weathering of soil clays on plant availability of native and added potassium and on clay mineral structure. *Soil Science Soc. of Am. Proc.* **32,** 341–5.

Daragan A. Yu. (1971) Decomposition of minerals containing iron by soil microorganisms. *Soviet Soil Sci.* **3,** 567–72.

Doetsch R.N. & Cook T.M. (1973) *Introduction to bacteria and their ecobiology.* Medical technical publishing Co Ltd., Lancaster.

Dommergues Y. & Duchaufour Ph. (1965) Etude comparative de la dégradation biologique des complexes organo-ferriques dans quelques types de sols. *Science du Sol* **1,** 43–59.

Dommergues Y., Jacq V., Balandreau J. & Combremont R. (1969) *Rhizospherical and spermatospherical sulfate reduction and rhizospherical nitrogen fixation in saline soils.* Proceedings of the third International Conference on the global Impacts of Applied Microbiology (Giam III), Bombay.

Dommergues Y. & Mangenot F. (1970) *Ecologie Microbienne du Sol.* Masson, Paris.

Domsch K.H. (1968) Microbial stimulation and inhibition of plant growth. *9th Int. Congr. Soil Sci., Adelaide* **3,** 455–63.

Duff R.B. & Webley D.M. (1959) 2-ketogluconic acid as a natural chelator produced by soil bacteria. *Chem. and Ind.* 1376–7.

Eckhardt F.E.W. (1978) Microorganisms and weathering of a sandstone monument. In *Environmental Biogeochemistry and Geomicrobiology*, vol. 2, *The Terrestrial Environment* (Ed. W.E.Krumbein), pp. 675–86. Ann Arbor Science Publishers, Ann Arbor.

Eckhardt F.E.W. (1979) Über die Einwirkung heterotropher Mikroorganismen auf die Zersetzung silikalischer Minerale. *Zeitschrift für Planz. Boden.* **142,** 434–45.

Ehrlich H.L. (1966) Reactions with manganese by bacteria from marine ferro-manganese nodules. *Develop. Ind. Microbiol.* **7,** 272–86.

Ehrlich H.L. (1978) Inorganic energy sources for chemolithotrophic and mixotrophic bacteria. *Geomicrobiology J.* **1,** 65–83.

Ehrlich H.L. (1981) *Geomicrobiology.* Marcel Dekker Inc., New York.

Ehrlich H.L. (1981) Microbial oxidation and reduction of manganese as aids in its migration in soil. Colloque International C.N.R.S. n° 303 *Migrations organo-minérales dans les sols tempérés*, Nancy, 209–13.

El Gibaly M.H., El Reweiny F.M., Abdel-Nasser M. & El Dahtory T.A. (1977) Studies on phosphate-solubilizing bacteria in soil and rhizosphere of different plants. Occurrence of bacteria acid producers and phosphate dissolvers. *Zentrabl. Bakt. Parasitkde Infek. Kr. Hyg.* **132,** 233–9.

Emery T. (1974) Biosynthesis and mechanism of action of hydroxamate type siderochromes. In *Microbial iron metabolism* (Ed. J.B.Neilands), pp. 107–23. Academic Press, New York.

Ennever J., Vogel J.J. & Streckfuss J.L. (1974) Calcification by *Escherichia coli. J. Bacteriol.* **119,** 1061–2.

Fisher J.R. (1966) *Bacterial leaching of Elliot lake uranium ore.* Annual Western Meeting, Winnipeg, October 1965, Transactions, vol. LXIX, 164–71.

Fisher W.R. & Ottow J.G.G. (1972) Decomposition of iron (III)-citrate in a well-aerated, aqueous solution by soil bacteria. *Zeit. Pflanz. Bodenk.* **131,** 243–53.

Flaig W. (1971) Organic compounds in soil. *Soil Science* **111,** 19–33.

Friedmann I. (1971) Light and scanning electron microscopy of the endolithic desert algal habitat. *Phycologia* **10,** 411–28.

Friedmann I. & Ocampo R. (1976) Endolithic blue-green algae in the dry valleys: primary producers in the Antarctic Desert ecosystem. *Science* **193,** 1247–9.

Gerretsen F.C. (1948) The influence of microorganisms on the phosphate intake by the plant. *Plant and Soil* **1,** 51–81.

Goni J., Gugalski T. & Sima M. (1973) Solubilisation du potassium de la muscovite par voie microbienne. *Bull. B.R.G.M.* **4,** 31–47.

Gordienko S., Glushchenko T. & Ivahno L. (1972) Decomposition of metal humic acid complexes by microorganisms. *Proceedings of the Symposium on Soil Microbiology* (Ed. J.Szegi), pp. 191–6. Akademiai Kiado, Budapest.

Gundersen K. (1960) An all-round soil percolator. *Science* **132,** 224–5.

Heinen W. (1974) Microbial interaction with non-physiological elements and the substitution of bio-elements. *Biosystems* **6,** 133–51.

Henderson M.E.K. & Duff R.B. (1963) The release of metallic and silicate ions from minerals, rocks and soils by fungal activity. *J. Soil Sci.* **14,** 236–46.

Hirsch P. (1968) Biology of budding bacteria. IV. Epicellular deposition of iron by aquatic budding bacteria. *Arch. Mikrobiol.* **60,** 201–16.

Holler H. (1967) The experimental formation of alunite-jarosite through the action of sulfuric acid on mineral and rocks. *Contrib. Miner. and Petrol.* **15,** 309–29.

Holobrady K., Bujdos G. & Dobis A. (1970) Beitrag zum Studium biologischer Mobilisierung des Kaliums aus Kalihaltigen alumosilikaten II. *Biologia* (Bratislava) **25,** 461–9.

Huang W.H. & Kiang W. (1972) Laboratory dissolution of plagioclase feldspaths in water and organic acids at room temperature. *Amer. Mineral.* **57,** 1849–59.

Hutner S.H. (1972) Inorganic nutrition. *Ann. Rev. Microbiol.* **26,** 313–46.

Ivarson K.C. (1973) Microbiological formation of basic ferric sulfates. *Can. J. Soil Sci.* **53,** 315–23.

Ivarson K.C., Ross G.J. & Miles M.M. (1978) Alterations of micas and feldspars during microbial formation of basic ferric sulfate in the laboratory. *Soil Sci. Soc. Am. J.* **42,** 518–24.

Iverson W.P. & Brinckman F.E. (1978) Microbial metabolism of heavy metals. *In Water Pollution Microbiology*, vol. 2 (Ed. R.Mitchell), pp. 201–32. John Wiley & Sons, New York.

Jackson T. & Voigt G.K. (1971) Biochemical weathering of calcium-bearing minerals by rhizosphere micro-organisms, and its influence on calcium accumulation in trees. *Plant and Soil* **35,** 655–8.

Jaton C. (1972) Alterations microbiologiques de l'église monolithe d'Aubeterre-sur-Dronne. *Rev. Ecol. Biol. Sol* **9,** 471–7.

Jones D. & Wilson M.J. (1980) *Microbiological weathering of minerals.* Abst. of the second international symposium on microbial ecology, pp. 87–8, 7–12 Sept., University of Warwick, Coventry, England.

Juang T.C. & Uehara G. (1968) Mica genesis in Hawaiian Soils. *Soil Sci. Soc. Am. Proc.* **32,** 31–5.

Juste C. (1965) Contribution à l'étude de la dynamique de l'aluminium dans les sols acides du Sud-Ouest atlantique: application à leur mise en valeur. Doctoral thesis in Engineering, Faculty of Science, University of Nancy.

Kabata-Pendias A. (1971) Pobieranie Mikroelementow przez koniczyne z roznych poziomow glebowych. *Pam. Pul.* **45,** 127–45.

Kauffmann J. (1960) Corrosion et protection des pierres calcaires des monuments. *Corrosion et anti-corrosion* **8,** 87–95.

Kaufman D.D. (1966) An inexpensive, positive pressure, soil perfusion system. *Weeds* **14,** 90–1.

Kee N.S. & Bloomfield C. (1962) The effect of flooding and aeration on the mobility of certain trace elements in soils. *Plant and Soil* **16,** 108–35.

Kiel H. (1977) Laugung von Kupferkarbonat und Kupfersilikat-Erzen mit heterotrophen Microorganismem. *Conference on bacterial leaching* (Ed. W.Schwartz), pp. 261–70. Verlag Chemie Publishers, New York.

Kobus J. (1962) The distribution of microorganisms mobilizing phosphorus in different soils. *Acta Microbiol. Polonica* **11,** 255–64.

Krumbein W.E. (1969) Über den Einfluss der Mikroflora auf die exogene Dynamik (Verwitterung und Krustenbildung). *Geol. Rdsch.* **58,** 333–63.

Krumbein W.E. (1971) Manganese oxidizing fungi and bacteria. *Naturwissenschaften* **58,** 56–7.

Krumbein W.E. (1972) Rôle des microorganismes dans la genèse, la diagenèse et la dégradation des roches en place. *Rev. Ecol. Biol. Sol* **9,** 283–319.

Krumbein W.E. & Altmann H.J. (1973) A new method for the detection and enumeration of manganese oxidizing and reducing microorganisms. *Helgoländer Wiss. Meeresunters.* **25,** 347–56.

Krumbein W.E. & Lange C. (1978) Decay of plaster, paintings and wall material of the interior of buildings via microbial activity. In *Environmental Biogeochemistry and Geomicrobiology*, vol. 2 (Ed. W.E.Krumbein), pp. 687–97. Ann Arbor Science, Ann Arbor.

Krumbein W.E. & Pochon J. (1964) Ecologie bactérienne des pierres altérées des monuments. *Ann. Inst. Pasteur* **107,** 724–32.

Kutuzova R.S. (1973) Possible ways of mineral weathering in alkaline soils. *Soviet Soil Sci.* **5,** 111–16.

Kuznetsov S.I., Ivanov M.V. & Lyalikova N.N. (1963) *Introduction to geological microbiology*. McGraw Hill, New York.

Laboureur P. (1981) Lixiviation microbienne et hydrobiométallurgie. *Annales des Mines*, 63–76.

Lebedeva E.V., Lyalikova N.N. & Bugelsky Yu.Yu. (1978) Participation of nitrifying bacteria in weathering of serpentinized ultrabasic rocks. *Mikrobiol.* **47,** 1101–7.

Lees H. & Quastel J.H. (1944) A new technique for the study of soil sterilization. *Chem. and Indus.* **26,** 238–9.

Lehman D.S. (1963) Chelates in soils. A symposium on some principles of chelations chemistry. *Soil Sci. Soc. Proc.*, **27,** 167–70.

Leleu M., Sarcia C. & Goni J. (1973) *Altération expérimentale de deux feldspaths naturels par voie microbiologique directe et simulée.* Adv. in organic geochemistry actes du 6ème Congrès International de Géochimie organique, 18–21 sept. 1973. Ruel-Malmaison, France, pp. 905–24. Editions Techniques.

Le Roux N.W. & Mehta K.B. (1978) Examination of a copper ore after leaching with bacteria. In *Metallurgical applications of bacterial leaching and related microbial phenomena* (Eds L.E.Murr, A.E.Torma & J.A.Brierley) pp. 463–76. Academic Press, New York.

Le Roux N.W., Wakerley D.S. & Perry V.F. (1978) Leaching of minerals using bacteria other than *Thiobacilli*. In *Metallurgical applications of bacteria leaching and related microbiological phenoma* (Eds L.E.Murr, A.E.Torma & J.A.Brierley), pp. 167–91. Academic Press, New York.

Leyval C. (1981) Etude préliminaire de l'altération des minéraux par les microflores rhizosphériques de graminées et d'éricacées. Doctoral Thesis, University of Nancy.

Leyval C., Chakly M. & Berthelin J. (1981) Rôle des microflores rhizosphériques (symbiotiques ou non symbiotiques) sur la solubilisation d'éléments minéraux: influence sur la nutrition et la croissance des vegetaux. Colloque international du C.N.R.S., n° 303 *Migrations organo-minérales dans les sols tempérés*, Nancy, 481–4.

Lipman J.G., McLean H. & Lint H.C. (1916) Sulfur oxidation in soils and its effect on the availability of mineral phosphates. *Soil Sci.* **1,** 533–9.

Louw H.A. & Webley D.M. (1958) A plate method for estimating the numbers of phosphate-dissolving and acid producing bacteria in soil. *Nature* **182,** 1317–18.

Louw H.A. & Webley D.M. (1959) The bacteriology of the root region of the oat plant grown under controlled pot culture conditions. *J. Appl. Bacteriol.* **22,** 216–26.

Lundgren D.G. & Silver M. (1980) Ore leaching by bacteria. *Ann. Rev. Microbiol.* **34,** 263–83.

MacRae I.C. & Edwards J.F. (1972) Adsorption of colloidal iron by bacteria. *Appl. Microbiol.* **24,** 819–23.

MacRae I.C., Edwards J.F. & Nour D. (1973) Utilization of iron gallate and other organic iron complexes by bacteria from water supplies. *Appl. Microbiol.* **25,** 991–5.

Macura J. (1961) Continuous flow method in soil microbiology. I. Apparatus. *Fol. Microbiol.* **6,** 328–34.

Macura J. & Malek I. (1958) Continuous flow method for the study of microbiological processes in soil sample. *Nature, London*, **182,** 1796.

Magne R., Berthelin J. & Dommergues Y. (1974) Solubilisation et insolubilisation de l'uranium des granites par des bactéries hétérotrophes. Formation of uranium ore deposits. I.A.E.A., Vienna, Austria, 73–88.

Mehta A.P., Torma A.E. & Murr L.E. (1978) Biodegradation of aluminium-bearing rocks by *Penicillium simplicissimum. I.R.C.S. Medical Science* **6,** 416.

Mojallali H. & Weed S.B. (1978) Weathering of micas by mycorrhizal soybean plants. *Soil Sci. Soc. Amer. J.* **42,** 367–72.

Mortensen J.L. (1963) Complexing of metals by soils organic matter. *Soil Sc. Soc. Proc.* **27,** 179–86.

Mortland M.M., Lawton K. & Uehara G. (1956) Alteration of biotite to vermiculite by plant growth. *Soil Science* **82,** 477–81.

Mouraret M. & Baldensperger J. (1977) Use of membrane filters for the enumeration of autotrophic Thiobacilli. *Microbial Ecology* **3,** 345–59.

Moureaux C. (1970) Montages pour percolation continue de sols ou de roches. *Biol. Sol* **12,** 14–19.

Moureaux C. (1972) Influence du facteur microbiologique sur la solubilisation d'éléments minéraux à partir d'un sol ferrallitique malgache et à partir de biotite en présence de litières tropicales (teck et Niaoulé). *Rev. Ecol. Biol. Sol* **9,** 539–47.

Müller G.& Förster I. (1963) The effect of microscopic soil fungi on nutrient release from primary mineral, as a contribution to biological weathering. *Zbl. Bakt.* **116,** 372–409.

Müller G. & Förster I. (1964) The effect of microscopic soil fungi on nutrient release from primary minerals as a contribution to biological weathering. *Zbl. Bakt.* **118,** 594–621.

Munch J.C., Hillebrand Th. & Ottow J.C.G. (1978) Transformations in the Feo/Fed ratio of pedogenic iron oxides affected by iron-reducing bacteria. *Can. J. Soil Sci.* **58,** 475–86.

Munch J.C. & Ottow J.C.G. (1977) Modelluntersuchungen zum Mechanisms der Bakterielen Eisen-reduktion in Hydromorphen Böden. *Z. Pfl. Bodenk.* **140,** 549–62.

Munier-Lamy C. (1980) Etude expérimentale du rôle des microorganismes hétérotrophes dans la géochimie du cuivre et de l'uranium. Doctorate in Soil Science, University of Nancy I.

Murr L.E. (1980) Theory and practice of copper sulphide leaching in dumps and *in situ. Minerals Sci. Engng* **12** (3), 121–89.

Murr L.E. & Berry V.K. (1976) Direct observations of selective attachment of bacteria on low grade sulfide, ores and other mineral surfaces. *Hydrometallurgy,* **2,** 11–24.

Ottow J.C.G. (1968) Evaluation of iron reducing bacteria in soil and the physiological mechanism of iron reduction in aerobacter aerogenes. *Z. Allg. Mikrobiol.* **8,** 441–3.

Ottow J.C.G. (1969) The distribution and differentiation of iron-reducing bacteria in gley soils. *Zbl. Bakt.* **123,** 600–15.

Ottow J.C.G. & Munch J.C. (1981) Role of bacterial enzymes in the reductive dissolution of amorphous and crystalline iron compounds. Colloque International, C.N.R.S. n° 303 *Migrations organo-minérales dans les sols tempérés,* Nancy, 189–97.

Pedro G. (1971) *Altération des roches.* Encyclopedia Universalis, vol. I, pp. 813–17. Encyclopedia Britannica, Paris.

Perlman D. (1965) Microbial production of metal organic compounds and complexes. In *Advances in Applied Microbiology,* vol. 7 (Ed. W.W.Umbreit), pp. 103–38. Academic Press, New York.

Peters W. & Warren R.A.J. (1970) The accumulation of phenolic acids and coproporphyrin by iron-deficient cultures of *Bacillus subtilis. Can. J. Microbiol.* **16,** 1179–85.

Pochon J. & Chalvignac M.A. (1964) Oxydation biologique du soufre et dégradation des sols de vignoble. Vè Symp. Int. Agrochim. *La zolfo in agricoltura,* Palermo, pp. 373–80.

Pochon J. & Jaton C. (1967) The role of microbiological agencies in the deterioration of stone. *Chemistry and Indust.* 1587–9.

Pochon J., Tardieux P., Lajudie J. & Charpentier M. (1960) Dégradation des temples d'Angkor et processus biologiques. *Ann. Inst. Pasteur* **98,** 457–61.

Powell C.L. & Daniel J. (1978) Mycorrhizal fungi stimulate uptake of soluble and insoluble phosphate fertilizer from a phosphate-deficient soil. *New Phytol.* **80,** 351–8.

Powell P.E., Cline G.R., Reid C.P.P. & Szaniszlo P.J. (1980) Occurrence of hydroxamate siderophore iron chelators in soils. *Nature* **287,** 833–4.

Quastel J.H. (1965) Soil metabolism. *Ann. Rev. Plant Physiol.* **16,** 217–40.

Raghu K. & MacRae I.C. (1966) Occurrence of phosphate-dissolving microorganisms in the rhizosphere of rice plants and in submerged soils. *J. Appl. Bacteriol.* **29,** 582–6.

Ralston D.B. & McBride R.P. (1976) Interaction of mineral phosphate dissolving microbes with red pine seedlings. *Plant and Soil* **45,** 493–507.

Razzaghe-Karimi M.H. (1976) Contribution à l'étude expérimentale des phénomènes d'altération en milieu organique acide. Thèse de Doctorat ès-Sciénces Naturelles, Université Pierre et Marie Curie, Paris VI.

Razzaghe-Karimi M.H. & Robert M. (1975) Altération des micas et géochimie de l'aluminium: rôle de la configuration de la molécule organique sur l'aptitude à la complexation. *C.R. Acad. Sci., Paris* **280,** 2645–8.

Ribeiro R.M., Moureaux C. & Novikoff A. (1976) Etude comparative de l'altération microbienne des différents minéraux constituants d'une diabase. *Cah. ORSTOM, sér. Pédol.* **14,** 161–8.

Robert M., Eyralde J., Berrier J. & Pelisonnier C. (1979) Illustration du rôle des êtres vivants dans l'altération et la pédogenèse à l'étage alpin et subalpin. *Annales de l'Université de Savoie* **4,** 23–5.

Robert M. & Razzaghe-Karimi M.H. (1975) Mise en évidence de deux types d'évolution minéralogique des micas trioctaédriques en présence d'acides organiques hydrosolubles. *C.R. Acad. Sci. Paris* **280,** 2175–8.

Rossi G. (1978) Potassium recovery through leucite bioleaching possibilities and limitations. In *Metallurgical applications of bacterial leaching and related microbiological phenomena* (Eds L.E.Murr, A.E.Torma & J.A.Brierley), pp. 297–319. Academic Press, New York.

Savostin P. (1976) A procedure for isolation of silicate and other microorganisms from tropical soils. *Soil Science* **122,** 154–8.

Schatz A. (1963) Soil microorganisms and soil chelation. The pedogenetic action of lichens and lichen acids. *Agricult. food Chemi.* **11,** 112–18.

Schatz A., Schatz V., Schalscha E. & Martin J.J. (1964) Soil organic matter as a natural chelating material. Part I. The chemistry of chelation. *Compost Sci.* **4,** 25–8.

Schnitzer M. (1978) Reactions of humic substances with minerals in the soil environment. In *Environmental biogeochemistry and geomicrobiology*, vol. 2 (Ed. W.E.Krumbein), pp. 639–47. Ann Arbor Science, Ann Arbor.

Schnitzer M. & Khan S.U. (1972) *Humic substances in the environment.* Marcel Dekker, New York.

Schnitzer M. & Skinner S.I.M. (1964) Organo-metallic interactions in soils: 3é properties of iron and aluminium organic matter complexes prepared in the laboratory and extracted from a soil. *Soil Sci.* **98,** 197–203.

Silverman M.P. & Ehrlich H.L. (1964) Microbial formation and degradation of minerals. *Adv. Appl. Microbiol.* **6,** 153–206.

Silverman M.P. & Lundgren D.G. (1959) Studies on the chemo-autotrophic iron

bacterium *Ferrobacillus ferrooxidans*. I. An improved medium and a harvesting procedure for securing high cell yields. *J. Bact.* **77,** 642–7.

Silverman M.P. & Munoz E.F. (1970) Fungal attack on rock: solubilization and altered infrared spectra. *Science* **169,** 985–7.

Silverman M.P. & Munoz E.F. (1971) Fungal leaching of titanium from rock. *Appl. Microbiol.* **22,** 923–5.

Smyk B. & Drzal M. (1964) Research on the influence of microorganisms on the development of karst phenomena. *Geographica polonica* **2,** 57–60.

Sperber J.I. (1958) The incidence of apatite-solubilizing organisms in the rhizosphere and soil. *Austral. J. Agric. Res.* **9,** 778–81.

Sperber J.I. & Sykes B.J. (1964) A perfusion apparatus with variable aeration. *Plant and Soil* **20,** 127–30.

Spyridakis D.E., Chesters G. & Wilde S.A. (1967) Kaolinisation of biotite as a result of coniferous and deciduous seedling growth. *Soil Sci. Soc. Amer. Proc.* **31,** 203–10.

Starkey R.L. (1966) Oxidation and reduction of sulfur compounds in soils. *Soil Science* **101,** 297–306.

Stevenson F.J. (1967) Organic acids in soil. *Soil Biochemistry,* vol. I (Eds A.D.McLaren & G.H.Peterson), pp. 119–46. Marcel Dekker Inc., New York.

Summers A.O. & Silver S. (1978) Microbial transformations of metals. *Ann. Rev. Microbiol.* **32,** 637–72.

Swaby R.J. & Sperber J. (1958) Phosphate dissolving microorganisms in the rhizosphere of legumes. *Nutrition of legumes.* Proc. Univ. Nottingham fifth Easter agric. Sci., 289–94 (CSIRO, Adelaide).

Tardieux-Roche A. (1966) Contribution à l'étude des interactions entre phosphates naturels et microflore du sol. *Ann. Agron.* **17,** 403–71, 479–528.

Tesic Z.P. & Todorovic M.S. (1958) Contribution à la connaissance des propriétés spécifiques des bactéries silicatées. *Zemlogiste i biljka* **8,** 238–40.

Torma A.E. (1976) Biodegradation of chalcopyrite. *Proc. 3rd Int. Biodegr. Symp.* (Eds J.M.Sharpley & A.M.Kaplan), pp. 937–46. Applied Science Publishers Ltd, London.

Torma A.E. & Guay R. (1976) Effect of particle size on the biodegradation of a sphalerite concentrate. *Natur. Can.* **103,** 133–8.

Torma A.E. & Sakaguchi M. (1978) Relation between the solubility product and the rate of metal sulfide oxidation by *Thiobacillus ferrooxidans*. *J. Ferment. Technol.* **56,** 173–8.

Torma A.E. & Tabi M. (1973) Mise en solution des métaux de minerais sulfurés par voie bactérienne. *L'ingénieur* **294,** 2–8.

Tsuchiya H.M., Trivedi N.C. & Schuler M.L. (1974) Microbial mutualism in ore leaching. *Biotechnol. Bioeng.* **16,** 991–5.

Tsyurupa I.G. (1964) Some data on complex products of microbial activity and autolysis with soil minerals. *Soviet Soil Science* **3,** 261–5.

Tuovinen O.H. & Carlson L. (1979) Jarosite in culture of iron-oxidizing Thiobacilli. *Geomicrobiology J.* **1,** 205–10.

Védy J.C. & Bruckert S. (1979) Les solutions du sol. Composition et signification pédogénétique. In *Pédologie,* T.2, *Constituants et propriétés du sol* (Eds M.Bonneau & B.Souchier), pp. 161–86. Masson, Paris.

Vincente Hernandez M.A. & Robert M. (1975) Transformation profonde des micas

sous l'action de l'acide galacturonique. Problème des spectites des podzols. *C.R. Acad. Sci., Paris* **281,** 523–6.

Wagner M. & Schwartz W. (1967) Geomikrobiologische Untersuchungen VIII. Über das Verhalten von Bakterien auf der Oberflüche von Gesteinen und Mineralieu und ihre Rolle bei der Vitterung. *Z. für Allg. Mikrobiol.* **7,** 33–52.

Webley D.M., Duff R.B. & Mitchell W.A. (1960) A plate method for studying the breakdown of synthetic and natural silicates by soil bacteria. *Nature , London* **188,** 766–7.

Webley D.M., Henderson E.K. & Taylor F. (1963) The microbiology of rocks and weathered stones. *J. Soil Sci.* **14,** 102–12.

Weed S.B., Davey C.B. & Cook M.G. (1969) Weathering of mica by fungi. *Soil Sci. Soc. Am. Proc.* **33,** 702–6.

Wildung R.E., Biggar J.W. & Chesters G. (1969) Soil perfusion under controlled conditions of soil moisture content and soil atmospheric composition. *Soil Sci. Soc. Amer. Proc.* **33,** 813–14.

Wood P.A. & MacRae I.C. (1972) Microbial activity in sandstone deterioration. *Int. Biodetn. Bull.* **8,** 25–7.

Wright S.J.L. & Clark C.G. (1969) Controlled soil perfusion with a multichannel peristaltic pump. *Weed Res.* **9,** 65–8.

Yoshida T. (1975) Microbial metabolism of flooded soils. In *Soil Biochemistry*, vol. 3 (Eds E.A.Paul & A.A.McLaren), pp. 83–122. Marcel Dekker Inc., New York.

Zajic J.E. (1969) *Microbial biogeochemistry*. Academic Press, New York.

CHAPTER 9
MICROORGANISMS AND DIAGENESIS OF SEDIMENTS

CHARLES D.CURTIS

9.1 THE DIAGENETIC ENVIRONMENT

9.1.1 Introduction

This chapter attempts to summarize what contribution microorganisms make to the processes which take place as uncompacted sediments are converted, during burial, into indurated sedimentary rocks. It is thus concerned with the diagenetic environment as a whole rather than specific processes or specific microorganisms.

Most is known about the relatively rapid chemical changes that take place at the sediment–water interface and in the uppermost few metres of sediment. Very much less is known about the possible involvement of microorganisms at much greater depths although many workers now agree that microbiological degrading of oil within reservoirs is probably quite commonplace. Sampling (contamination) problems make positive proof difficult but recent work with specific organic compounds of known affinity looks promising in this context.

The early diagenetic processes referred to above are very much the concern of chemical oceanographers who are particularly interested in the flux of chemical species across the sediment–water interface. Mass transfer across this boundary is one very important step in the geochemical cycles of elements as well as being in part responsible for the steady-state composition of depositional water masses. These topics are treated more comprehensively elsewhere in the book (cf. Chapters 3, 4 and 7).

Geologists are concerned with the end-product of diagenesis: sedimentary rocks. Again certain aspects are treated elsewhere, notably fossil fuels and metal concentration. Many common sedimentary rocks, however, are significantly modified during burial diagenesis by microorganisms. Sometimes the effects are very obvious; as in the case of concretions. In other cases relatively subtle changes in sediment porosity and permeability may influence pore fluid migration and confinement and so make the difference between an oil reservoir and a barren sandstone. Not a lot is known about the importance of microorganism involvement relative to inorganic reactions at the considerable depths where some cementation takes place: this is an area requiring urgent study.

Not unnaturally most of this chapter will be devoted to a discussion of early diagenetic processes in which microorganisms are certainly responsible for important organic matter degradation and mineral precipitation and dissolution reactions.

9.1.2 Temperature and pressure variation

Heat flow within the earth's crust causes temperature gradients within buried sediment columns (or vice versa). With burial, therefore, sediments encounter environments of successively higher temperature and pressure. The term *diagenesis* is used to describe all the changes that take place from the time of deposition until the temperatures and pressures normally associated with the lowest grades of *metamorphism* are reached. It is not easy to agree on a precise definition of the boundary between the two but temperatures around 120 °C are not too unreasonable: above this temperature lies the supermature stage of hydrocarbon generation and very significant mineral recrystallization starts to occur. A useful review of thermal indicators during burial diagenesis was recently presented by Heroux *et al.* (1979). It also seems unlikely that microbiological processes are important at the upper end of the diagenetic temperature range.

Heat flow varies from point to point on the earth's crust and thermal gradients tend to fall in the range of 10–40 °C km^{-1}. This means that temperatures below 50 °C can be anticipated at depths down to approximately 5 km in areas of very low thermal gradient. In other locations this temperature will be encountered at 1 km depth.

Pressures are generally hydrostatic within compacting sediment columns: particularly during the earlier stages. Good pore interconnection is the norm with relatively high porosities. Permeability, however, does tend to vary markedly between different sediment types with complicated consequences for porewater migration. This will be treated below.

This brief analysis suggests that microbiological activity is not debarred from many diagenetic environments for reasons of excessive temperature or pressure. This is certainly true up to depths very much greater than are normally associated with microbiological activity in sediments.

9.1.3 Physical compaction

The porewater content of sediments and its variation with depth have been studied in great detail—not least because of its relevance to petroleum migration and accumulation. Throughout burial there is more or less continuous expulsion of porewater with resultant sediment compaction. At deposition, muddy sediments contain something like 80% porewater. By the time these have been buried to a depth of 1 km this has been reduced to about 30% and probably to less than 10% at 5 km depth, although an uncertain but significant fraction is structurally bound to mineral matter at these burial depths. Compaction of fine-grained sediments has been reviewed by Riecke and Chilingarian (1974).

Coarser-grained sediments tend to contain rather less pore space at deposition and exhibit greater resistance to compaction. This is mostly for simple physical reasons: sand particles are sub-spherical and pack relatively neatly at deposition. Physical compaction adjusts the packing patterns but significant pore space (30% or more) is left even within perfectly sorted and packed spheres. Clay particles, on the other hand, are small and very thin platelets which initially pack, for electrostatic charge neutralization, in a very open 'edge-to-face' pattern. With burial and increasing overburden pressure, these forces are overcome and flakes rotate to give space-conservative 'face-to-face' structures.

The degree of compaction is clearly a function of burial depth, as is pore volume. Sedimentary rocks are said to be under-compacted or over-compacted depending on their departure from the normal trend. Such departures can give important evidence of depositional and tectonic history.

During compaction, vast volumes of porewaters are squeezed out from all types of sediment and must find their way out to the depositional environment. Much current research activity is focused in this area because migration pathways probably modify sediment permeability and porosity and thereby may radically affect liquid hydrocarbon location. It seems likely that expressed porewaters migrate laterally along higher permeability sediment units and upwards wherever possible. Often the pattern is from basin centre to margin as well as upwards. Emergence to depositional waters is obvious when porewaters are charged with gases, most commonly methane (Hedberg 1974).

This last example makes an important point. Pore fluids carry the products of microbiological processes (in this case methanogenesis) away from the site of their production. They may equally transport nutrients produced by essentially inorganic means at greater depth and thereby support microbiological communities. It is essential to appreciate the very open nature of diagenetic environments and to realize that conditions for the maintenance of microbiological activity could exist at very great depth within sediment columns. It is the shallow burial environments which have been studied extensively—for obvious reasons.

9.1.4 Concretions

Perhaps most striking of all geological records of microbiological activity are *concretions*. These are found in sedimentary rocks of all kinds and range in shape from ellipsoidal through flattened forms via coalescence into continuous sheets. Thin sections show that they include the same detrital sediment grains as the enclosing sedimentary rocks themselves but in a much less compacted state. Most people accept that cements, usually carbonate or sulphide minerals, were precipitated within pore space at an early stage of compaction; effectively replacing porewater. Further compaction was then prevented.

The evidence for their microbiological derivation most commonly cited is the stable isotope composition of carbon, oxygen and sulphur atoms in the cement. Values wildly different from 'normal' carbonate and sulphide minerals are often found and these are readily linked with fractionations typical of sulphate reduction and fermentation.

Why they are important as well as striking is that they preserve evidence of compactional history and, most important of all, the probable depth at which cementation occurred within the sediment column. Figure 9.1 illustrates some of these points. Two obvious elements of fabric within sediments are clay fabric (muds/mudrocks) and 'bedding' or time planes picked out by some distinguishable layer of sediment (all sediments). Both change with depth and can be measured to estimate compaction (Oertel & Curtis 1972). Concretions formed rapidly and early during compaction (Fig. 9.1(a)) will show bedding surfaces preserved at a much greater vertical separation than surrounding sediments—the latter are invariably 'wrapped around' the concretions. Concretions precipitated at greater burial depth will show less compaction relative to surrounding sediments (also a lesser cement/detrital sediment ratio). Concretions slowly precipitated over a lengthy timespan covering significant compaction will exhibit bedding deflection *within* the structure (Fig. 9.1(b)). All these structures have been described in the literature and do offer significant help with diagenetic environment description.

The intimate association of organic matter and carbonate cements with isotopic composition suggesting direct derivation was described by Galimov and Girin (1968). Since this classic paper, numerous studies of sequential cementation within concretions and attempts to link observations with microrganism activity have been reported (Curtis *et al.* 1971, Sass & Kolodny 1972, Dickson & Barber 1976, Hudson 1978). Raiswell (1976) specifically attempted to reconstruct microbiological processes and controls and, more recently (Raiswell & White 1978), to look for evidence of nutrient supply via migrating porewaters.

Early, rapidly precipitated concretions are often composed of calcite ($CaCO_3$) with, less often, pyritic (FeS_2) rims. It is by no means uncommon to find beautifully preserved fossils within them—even soft structures are preserved in uncompressed conditions. It has been suggested that organic

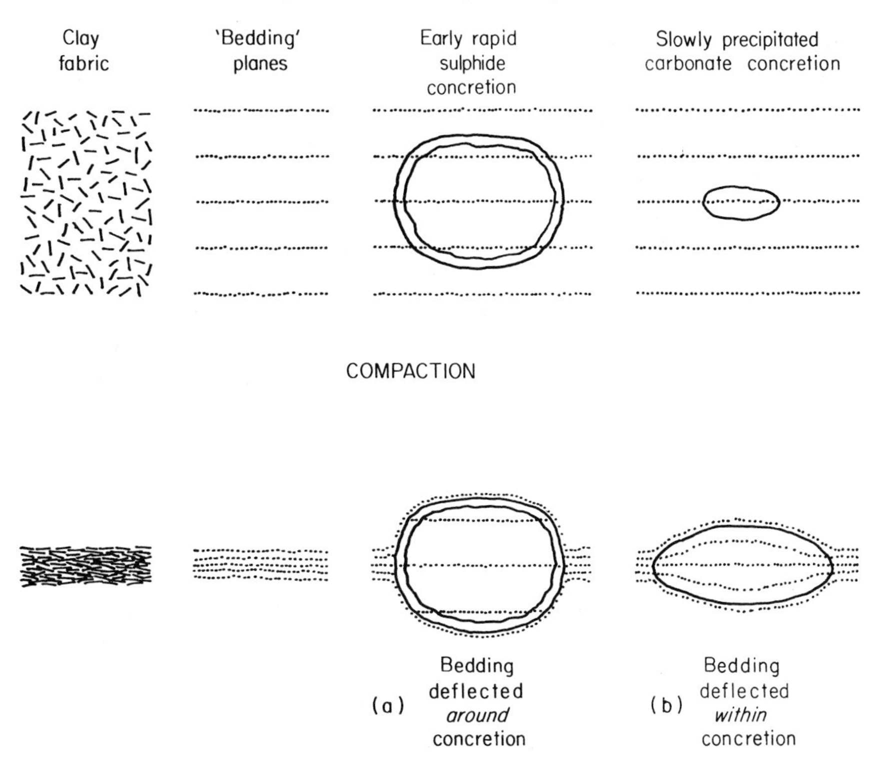

Fig. 9.1. Sediment compaction and concretion formation. (a) Growth complete before significant compaction: bedding deflected around concretion ('body' often calcite with pyrite rim). (b) Growth during compaction: bedding deflected within concretion—fossil 'nucleii' rare, suggesting nutrient supply from migrating porewaters?

decay supports localized microorganism activity, the products of which cause cement precipitation. Berner (1968) investigated this proposition in the laboratory and suggested that calcium salts of fatty acids were probably responsible for initial cementation; subsequently to be replaced by calcite. Later concretions, as inferred from structures, seldom preserve evidence suggesting that microorganism nutrient supply was localized in the form of deceased macroorganisms. Here, migrating porewaters carrying soluble nutrients may be responsible. What is clear is that such suggestions are really speculative: there is much to be said for further careful study of concretions and the evidence that they provide about the environment.

9.1.5 Variations in sediment composition and depositional environment

Concretions and other evidence of extensive microbiological activity are found in many different types of sediment but their size, composition and frequency of occurrence do depend on host sediment composition and depositional environment. It is, therefore, necessary to recognize these controls on microorganism activity during diagenesis.

At the simplest possible level it is useful to acknowledge shallow water depositional environments on land (lacustrine) and in the sea (shelf) as well as deep-water (pelagic) situations. Mineral matter is deposited directly from depositional water as carbonate and silica, usually as skeletal material. This is mixed with sediment derived from the land surface by erosion and transport of soil profiles. On a global scale this component is greater than that due to chemical precipitation although shallow-water carbonate sediments and pelagic oozes in which the chemical fraction dominates are by no means uncommon.

Coarse detrital material is separated from fine by water currents of different velocity and particle-transporting capacity. Now, and throughout most of geological time, the most common sediment type consists of mud and silt deposited in relatively near-shore marine and estuarine situations. Organic detritus from the land surface and that from productivity within the water column tend to accumulate to high levels in such sedimentary environments. In this chapter most attention will be paid to these particular situations. The vast variety of sedimentary environments has been recently and well reviewed by Reading (1978). Obviously it is quite beyond the scope of this chapter to attempt any kind of comprehensive cover of diagenesis in different sediment types.

9.1.6 Porewater

In the last section no mention was made of the most important (in volume terms) sediment component of all at deposition: porewater. Most sediments contain something between 50 and 90%. In the case of seawater which

Table 9.1. Principal solutes in seawater and 'average' river water (after Garrels & Mackenzie 1972).

	River water		Sea water
Solute	Concentration mmol l^{-1}	Solute	Concentration mmol l^{-1}
HCO_3^-	0.958	Cl^-	535.2
Ca^{2+}	0.375	Na^+	456.2
Na^+	0.270	Mg^{2+}	54.2
Cl^-	0.220	SO_4^{2-}	27.6

includes 3.5% by weight of dissolved salts, a sizeable proportion of the total sediment mineral matter is included in the aqueous phase.

Fresh water and seawater differ greatly in the solutes they contain as well as in their total concentration. Table 9.1 lists the four most important solutes in seawater and 'average' river water. Normal marine sediments thus include highly saline porewaters with very significant chloride and sulphate contents, whereas the most concentrated anion in fresh waters is bicarbonate. Diagenetic reactions between mineral particles and porewaters, microorganism metabolism and mass transfer between porewaters and overlying depositional waters cause systematic changes in porewater composition with depth. These porewater solute concentration profiles have provided some of the most telling clues in the search to elucidate reaction mechanisms, rates and controls.

In the early stages of diagenesis, connection between porewater and overlying water is good. Convective mass transport, of course, is virtually eliminated from porewaters but diffusive transfer is modified only by the tortuousness of pore systems, which is not very great in sediments containing more than 70% porewater. Establishment of concentration differences across the sediment–water interface instigates diffusive mass transfer and sediments close to the sediment–water interface are effectively 'open' with respect to the depositional water solute reservoir. Modification to sediment composition can thus be very extensive indeed *provided bottom water circulation* is effective. Oxygen supply and sulphate supply in marine systems are critical diagenetic controls: both limit the metabolic rate of important groups of microorganisms.

9.1.7 Chemical reactivity

Some mention of compositional variation was made in section 9.1.5. The starting place for detrital sediments is soils: themselves the end-point of processes with significant microorganism involvement (see Chapter 8). The coarse fraction, ultimately to dominate sandy sediments, is mostly composed of residual mineral particles from bedrock which have survived weathering processes intact. They are thus very resistant to aqueous solutions containing oxygen, carbon dioxide and organic acids. By the same measure they can be anticipated to be relatively inert in shallow burial environments.

Soil fine fractions include clay minerals and 'sesquioxides'—amorphous or crystalline hydrated aluminium and iron oxides. Both groups are stable in soil systems although relatively reactive (with a very large surface area). These constituents find their way to be deposited as mud; along with degraded organic residues from the soils and organic input from depositional water productivity. Detailed description of the chemical constitution of these organic residues is clearly beyond the scope of this chapter. Variations between different sedimentary environments are important here, however, in influencing 'substrate quality' for microbial populations.

At present, most soil systems are adequately supplied with molecular oxygen. This means that mineral phases (notably sesquioxides) containing iron are in the oxidized (ferric) form. Soils commonly contain several percent iron as Fe^{3+} hydrated oxides. Once sedimented as mud, this material comes into intimate contact with reduced carbon compounds in an oxygen-free environment. At deposition, therefore, mud is a fundamentally unstable assemblage and diagenetic reactions are to be anticipated irrespective of microbiological involvement.

Carbonate sediments may be coarse- or fine-grained: the latter usually containing most organic matter. The very specific dependence of carbonate mineral solubility on carbon dioxide partial pressure suggests at once that diagenetic reactions involving dissolution and precipitation are very likely. This is in fact the case: something between 30 and 50% of many limestones is made up from calcium carbonate precipitated in pore space after deposition. Dickson (1979) has made the point that something over one-third of the height of the Canadian Rockies (where they are limestone!) can be attributed directly to diagenetic precipitation. It is a far from insignificant process.

9.2 CHEMICAL AND MINERALOGICAL REACTIONS SUBJECT TO MICROORGANISM CONTROL

9.2.1 Introduction

The first important environmental boundary to be considered must obviously be that between oxygen-containing and anoxic waters. For most sedimentary environments this occurs just below the sediment–water interface. Anoxic conditions develop as soon as oxygen supply falls short of uptake by microorganism respiration or by reduced carbon compounds. Within sediments oxygen supply is limited by diffusion rather than by convection: very rapid exhaustion is to be expected downwards from the sediment–water interface in organic-rich muds.

Anoxic conditions can develop, of course, within the depositional water column. In such situations molecular oxygen is present in insignificant amounts and aerobic biological activity makes no contribution to diagenesis. Anaerobic conditions can develop in a variety of situations, mostly determined by basin geometry and current strength. In a recent review, Demaison and Moore (1980) recognized four present-day anoxic depositional basin types:

(1) large anoxic lakes (e.g. Lake Tanganyika);
(2) anoxic silled basins (e.g. Black Sea);
(3) anoxic layers caused by upwelling (e.g. Peru coastal upwelling);
(4) open ocean anoxic layers (e.g. intermediate depths in North-East Pacific).

In former geological times it is very likely that different climatic distribution patterns (especially absence of icecaps), different ocean circulation patterns and lower atmospheric oxygen levels all would contribute to more widespread anoxic water masses than occur today.

In order to examine diagenetic environments as generally as possible, however, it is logical to start where oxygen supply is good and organic input low: under these conditions the zone of influence of molecular oxygen in sediments is expanded.

9.2.2 Suboxic diagenesis: zonal sequence 1

In relatively slowly deposited pelagic sediments from the eastern equatorial Atlantic, the influence of molecular oxygen can be traced down to depths approaching 1 m. Porewater solute profiles are complicated but sulphate depletion (i.e. presence of sulphate-reducing bacteria) is not detectable. The recent paper by Froelich *et al.* (1979) describes a number of gravity cores taken from these sediments and presents a detailed analysis of the chemical changes that take place. Porewater data for total carbonate, sulphate, pH, nitrate, ammonia, phosphate and dissolved manganese and iron were obtained to depths of about 80 cm. Reproducible solute trends strongly suggested that several important redox reactions were taking place; often over characteristic (relative) depth intervals in which one reaction was prominent.

Figure 9.2 is modified and simplified from Froelich *et al.* (1979) to demonstrate how porewater solute profiles can provide good evidence of solute production, removal and vertical transfer. The uppermost part of each profile was characterized by respiration: dissolved molecular oxygen being consumed and bicarbonate, nitrate and phosphate being added to porewater:

$$(CH_2O)_{106}(NH_3)_{16}(H_3PO_4) + 138\ O_2 \rightarrow 106\ CO_2 + 16\ HNO_3 + H_3PO_4 + 122\ H_2O$$

$$\Delta G^0_r = -3190\ \text{kJ mol}^{-1}\ \text{glucose.} \qquad (9.1)$$

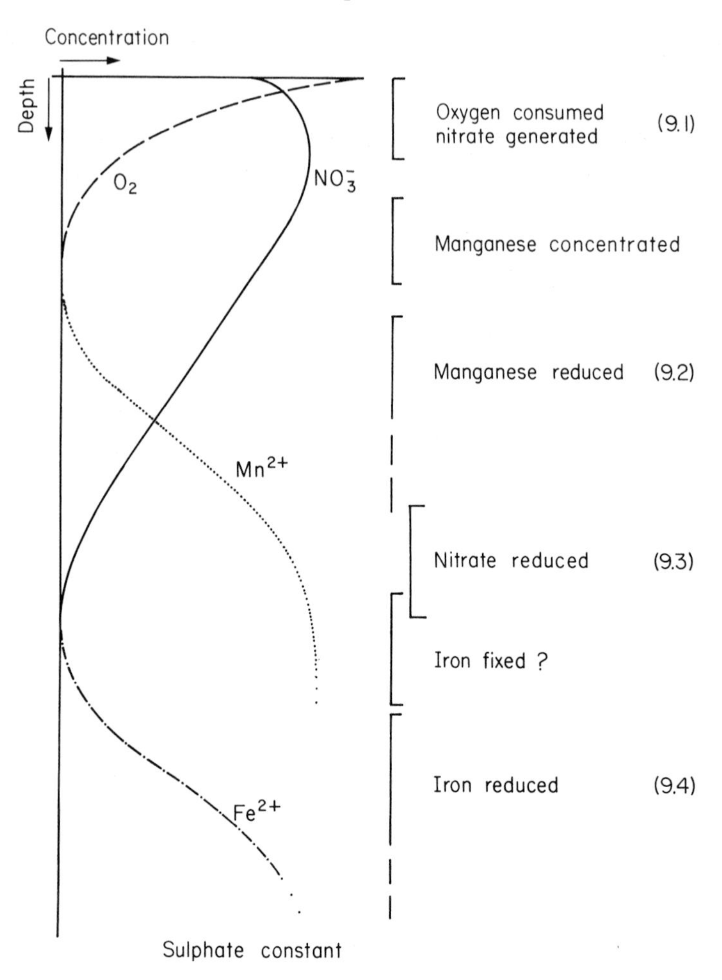

Fig. 9.2. Idealized porewater profiles: eastern equatorial Atlantic (after Froelich *et al.* 1979). (Numbers in brackets refer to equations in the text.)

Nitrate and bicarbonate produced in this reaction diffuse up and down, in the latter case accompanied by any molecular oxygen having escaped the attention of oxygen respirers. Manganous manganese, diffusing upwards from reactions taking place deeper in the profile, fixed the molecular oxygen by precipitation of oxides within sediment pore space. Whether or not microorganisms are involved directly in redox reactions involving manganese and iron in sediments remains to be answered: clearly oxidative precipitation does not *require* such involvement.

Beneath the manganese precipitation horizon evidence for manganese reduction was plentiful. Formulation of this redox reaction in terms similar to those of equation 9.1 showed it to be only slightly less favourable than molecular oxygen reduction:

$$(CH_2O)_{106}(NH_3)_{16}(H_3PO_4) + 236\ MnO_2 + 472\ H^+ \rightarrow 236\ Mn^{2+} + 106\ CO_2 + 8\ N_2 + H_3PO_4 + 366\ H_2O$$
$$\Delta G^0_r = -3090\ \text{kJ mol}^{-1}\ \text{(birnessite)} \qquad (9.2)$$

Successively deeper zones involve nitrate reduction and ferric oxide reduction: each with a somewhat smaller energy liberation:

$$(CH_2O)_{106}(NH_3)_{16}(H_3PO_4) + 94.4\ HNO_3 \rightarrow 106\ CO_2 + 55.2\ N_2 + H_3PO_4 + 177.2\ H_2O$$
$$\Delta G^0_r = -3030\ \text{kJ mol}^{-1} \qquad (9.3)$$

$$(CH_2O)_{106}(NH_3)_{16}(H_3PO_4) + 212\ Fe_2O_3 + 848\ H^+ \rightarrow 424\ Fe^{2+} + 106\ CO_2 + 16\ NH_3 + H_3PO_4 + 530\ H_2O$$
$$\Delta G^0_r = -1410\ \text{kJ mol}^{-1}\ \text{(hematite)} \qquad (9.4)$$

Thus, Froelich *et al.* (1979) were able to show that oxidants were consumed in order of decreasing energy production per mole; each of the above reactions being much more energetic than sulphate reduction or methane production calculated on the same basis:

$$(CH_2O)_{106}(NH_3)_{16}(H_3PO_4) + 53\ SO_4^{2-} \rightarrow 106\ CO_2 + 16\ NH_3 + 53\ S^{2-} + H_3PO_4 + 106\ H_2O$$
$$\Delta G^0_r = -380\ \text{kJ mol}^{-1} \qquad (9.5)$$

$$(CH_2O)_{106}(NH_3)_{16}(H_3PO_4) \rightarrow 53\ CO_2 + 53\ CH_4 + 16\ NH_3 + H_3PO_4$$
$$\Delta G^0_r = -350\ \text{kJ mol}^{-1} \qquad (9.6)$$

This study demonstrates that in pelagic sediments, degradation of organic matter by redox reactions involving mineral species (iron and manganese oxides) occurs in close association with metabolic reactions utilizing dissolved species as electron acceptors (O_2, NO_3^-). There appears to be a rather clearly defined succession of 'zones' in which one reaction takes precedence over the others and this correlates rather simply with net energy release.

In the past (e.g. Claypool & Kaplan 1974), such a succession has been linked with the competitive advantage offered to specific microorganisms by more efficient metabolism. The clear demonstration of oxidation of organic matter by ferric and manganic mineral phases at the appropriate point in the 'microbial metabolism ecological succession' raises very interesting questions about the direct involvement of bacteria in these reactions.

9.2.3 Diagnesis in 'near-shore' sediments: zonal sequence 2

Moving from pelagic environments towards the sources of terrestrial detritus, there is a general increase in sedimentation rate and a shallowing of depositional water. Except in relatively rare situations with severely restricted water circulation, bottom waters are oxygenated. In these more rapidly deposited 'near-shore' sediments, however, anoxic conditions are established within porewaters just below the sediment–water interface and diagenesis is characterized by anaerobic degradation of organic matter and the response of detrital mineral species to metabolic products.

As in the case of 'suboxic' diagenesis, rather distinctive depth zones can be recognized which are best related to specific microbial ecosystems. Numerous investigations of shallow, marine sediment cores have been undertaken throughout the world. A useful summary of more important findings is to be found in Claypool and Kaplan (1974). They established a sequence based on a shallow aerobic zone, which varies in thickness from a few mm to a few cm depending on the activity of burrowing organisms, followed by a zone, some metres thick, dominated by the activities of sulphate-reducing bacteria. Beneath the sulphate reduction zone, organic degradation tends to include methane as one important product. Methanogenesis appears to commence as soon as sulphate reduction is complete (some metres below the sediment–water interface) and continue to considerable depths. Experimental difficulties associated with bringing deep sediment cores to the surface without contamination or gas loss are considerable: it is very much harder still to prove the presence of microbial metabolism in such material.

The metabolic activity of sulphate-reducing bacteria has been measured directly in near-shore muddy sediments by Goldhaber *et al.* (1977). Sediment gravity and box cores were taken at different seasons and a wide range of porewater and mineral properties established. Sediment aliquots from different depths were incubated under anaerobic conditions and sulphate reduction rates equated with porewater solute profiles. The effects of burrowing organisms irrigating the uppermost sediment layers with oxygenated waters and their marked seasonal dependence were documented. Beneath this zone of bioturbation, the sulphate reduction zone was shown to be maintained by sulphate diffusion alone.

Recent studies of methane production have established metabolic rates for this zone also and have demonstrated mechanisms of gas migration in shallow marine sediments (Martens & Klump 1980). Methane generated below the sulphate reduction zone migrates upwards by diffusion and gas bubble rise. The latter mechanism, although seasonal, enhances the former by sediment texture modification. The common observation that methane is not found in sediments until virtually all sulphate has been reduced (see Martens & Berner 1974 for discussion) may thus be obscured by mixing. These authors also suggest the possibility that methane may be a substrate for sulphate-reducing bacteria: a suggestion subsequently favoured by Murray *et al.* (1978). Experimental investigation of the relationship between sulphate reduction and methanogenesis in marine sediments was undertaken by Oremland and Taylor (1978). They showed that the two processes were *not* mutually exclusive although sulphate reduction predominates when sulphate is available. It was suggested that both sulphate reducers and fermenters might generate hydrogen from organic matter in the absence of sulphate: methanogenic bacteria then utilizing this and reducing bicarbonate.

Methanogenesis in freshwater sediments has been more extensively investigated than that in the more complicated and less readily accessible marine situation. Methanogenic bacteria as a group were reviewed relatively recently by Zeikus (1977). On a more geochemical theme, Zeikus and Winfrey (1976) cultured methanogens from a freshwater lake in Wisconsin. They established certain metabolic systems and certainly demonstrated the complexity of natural populations as well as their strong response to environmental controls. The most common methanogens were metabolically active over the temperature range 4–45 °C and a chemolithotrophic species was isolated in pure culture which showed a temperature optimum above 30 °C and possessed simple nutritional requirements.

It seems that there is every likelihood that complex methanogen populations also inhabit marine sediment porewaters and that, by analogy with the work of Zeikus and Winfrey (1976), certain population components may be capable of metabolism at very considerable depths. A difficult question to answer, however, is what kind of substrate would be utilized by such bacteria or bacterial communities?

In spite of the uncertainties hinted at above, it is well established that the zonal sequence suggested by Claypool and Kaplan (1974) is representative of microbial activity taking place in shallow marine sediments beneath oxygenated waters. In stagnant basins or beneath anoxic waters generally, sulphate reduction takes place immediately beneath the sediment–water interface and may occur above it.

Methanogenesis commences as soon as sulphate concentrations have been reduced to low levels (or once anoxic conditions are established in freshwater sediments). The zonal sequence and its simplest consequences for porewater composition is illustrated in Table 9.2. Overall reactions are approximated to by equations 9.5 and 9.6 (see section 9.2.2) and continue the general trend of zonal stratification with decreasing molar (glucose) free energy release, much as discussed by Claypool and Kaplan (1974).

Table 9.2. Principal simple solutes derived from or utilized in microbial metabolism: 'open' marine sedimentary environments.

		Simple aqueous solutes involved	
Oxgen status	Dominant overall reaction†	Removed from porewater	Added to porewater
Oxic	Aerobic respiration	O_2	HCO_3^- NO_3^- H_3PO_4 H+
Anoxic	Sulphate reduction	SO_4^{2-}	HCO_3^- HS^- H_3PO_4 H+ (NH_3/N_2)
	Methane formation	HCO_3^-*	HCO_3^-* CH_4 (H_3PO_4) (NH_3/N_2)

* Probably net *production* of HCO_3 even though bicarbonate reduction is one plausible mechanism of methane formation.
† Organic matter destroyed in each case.

9.2.4 Soluble organic matter in sediment porewaters

The solutes listed in Table 9.2 do not include dissolved organic matter generated during microbial metabolism. Nissenbaum *et al.* (1972) analysed interstitial porewater in reducing sediments from Saanich Inlet, British Columbia. They found up to 150 mg l^{-1} dissolved organic carbon with high oxygen content and average $^{13}C/^{12}C$ ratio lying between -20 and $-21‰$ This material contained high molecular weight polymers of amino acids and carbohydrates and its concentration increased with depth.

It was suggested that the high molecular weight material formed by combination of low molecular weight metabolic products (plankton decomposition). It was also suggested that further polymerization would lead to the formation of humic and fulvic acids which would then be incorporated into the sediment.

A more extensive investigation of dissolved organic matter in porewaters from suboxic sulphate reduction and methanogenesis zone sediments was carried out by Krom and Sholkovitz (1977). They found that low molecular weight compounds were present at concentrations close to 10 mg l^{-1} in both oxygen-containing and anoxic porewaters and showed only slight variation with depth. In anoxic porewaters, however, the *total* dissolved organic matter increased systematically to values around 70 mg l^{-1} at 80 cm: the additional component being entirely in the high molecular weight range.

This work confirms the general pathway for humification proposed by Nissenbaum *et al.* (1972) and demonstrates that dissolved organic matter can offer important clues as to the nature of metabolic pathways in different diagenetic zones. It seems important to develop studies of dissolved organic matter further: very few investigations of diagenetic processes record or refer to species other than the simple 'inorganic' ions included in Table 9.2. It is likely that precise characterization of microorganism involvement during diagenesis will only follow detailed 'fingerprinting' of organic molecules present in porewaters and the establishing of links between these and biochemical pathways found by laboratory experiment.

9.2.5 Modification of organic matter: oil-generating potential

The microbial degradation that takes place in shallow diagenetic zones produces large quantities of bicarbonate and methane (Table 9.2). Unlike low molecular weight products which may be retained in the sediments after polymerization, bicarbonate and methane represent organic matter 'destroyed'. The oil-generating potential of the sediment is thus so much the poorer.

The extent to which microbiological processes do affect 'kerogen' composition will affect both the amount and type of liquid hydrocarbons produced when sediments are buried to sufficient depth for thermally induced cracking to occur. Considerations of this kind are of concern to the exploration geologist or geochemist (Tissot *et al.* 1974, Saxby 1977). The most obvious variable is the oxygen status of the depositional water body: beneath anoxic water masses, neither oxic microbial degradation nor feeding by benthic scavengers takes place. Demaison and Moore (1980) thus associate anoxic depositional environments with good source potential and discuss the reasons for the occurrence of anoxic water bodies at some length. Sedimentation rate is the other principal environmental control. Curtis (1977) and Coleman *et al.* (1979) show how rate of sediment burial (i.e. sedimentation rate) simply limits the time period which any sediment unit spends within a given microbial diagenesis zone. Rapid burial minimizes degradation by microorganisms and optimizes liquid hydrocarbon generation at depth.

Rate of sedimentation control of microbial activity has also been the subject of several recent studies in shallow marine cores. Goldhaber and Kaplan (1975), Berner (1977) and Toth and Lerman (1977) all showed that the rate of sulphate reduction in marine sediment porewaters decreased with decreasing depositional rate. It seems clear that in slow depositional rate situations, organic matter is degraded more extensively by aerobic organisms and becomes less attractive (more 'refractory') as a substrate for the sulphate-reducing ecosystem. By the same token, the total amount available and its reactivity in thermal cracking reactions will be considerably modified.

9.2.6 Mineralogical transformation 1: present-day sediments

The influence exerted by microorganisms on mineral assemblages during diagenesis arises largely from introduction of anions (especially bicarbonate, hydrosulphide, phosphate and organic) and elimination of molecular oxygen. Direct involvement of bacteria in mineral precipitation reactions may also be important, but is dealt with elsewhere in the volume.

The most important evidence for microorganism involvement during anoxic diagenesis stems from stable isotope studies: carbon, nitrogen, oxygen and sulphur. The isotopic composition of bicarbonate, sulphide, sulphate and methane within sediment porewaters where microbial metabolism is active differs markedly from the same species present in the oceanic reservoir or produced by inorganic means. Studies such as those described in Presley and Kaplan (1968), Presley *et al.* (1973) and Claypool and Kaplan (1974) have established links between stable isotope ratios, metabolic processes and reaction rates; for example, sulphate depletion involves the selective precipitation of ^{32}S relative to ^{34}S, methanogenesis produces methane enriched in ^{12}C and bicarbonate enriched in ^{13}C relative to substrate organic matter. Sulphate reduction produces bicarbonate little different from substrate organic matter but much enriched in ^{12}C relative to marine reservoir bicarbonate (and normal carbonate sediments) on account of the selective incorporation of the light isotope during photosynthesis. Availability of this information has assisted greatly in interpretation of the history of early diagenetic minerals and has proved invaluable in studies of ancient sediments.

Within anoxic sediments, the common presence of hydrated iron oxides means that the first mineralogical reaction of note is the formation of iron sulphides by reduction of iron and precipitation with microbiologically produced hydrosulphide. Although numerous workers have been involved in diagenetic sulphide mineral studies, the work of R.A.Berner is preeminent. The overall picture was adequately summarized by Berner (1970) as a sequence of events.

"1. Organic matter derived from dead marine organisms and iron contained within or adsorbed upon fine grained detrital minerals is

deposited in relatively quiet water where the light particles may settle out.
2. Upon removal of dissolved oxygen from associated waters by aerobic metabolism of the organic matter, anaerobic bacterial sulfate reduction occurs within sediments and more rarely in waters overlying the sediments. As a result of sulfate reduction, H_2S is formed.
3. Dissolved H_2S reacts immediately with the most reactive forms of iron present to form black non-crystalline FeS. If H_2S formation ceases at this point and no extra elemental sulfur is added to the sediment, no further reactions, other than the crystallization of FeS to various iron monosulfides, occur. The result is greigite, mackinawite, or pyrrhotite and little or no pyrite (or marcasite).
4. If continued sulfate reduction takes place, the H_2S concentration in the sediment increases which brings about the conversion of additional detrital iron to FeS. The amount of FeS (and ultimately pyrite) that can form is limited by the rate of production and concentration of H_2S and/or by the amount and reactivity of detrital iron.
5. Simultaneously with step 4 the concentration of sulfate in the sediment pore water falls and additional sulfate is made available for reduction by diffusion down the concentration gradient from the overlying water. Likewise, the produced H_2S builds up in the pore water and begins to diffuse out of the sediment. The concentration of available (metabolizable) organic matter controls the rate of sulfate reduction and consequently, the concentration of H_2S that can be maintained in the presence of loss by diffusion.
6. Some of the H_2S is oxidized, either inorganically or by sulfur oxidizing bacteria, to elemental sulfur. Under aerobic bottom conditions considerable elemental sulfur is formed by the reduction of FeS and H_2S with dissolved oxygen which is stirred into the sediment from the overlying water by burrowing organisms or by storm waves and currents. Part of the elemental sulfur is subsequently oxidised by bacteria to dissolved sulfate.
7. The remaining elemental sulfur slowly reacts with FeS in the sediment to form pyrite which crystallizes as minute framboidal microspheres. The time for complete transformation of FeS to FeS_2, in the presence of abundant H_2S and elemental sulfur, is of the order of several years."

More recent investigations (Berner 1974, Berner *et al.* 1979) have confirmed these general conclusions and extended the work towards establishing definite links between sulphide mineralogy and depositional environment, via the microbiological pathways involved and environmental controls on those pathways. It should be noted that transformations between different sulphide minerals during early diagenesis involve sulphur in zero valence state and as polysulphide ions: sulphide-oxidizing bacteria almost certainly play some part.

Beneath oxygen-containing waters, high oxidation state compounds of iron and manganese occur at, or close to the sediment–water interface. These, however, are almost invariably destroyed by reduction at greater depth (see section 9.2.2). Increased activities of iron and manganese are thus to be anticipated together with higher hydrosulphide, bicarbonate and phosphate levels. Only one possible combination (iron sulphides) has been considered, and although this is the most common one in marine sediments and shallow burial depths, other combinations are precipitated. Suess (1979) studied early diagenetic minerals formed in an anoxic Baltic basin. Using a wide variety of modern and classical mineralogical and chemical techniques, he was able to identify siderite $FeCO_3$, calcic rhodocrosite, $MnCO_3$ and manganese sulphide MnS. On the basis of chemical data alone he identified amorphous iron sulphide FeS, a manganese phosphate and a calcic iron phosphate. In these sediments the diagenetic minerals constituted a sizeable proportion of the total sediment yet it was difficult to characterize each mineral fully.

It is clear from the above that numerous early diagenetic phases can form from metal cations and metabolic products of microbial activity. As yet, however, only the iron sulphide story has been fully elucidated: much more information is needed about other early diagenetic minerals formed under anoxic conditions. The fact that many sediment profiles are distinctly zoned in terms of anion concentration suggests that some of the minerals described by Suess (1979) may be unstable and transitory: like the oxidized phases beneath oxygenated depositional waters.

Significant cementation by carbonates and oxides has also been reported for freshwater sediment sequences (Ho & Coleman 1969, Whelan & Roberts 1973). In general terms, diagenetic processes are as expected by analogy with marine diagenesis in the absence of sulphate. It is interesting to note that concretion formation is common; unlike in most marine diagenetic sequences.

Within carbonate sediments, diagenesis inevitably must follow somewhat different pathways, simply on account of the fact that thermodynamically unstable (biogenically precipitated) and relatively soluble minerals form the bulk of the sediment at deposition. Net dissolution, precipitation or recrystallization to more stable phases is bound to be much influenced by introduction of bacterial metabolic products such as bicarbonate or organic acids. Phosphate input almost certainly is responsible for phosphorite formation (Price & Calvert 1978).

Ancient limestones exhibit a bewildering diversity of textures, very many of which have been attributed to diagenetic processes. Bathurst (1975) has provided a comprehensive and analytical review of the extremely large research output in this area: so large, in fact, that any kind of review would be inappropriate here. More recently, Hudson (1977) has reviewed the evidence provided by stable isotope investigations and these establish the importance of microbial ecosystems in carbonate sediment. Although the bulk of diagenetic cement in limestone is probably derived from dissolution of less stable and reprecipitation as more stable carbonate minerals some cements clearly show carbon derived by oxidation of organic matter. Evidence of oxidation by aerobic organisms, by sulphate reducers and in methanogens has been obtained and certain extremely 'light' cements have been attributed to direct bacterial oxidation of methane. In modern carbonate sediments, like all other types considered here, it is plain that microbial metabolism contributes significantly to early diagenetic modification.

9.2.7 Mineralogical transformation 2: evidence from the geological record

Post-depositional alteration of all types of sediment is seen when ancient sedimentary rocks are examined in detail. Some mention has already been made of carbonate diagenesis: mudrocks and sandstones also exhibit diagenetic alteration. Porosity and permeability preserved at depth usually owe their establishment and/or preservation to diagenetic processes. This fact has not been missed by those geologists concerned with petroleum reservoirs.

The problem in so far as this book is concerned is separating the consequences of microbial metabolism from diagenetic reactions that undoubtedly proceed via inorganic mechanisms. Some of these, such as thermally induced decarboxylation of organic matter, introduce bicarbonate to porewater systems with an isotopic composition not dissimilar to that produced by sulphate reduction and aerobic respiration. Carbonate minerals precipitated or reprecipitated in consequence will have carbon isotope ratios indicative of derivation from organic matter yet uninfluenced by microbial processes.

In Curtis (1977), an attempt was made to present a general model for diagenesis in which both 'biogeochemical' and 'thermal' depth zones were included and the effects of variables such as sedimentation rate evaluated. Subsequently certain aspects of this model were tested by Irwin *et al.* (1977) and the influence of both thermally induced and microbiological reactions confirmed in the same sediment sequence. It proved possible to separate very early bicarbonate generated close to the sediment–water interface from that produced thermally by consideration of oxygen isotope ratios, which show strong temperature dependence. This is not always the case because input of meteoric waters of very different isotopic composition from marine waters can more than negate any temperature effect. An excellent discussion of these factors is included with Hudson's (1978) detailed study of carbonate concentrations in Jurassic clays of central England; both microbial and 'inorganic' processes are considered.

Friedman and Murata (1979) also discuss the development of dolomites in the Miocene Monterey Shale in terms of microbial activity and environmental controls reflected in carbon and oxygen isotope ratios. Sulphate reduction and methanogenesis are identified as important processes leading to carbonate precipitation.

Dickson and Coleman (1980) found that oxygen and carbon isotope ratio determinations for a sequence of pore-filling cements in a limestone proved 'inorganic' diagenesis; no carbonate indicative of microbial influence was detected. Collectively, these studies demonstrate that stable isotope studies, when linked with chemical data and careful petrographic observations, can separate the influence of purely microbiological processes from the quantitatively more important 'abiological' reactions which tend to occur at greater depths and temperatures.

As was mentioned in section 9.1.4, the most striking examples of microbial diagenesis are to be found in concretions. These occur in all types of sediment and frequently exhibit stable isotope ratios indicative of microbially induced sulphide, carbonate and phosphate precipitation. Iron-rich carbonate concretions have been shown to exhibit systematic chemical variation from core to margin, precisely mirroring carbon isotope variations (Curtis *et al.* 1975).

Careful chemical and isotopic studies of sequential cements in concretions, septaria and pore-cavities in sandstones and limestones have revealed much about diagenesis in ancient sediments. It seems clear that the influence of microorganisms active in the uppermost layer of all types of sediment is to be found preserved in ancient sedimentary rocks throughout the Geological Column.

Since the most primitive living organisms could well have inhabited the uppermost sediment layer environment, it is interesting to speculate that the earliest sediments known were probably modified by microbial diagenesis.

9.3 UNANSWERED QUESTIONS

This chapter has demonstrated that microbial processes of diverse kinds are important contributors to diagenetic alteration that takes place in shallowly buried sediments. It has also been shown that mineralogical precipitates and organic degradation resulting from microbial metabolism may profoundly modify sediments and influence markedly the properties of the resulting sedimentary rocks. This is obviously true for petroleum-source rocks and, if porosity is reduced by mineral cement, for petroleum reservoirs.

Although a case has been made for recognizing an orderly progression of distinctive diagenetic zones through which sediments pass on burial, much remains to be answered about the nutritional requirements of the micro-organisms active in those zones. Clearly each represents a microbial ecosystem with nutritional chain rather than some single organism and it is important to learn more about nutritional requirements. It is easy to appreciate that downward diffusion of molecular oxygen and dissolved sulphate (considered in their kinetic interaction with burial) limits the activities of aerobic respirers and sulphate reducers respectively. However, what limitations are imposed by the composition of the organic substrate deposited with the sediments? It is obvious that such a limitation applies to sulphate reduction. Numerous observations of links between sulphate reduction rate, sedimentation rate and the 'refractory' character of organic matter have been made.

One problem would seem to be that laboratory methods of micro-organism culture tend to accelerate metabolic rates whereas the geologist wishes to know about the possibility of much slower processes which could, conceivably, operate via different metabolic pathways. It is certainly important to know if fermentation, methanogenesis or sulphate reduction can take place at depths of some kilometres. Sulphate can be supplied to deep environments by circulating groundwaters; substrate quality would seem to be the limiting factor. Once again, some approach to these questions is made elsewhere in the volume.

One topic not discussed under the heading of mineralogical transformations was that of mineral dissolution in response to the introduction of organic acids and carbon dioxide. Both are capable of dissolution of silicates as well as carbonates as is evidenced by studies of chemical weathering. Organic acids also contribute to metal mobilization by chelation. Petroleum geologists are fairly sure that the porosity of reservoir sedimentary rocks is enhanced by dissolution at considerable depths. At present it seems likely that thermally induced organic degradation reactions are responsible but microbiological processes would certainly be important if they did occur at depth. Once again the depth limit for microbial activity within buried sediments emerges as an unknown of great significance.

References

Bathurst R.G.C. (1975) *Carbonate sediments and their Diagenesis*. 2nd ed. Elsevier, Amsterdam.

Berner R.A. (1968) Calcium carbonate concretions formed by the decomposition of organic matter. *Science* **159,** 195–7.

Berner R.A. (1970) Sedimentary pyrite formation. *Am. J. Sci.* **268,** 1–23.

Berner R.A. (1974) Iron sulphides in Pleistocene deep Black Sea sediments and their palaeo-oceanographic significance. *Am. Ass. Pet. Geol.* Memoir **20,** 524–31.

Berner R.A. (1977) Sulphate reduction and the rate of deposition of marine sediments. *Earth Planet. Sci. Lett.* **37,** 492–8.

Berner R.A. (1977) Sulphate reduction and the rate of deposition of marine sediments. *Earth Planet. Sci. Lett.* **37,** 492–8.

Berner R.A., Baldwin T. & Holdren G.R. (1979) Authigenic iron sulphides as palaeosalinity indicators. *J. Sediment. Petrol.* **49,** 1345–50.

Claypool G.E. & Kaplan I.R. (1974) The origin and distribution of methane in marine sediments. In *Natural Gases in Marine Sediments* (Ed. I.R.Kaplan) pp. 99–139. Plenum Press, New York.

Coleman M.L., Curtis C.D. & Irwin H. (1979) Burial rate, a key to source and reservior potential. *World Oil,* 83–88.

Curtis C.D. (1977) Sedimentary geochemistry: Environments and processes dominated by involvement of an aqueous phase. *Phil. Trans. R. Soc. Lond.* **A286,** 353–72.

Curtis C.D., Pearson M.J. & Somogyi V.A. (1975) Mineralogy, chemistry and origin of a concretionary siderite sheet (clay-ironstone band) in the Westphalian of Yorkshire. *Mineralog. Mag. London* **40,** 385–93.

Curtis C.D., Petrowski C. & Oertel G. (1971) Stable carbon isotope ratios in carbonate concretions: A clue to place and time of formation. *Nature, London* **235,** 98–100.

Demaison G.J. & Moore G.T. (1980) Anoxic environments and oil source bed genesis. *Organic Geochemistry* **2,** 9–31.

Dickson J.A.D. (1979) Personal communication.

Dickson J.A.D. & Barber C. (1976) Petrography, chemistry and origin of early diagenetic concentrations in the Lower Carboniferous of the Isle of Man. *Sedimentology* **23,** 189–211.

Dickson J.A.D. & Coleman M.L. (1980) Changes in carbon and oxygen isotope composition during limestone diagenesis. *Sedimentology* **27,** 107–18.

Friedman I. & Murata K.J. (1979) Origin of dolomite in Miocene Monterey shale and related formations in the Temblor Range, California. *Geochim. Cosmochim. Acta* **43,** 1357–65.

Froelich P.N., Klinkhammer G.P., Bender M.L., Luedtke N.A., Heath G.R., Cullen D., Daupin P., Hammond D., Hartman B. & Maynard V. (1979) Early oxidation of organic matter in pelagic sediments of the eastern equatorial Atlantic: suboxic diagenesis. *Geochim. Cosmochim. Acta* **43,** 1075–90.

Galimov E.M. & Girin Yu P. (1968) Variation in the isotopic composition of carbon during the formation of carbonate concretions. *Geokhimiya* **2,** 228–33.

Garrels R.M. & Mackenzie F.T. (1971) *Evolution of sedimentary rocks.* W.W.Norton, New York.

Goldhaber M.B., Aller R.C., Cochran J.K., Rosenfeld J.K., Martens C.S. & Berner R.A. (1977) Sulphate reduction, diffusion, and bioturbation in Long Island Sound sediments: Report of the FOAM group. *Am. J. Sci.* **277,** 193–237.

Goldhaber M.B. & Kaplan I.R. (1975) Controls and consequences of sulphate reduction rates in recent marine sediments. *Soil Sci.* **11,** 42–55.

Hedberg H.D. (1974) Relation of methane generation to undercompacted shales, shale diapirs, and mud volcanoes. *Bull. Am. Assoc. Pet. Geo.* **58,** 661–73.

Heroux Y., Chagnon A. & Bertrand R. (1979) Compilation and correlation of major thermal maturation indicators. *Bull. Am. Assoc. Pet. Geol.* **63,** 2128–44.

Ho C. & Coleman J.M. (1969) Consolidation and cementation of recent sediments in the Atchafalaya Basin. *Geol. Soc. Am. Bull.* **80,** 183–92.

Hudson J.D. (1977) Stable isotopes and limestone lithification. *J. Geol. Soc. Lond.* **133,** 637–60.

Hudson J.D. (1978) Concretions, isotopes and the diagenetic history of the Oxford Clay (Jurassic) of central England. *Sedimentology* **25,** 339–70.

Irwin H., Curtis C.D. & Coleman M.L. (1977) Isotopic evidence for source of diagenetic carbonates formed during burial of organic-rich sediments. *Nature, London* **269,** 209–13.

Krom M.D. & Sholkovitz E.R. (1977) Nature and reaction of dissolved organic matter in the interstitial waters of marine sediments. *Geochim. Cosmochim. Acta* **41,** 1565–1573.

Martens C.S. & Berner R.A. (1974) Methane production in the interstitial waters of sulphate-depleted marine sediments. *Science* **185,** 1167–9.

Martens C.S. & Klump J.V. (1980) Biogeochemical cycling in an organic-rich coastal marine basin—I. Methane sediment–water exchange processes. *Geochim. Cosmochim. Acta* **44,** 471–90.

Murray J.W., Grundmanis V. & Smethie W.M. (1978) Interstitial water chemistry in the sediments of Saanich Inlet. *Geochim. Cosmochim. Acta* **42,** 1011–26.

Nissenbaum A., Baedecker M.J. & Kaplan I.R. (1972) Studies of dissolved organic matter from interstitial water of a reducing marine fjord. In *Advances in Organic Geochemistry 1971* pp. 427–40. Pergamon Press, Oxford.

Oertel G. & Curtis C.D. (1972) Clay-ironstone preserving fabrics due to progressive compaction. *Geol. Soc. Am. Bull.* **83,** 2597–606.

Oremland R.S. & Taylor B.F. (1978) Sulfate reduction and methanogenesis in marine sediments. *Geochim. Cosmochim. Acta* **42,** 209–14.

Pamatmat M.M. & Bause K. (1969) Oxygen consumption by the seabed. II. *In-situ* measurements to a depth of 180 m. *Limnol. Oceanogr.* **14,** 250–9.

Presley B.J., Culp J., Petrowski C. & Kaplan I.P. (1973) In *Initial Reports of the Deep Sea Drilling Project*, vol. 11, pp. 805–10, U.S. Government Printing Office, Washington.

Presley B.J. & Kaplan I.R. (1968) Changes in dissolved sulphate, calcium and carbonate from interstitial water of near-shore sediments. *Geochim. et Cosmochim. Acta* **32,** 1037–48.

Price N.B. & Calvert S.E. (1978). The geochemistry of phosphorites from the Namibian shelf. *Chem. Geol.* **23,** 151–70.

Raiswell R. (1976) The microbiological formation of carbonate concretions in the Upper Lias of N.E. England. *Chem. Geol.* **18,** 227–44.

Raiswell R. & White N.J.M. (1978) Spatial aspects of concretionary growth in the Upper Lias of N.E. England. *Sed. Geol.* **20,** 291–300.

Reading H.G. (1978) *Sedimentary Environments and Facies*, Blackwell Scientific Publications, Oxford.

Riecke H.H. & Chilingarian G.V. (1974) *Compaction of argillaceous sediments.* Elsevier, Amsterdam.

Sass E. & Kolodny Y. (1972) Stable isotopes, chemistry and petrology of carbonate concretions (Mishash Formation, Israel). *Chem. Geol.* **10,** 261–86.

Saxby J.D. (1977) Oil-generating potential of organic matter in sediments under natural conditions. *J. Geochem. Expl.* **7,** 373–82.

Suess E. (1979) Mineral phases formed in anoxic sediments by microbial decomposition of organic matter. *Geochim. Cosmochim. Acta* **43,** 339–52.

Tissot B., Durand B., Espitalie J. & Combaz A. (1974) Influence of nature and diagenesis of organic matter in the formation of petroleum. *Am. Ass. Pet. Geol. Bull.* **58,** 499–506.

Toth D.J. & Lerman A. (1977) Organic matter reactivity and sedimentation rates in the ocean. *Am. J. Sci.* **277,** 465–85.

Whelan T. & Roberts H.H. (1973) Carbon isotope composition of diagenetic carbonate nodules from freshwater swamp sediments. *J. Sediment. Petrol.* **43,** 54–8.

Zeikus J.G. (1977) The biology of methanogenic bacteria. *Bacteriol. Rev.* **41,** 514–37.

Zeikus J.G. & Winfrey M.R. (1976) Temperature limitation of methanogenesis in aquatic sediments. *Appl. Environmental Microbiol.* **31,** 99–107.

CHAPTER 10
ANCIENT MICROBIAL ECOSYSTEMS

ANDREW H.KNOLL AND STANLEY M.AWRAMIK

> The leading idea which is present in all our searches and which accompanies every fresh observation, the sound to which the ear of the student of Nature seems continually echoed from every part of her works is—Time!–Time!–Time!
>
> G.P.Scrope (1858)

10.1 INTRODUCTION

The microbial processes and microbially mediated element cycles discussed in previous chapters are observed within the framework of ecological time; however, microbial ecosystems are ancient, perhaps nearly as old as the earth itself. The palaeontological record provides the opportunity to view these processes from the perspective of geological time and in so doing furnishes important data on the nature and tempo of microbial evolution.

Micropalaeontology differs from neontological microbiology in that metabolism cannot be measured directly, but must be inferred from the morphology and spatial distribution of fossil remains. Given the structural simplicity and physiological complexity of most prokaryotes, this inferential methodology has obvious limitations. None the less, the geological record does permit the construction of a reasonably detailed picture of microbial evolution. Although both prokaryotes and eukaryotic microorganisms continue to evolve on the present-day earth, the discussion in this chapter will be limited to Precambrian microbial evolution, because it was during this long period prior to the appearance of multicellular plants and animals that the fundamental elements of microbial ecosystems became established.

10.2 THE NATURE OF THE DATA

10.2.1 Microfossils

The best data on the nature of ancient microorganisms naturally are provided by the organically preserved remains of the organisms themselves; however, the fossil record is not simply an unabridged compendium of past life and cannot be read as such. The vast majority of microorganisms that have populated the earth through time have left no direct record. Cellular remains are generally obliterated rather quickly after death by autolysis and/or bacterial action; therefore, fossilization requires as a first step that post-mortem decomposition be arrested before it can proceed to completion. Since the likelihood of this happening varies from one environment to another, preservation processes introduce a significant degree of ecological bias into the fossil record. Degradation is strongly retarded by the imposition of an anaerobic environment. Rapid burial in organic-rich sediments can result in the depletion of ambient O_2, and cyanobacterial remains in stromatolites can also pass quickly into the anaerobic zone as the mat community builds upward. On the other hand, good microbial preservation is unlikely if porewaters are well aerated during early diagenesis.

Microbial decay can proceed in oxygen-free environments if several bacterial taxa are able to act in succession on cellular remains (Doemel & Brock 1977, Golubic & Barghoorn 1977). This anaerobic sequence is most readily broken by environmental conditions inimical to one or more members of the chain. Lowered pH is one possible environmental barrier, but in marine environments the condition most likely to stop anaerobic decomposition is hypersalinity. Highly saline groundwaters percolating through anaerobic cyanobacterial peats have preserved cellular remains in Holocene microbial mats from Abu Dhabi for more than 8000 years (Golubic 1973). The environmental bias introduced by the coincidence of anaerobic conditions and above-normal salinity should favour the preservation of intertidal to shallow subtidal microbial communities in areas of limited water circulation, and it is not surprising that such assemblages are well documented in the Precambrian fossil record.

Long-term preservation is often affected by the silification of partially degraded microbial assemblages. Relative to most other types of sedimentary rock, chert (cryptocrystalline SiO_2) is incompressible, impermeable, and difficult to shear or recrystallize. Thus, it shields delicate microfossils from the effects of moderately elevated temperature and pressure. If deposited sufficiently early, either as a primary sediment (as in the Gunflint Iron Formation, Canada, Barghoorn & Tyler 1965; and the Barney Creek Formation, Australia, J.H.Oehler 1977) or as an early diagenetic replacement of carbonate, silica can itself arrest decomposition or preserve remains not necessarily subjected to environmental extremes such as high salinity. It is also becoming increasingly apparent that phosphorites and fine mudstones can preserve ancient microbes in exquisite detail (Diver 1980).

Preservational biases do not operate only at the environmental level; within a single community there can be selective preservation of taxa. Nowhere is this more clearly demonstrated than in fossilized microbial mat systems. The cyanobacterial components are often well preserved, but associated photosynthetic bacteria and anaerobic heterotrophs, which are fundamental components of modern mat ecosystems, are not in evidence (Awramik *et al.* 1978). In part, this may occur because the blue-green populations, which are fully exposed to the rigours of the external environment, are structurally more robust than the organisms inhabiting the lower, more protected regions of the mat. Differential preservation may also result from basic differences in the biochemical composition of organisms; some biomolecules are more easily broken down than others.

Subtle differences in micro-environments of diagenesis can cause substantial variation in the quality of preservation among individual fossils within a single population. This is well illustrated by a chroococcalean cyanobacterial assemblage preserved in silicified microbial stromatolites from the 800–900 million-year-old Bitter Springs Formation, Australia. Here variations in preservation have introduced an impressive amount of post-mortem morphological variation into a population that in life was probably rather homogeneous (Knoll & Golubic 1979).

Finally, the biochemical bias mentioned above with respect to different taxa also occasions selective preservation of the component parts of individual microbes. Extracellular cyanobacterial sheaths and envelopes, for example, are relatively decay resistant and are frequently encountered in the fossil record sans cells.

The various levels of preservational bias discussed in the preceding paragraphs are not presented as a discouraging note. Indeed, the many microfossils that have survived since the Precambrian provide numerous insights into the nature of early biology. It simply is important to understand both the strengths and the limitations of the data used in the interpretation of ancient ecosystems.

10.2.2 Stromatolites

Stromatolites are the sedimentary record of distinctive microbial communities which trap, bind, and/or precipitate sediments. This dual biological and geological nature is important because long after the organic remains of the microbial assemblage have disappeared, sedimentary structures remain to document the former activity of microorganisms. Modern stromatolitic systems vary greatly in the taxonomy of the constituent microbiota but, in terms of ecosystem structure, they tend to be quite uniform (Golubic 1976). Microbial mats are, in general, vertically differentiated along micro-environmental gradients (Golubic 1973, Awramik *et al.* 1978). The top layer consists of cyanobacteria which, sometimes with eukaryotic or bacterial* associates, form coherent mat surfaces and trap and bind sediments. Diversity within this layer may be high, but usually one or a few mat-building blue-greens dominate. A second layer consisting predominantly of taxonomically distict cyanobacteria may exist in the region of reduced light intensity (but still aerobic) directly beneath the surface community. Still farther below the mat surface, where light can still penetrate but oxygen is depleted, photosynthetic bacteria thrive. Below the level of effective light penetration, anaerobic heterotrophs and sulphate-reducing bacteria are found. The vertical distribution of microbes within the mat ecosystem may be more complex than this (Krumbein *et al.* 1977, 1979), but the general pattern of green/purple-pink/black (cyanobacteria, purple photosynthetic bacteria, heterotrophs) appears to be characteristic of most, if not all, marine stromatolitic ecosystems. This uniformity of pattern enables one to infer certain properties of ancient microbial ecosystems from the mere presence of stromatolites, even though actual organic remains are not preserved.

The overall shape, or macrostructure, of stromatolites is strongly influenced by environment, but microstructure—the texture of the stromatolite as seen at the microscopic level—appears to be controlled by the mat-building microorganisms (Gebelein 1974, Semikhatov *et al.* 1979). Thus, one can even infer patterns of community distribution without microfossils.

10.2.3 Sedimentary geochemical evidence

Organisms interact physiologically with their ambient environment, often leaving a geochemical record of metabolic activity. The geochemical evidence observable in Precambrian sedimentary rocks can be used to draw conclusions about early biological processes. In effect, metabolic causes are inferred from geological effects.

Certain biological processes fractionate isotopes; for example, in the fixation of CO_2 for photosynthesis ^{12}C is incorporated preferentially to ^{13}C with the result that $^{13}C/^{12}C$ values for photosynthetically produced organic matter differ measurably from the same ratio in the atmosphere or ocean (Craig 1953, Park & Epstein 1960). Since the isotopes ^{12}C and ^{13}C are both stable, their ratio in buried organic matter does not change through time (in the absence of metamorphism). Thus, ancient photosynthetic material can be identified isotopically (Hoering 1967, Oehler *et al.* 1972, Eichmann &

* We recognize the fundamentally bacterial nature of the 'blue-green algae'; however, because of the prevalence of cyanobacterial remains in the Precambrian fossil record, we have found it useful for purposes of discussion to adopt the convention of using the term 'bacteria' to indicate all monerans *except* the cyanobacteria.

Schidlowski 1975). Stable carbon isotope ratios for paired samples of carbonate carbon and organic carbon are remarkably constant through the geological column, suggesting that photosynthesis is a very ancient process (Eichmann & Schidlowski 1975, Schidlowski *et al.* 1979). It must be noted, however, that both aerobic and anaerobic photoautorophs fractionate carbon isotopes, as do methanogens. It is also possible that fractionation associated with abiotic organic syntheses contributed to the ratios measured for some older Archaean sedimentary rocks.

Similarly, sulphate-reducing bacteria fractionate S isotopes. Although the degree of fractionation varies widely as a function of metabolic activity, itself controlled by nutrient supply and other environmental parameters, the ancient activity of biological sulphate reducers can also be inferred isotopically (Goodwin *et al.* 1976, Monster *et al.* 1979).

Weathering profiles and certain sedimentary mineral deposits also contribute evidence of ancient microbial metabolism. Oxidized weathering rinds on basalts and flecks of hematite in clastic rocks (red beds) indicate the presence of oxygen in the atmosphere and, by inference, the presumable activity of aerobic photoautotrophs. Diagenetic pyrite found in intimate association with organic-rich shales may be related to the activity of sulphate-reducing bacteria.

Precambrian sedimentary rocks contain a wide variety of biological molecules including amino acids, porphyrins, and isoprenoid alkanes. Some of these compounds may be true chemical 'fossils'; however, many are contaminants introduced into the rock well after its deposition. The question of contamination has bedevilled attempts at the interpretation of organic geochemical analyses because extractable compounds are usually found in minute quantities. McKirdy (1974) has reviewed the extensive literature on Precambrian organic molecules, noting both the pitfalls and the promise of such evidence. Analyses of both extractable (Hoering 1978) and insoluble kerogen (Sklarew & Nagy 1979) fractions of ancient sedimentary organic matter underscore the potential of organic geochemistry for the interpretation of microbial evolution; however, in the present state of research, molecular 'fossils' cannot be used systematically in the evaluation of early biotas.

10.3 STROMATOLITIC ECOSYSTEMS

Silicified stromatolites in Precambrian sedimentary rocks contain the most detailed information available for the reconstruction of ancient microbial ecosystems. This obtains for several reasons:

(1) preserved stromatolitic microfossil assemblages are large and often diverse;

(2) most of the microfossils found in Proterozoic cherts have morphological analogues among the extant cyanobacterial biota;
(3) the organisms are preserved *in situ* and spatial relationships are retained;
(4) modern stromatolitic ecosystems are well known;
(5) ancient stromatolitic microbiotas can be characterized well in terms of their physical environment of growth;
(6) these assemblages have been extensively studied.

Stromatolites are widely distributed in the geological record and are particularly abundant in Proterozoic carbonates. It is evident that, prior to the rise of the Metazoa, microbial mat communities were extremely important elements of many shallow marine environments. Stromatolites containing silicified microorganisms constitute only a very small subset of the total microbial mat record, but they serve to illuminate the nature and evolution of stromatolitic ecosystems.

Among the oldest known stromatolitic microfossils are assemblages preserved in silicified mammillate mats from the 2000 million-year-old Belcher Supergroup in Hudson Bay, Canada (Hofmann 1976). Associated red beds, mud-cracked dolomite, and evaporite casts indicate that the Belcher microbes grew in tidal flats similar to those bordering the present-day Persian Gulf. The environmental correspondence between the Persian Gulf sabkhas and the Belcher tidal flats is paralleled by a close morphological similarity in their respective stromatolites *and* in the mat-building cyanobacterial communities responsible for the accretion of the stromatolites (Golubic & Hofmann 1976). The principal mat-builder of the Belcher assemblage was a coccoid cyanobacterium (Fig. 10.1 (c)) morphologically identical to the modern blue-green *Entophysalis*, the dominant mat-building organism in the lower intertidal zone of Persian Gulf tidal flats in Abu Dhabi. Diverse other cyanobacterial filaments and unicells are found in various associations in the Belcher cherts, and these are also modern in aspect (Hofmann 1976). The Belcher fossils, then, suggest that 2000 million years ago, there existed stromatolitic ecosystems that can be considered modern, not only structurally, but also in taxonomy.

The correspondence between entophysalid-dominated mat communities and tidal flats in arid to semi-arid climates can also be seen in the 1500 million-year-old Balbirini Dolomite of Australia (D.Z.Oehler 1978), the 800–900 million-year-old Bitter Springs Formation (Knoll & Golubic 1979), and the *ca.* 700 million-year-old Narssârssuk Formation of northwestern Greenland (Strother *et al.* 1983; and see Fig. 10.1 (a,b,d,e)), allowing one to document the persistence of a single environmentally specific community through immense periods of time. Evidently, entophysalid cyanobacteria are optimally designed for existence in the rigorous arid

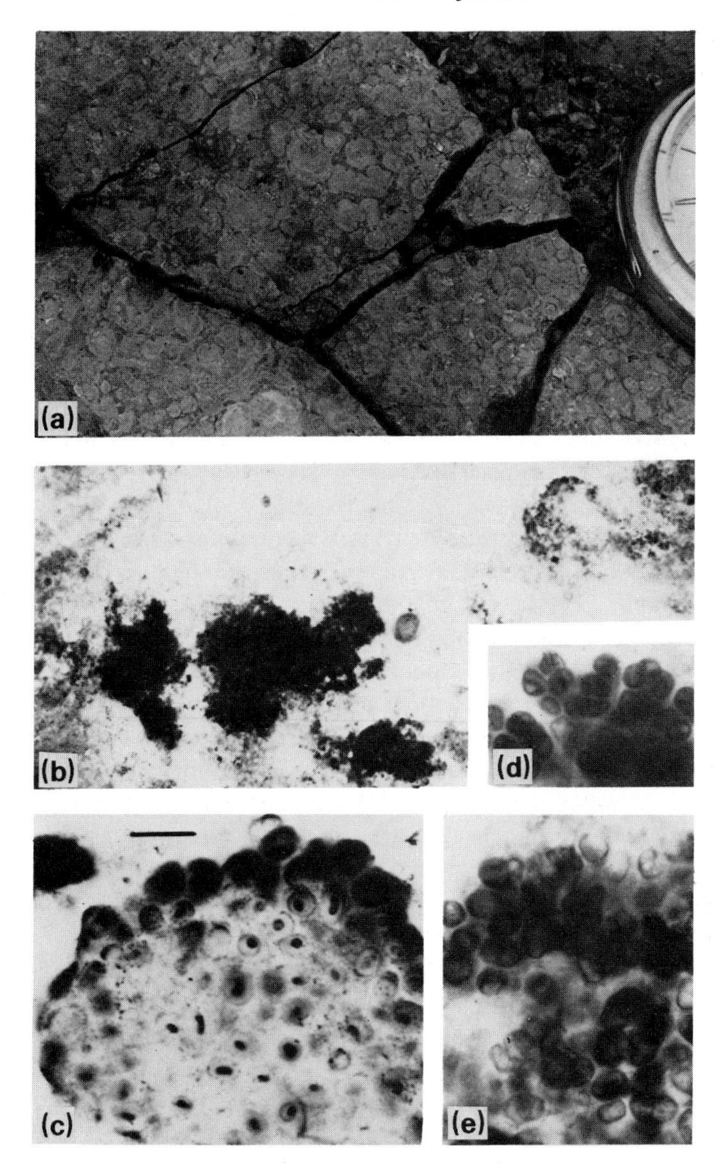

Fig. 10.1. (a) Small mammillate stromatolites built by entophysalid cyanobacteria from the late Precambrian Narssârssuk Formation, Greenland (× 1). (b) Low-power microscopic view of a silicified portion of (a), showing microstructural fabric of mat-building cyanobacteria (× 76). (d, e) Constituent entophysalid cyanobacteria (probable cells in (d) and external envelopes in (e)) making up the fabric in (b) (× 1000). (c) Mat-building entophysalid cyanobacteria (*Eoentophysalis belcherensis* Hofmann) from the 2000 million-year-old Belcher Supergroup, Canada. The bar in (c) = 10 μm for (c)–(e) and 130 μm for (b). ((c) reproduced by courtesy of H. Hofmann.)

intertidal zone; in the 2000 million years (at least) that this community has been in existence, evolution has produced no superior competitor.

Microfossiliferous stromatolites of the Gunflint Iron Formation, Ontario, document a completely different type of Early Proterozoic microbial mat ecosystem. The Gunflint stromatolites (Fig. 10.2 (a)) accreted on a shallow marine platform similar in many ways to the modern Bahama Banks, but characterized geochemically by an abundance of iron. This particular environmental combination appears to have been relatively widespread 2000 million years ago, but is unknown on the modern earth. The microfossils of the Gunflint stromatolites are beautifully preserved and have been studied extensively for more than 25 years (Tyler & Barghoorn 1954, Barghoorn & Tyler 1965, Cloud 1965, Awramik & Barghoorn 1977, Knoll *et al.* 1978).

The preserved microbial assemblage is essentially bispecific; thin (1–2 μm cross-sectional diameter) filaments and small (3–15 μm) coccoidal unicells comprise more than 99% of the microfossils found within the stromatolitic cherts (Fig. 10.2 (b)). These taxa are often considered to be cyanobacterial, a conclusion supported by the general morphology of the remains (the filaments and spheroids are similar in general organization to the simple forms of the modern cyanobacteria), as well as by the macrostructural and microstructural development of the stromatolites (Awramik & Semikhatov 1979). The stable carbon isotope ratios of kerogen contained in the cherts are also consistent with a cyanobacterial interpretation of the microfossils (Hoering 1967, Barghoorn *et al.* 1977). On the other hand, the close morphological similarity between the Gunflint filaments and modern iron bacteria was noted a number of years ago (Cloud 1965) and, indeed, the Gunflint filaments are quite as similar morphologically to extant *Sphaerotilus/Leptothrix* type iron bacteria as they are to any cyanobacterium. This demonstrates the limitations of drawing physiological inferences from form. The prevalence of the Gunflint microbes in an iron-rich environment and their absence from iron-poor areas such as the Belcher stromatolites are also consistent with the hypothesis of iron bacterial affinities for the dominant Gunflint fossils. (It is also possible that differences between Gunflint and Belcher assemblages reflect other or additional environmental differences, in particular the subtidal setting of the Gunflint stromatolites.) Other taxa that occur sporadically in the Gunflint stromatolites include both bacteria and cyanobacteria, as well as several problematic microbes.

We cannot at this juncture resolve unequivocally the question of the systematic position of the major elements of the Gunflint mat-building community; however, we can draw two important conclusions about this assemblage. First, the microorganisms preserved in the Gunflint stromato-

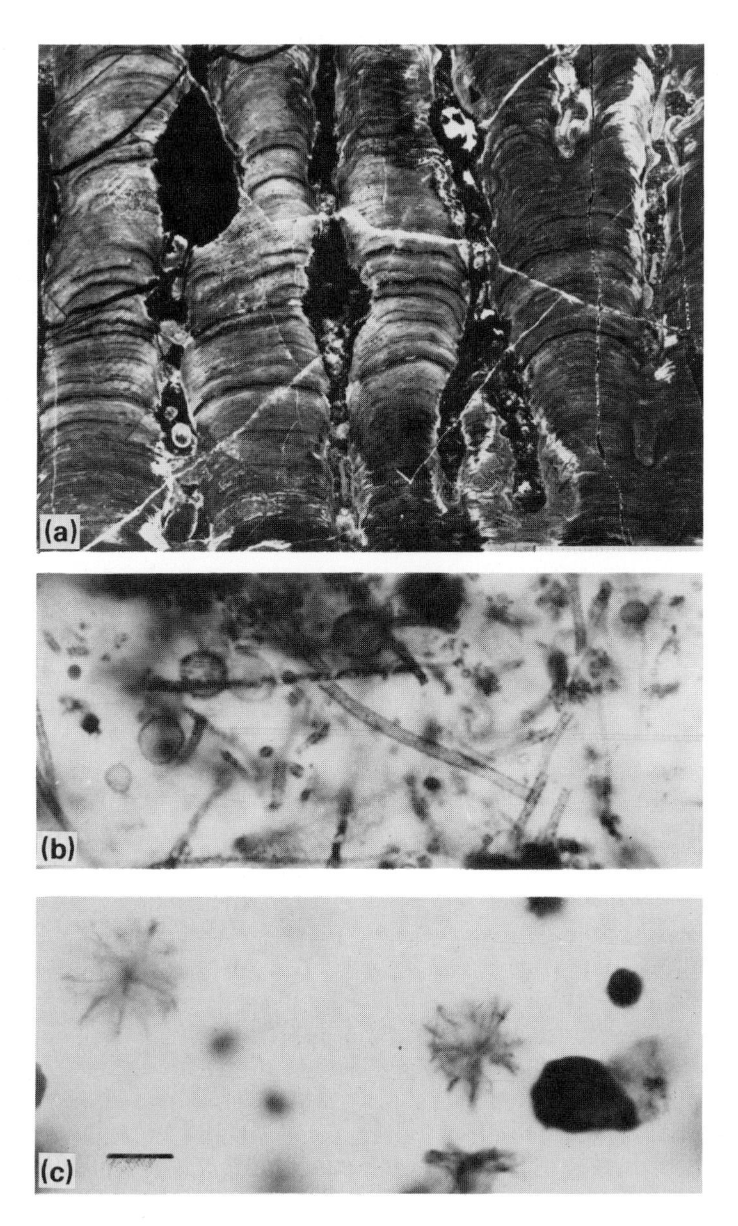

Fig. 10.2. (a) Polished section of columnar stromatolites from the 2000 million-year-old Gunflint Iron Formation, Canada (× 0.85). (b) Stromatolitic microbes from the Gunflint Iron Formation, Canada—filaments = *Gunflintia minuta* Barghoorn, unicells = *Huroniospora* sp. (× 1000). (c) Trichospheric bacteria (*Eoastrion simplex* Barghoorn) from distal non-stromatolitic cherts of the Gunflint Iron Formation (× 1000). The bar in (c) = 10 μm for (b) and (c).

lites are modern in aspect (questions of affinities revolve about physiologically modern cyanobacteria and iron bacteria, not primitive, evolutionarily transitional forms). Second, the Gunflint assemblage is ecologically distinct from that of the Belcher Supergroup, an indication that diverse microbial mat ecosystems existed 2000 million years ago.

The diversity of mat-building microbial associations is further attested to by the variety of stromatolite morphologies present in Early Proterozoic rocks (Donaldson 1976, Bertrand-Sarfati & Eriksson 1977, Hofmann 1977, Semikhatov 1978). Although the great majority of these stromatolites do not contain preserved microfossils, their differing microstructural organizations suggest that stromatolite diversity is indeed a reflection of microbial community diversity. Evidently, the environmental heterogeneity of the Early Proterozoic sea floor permitted the differentiation of many distinct mat-building communities.

Microfossils are widely distributed in younger Proterozoic stromatolites (Fig. 10.3), and like the Belcher assemblage they reveal a picture of mat building by modern cyanobacterial communities (Schopf 1978b); for example, in the Bitter Springs Formation (Schopf 1968, Schopf & Blacic 1971), distinct mat-building associations are variously dominated by *Phormidium/Lyngbya*-like filaments, sheathless oscillatorians, or entophysalidalean colonies (Knoll & Golubic 1979, D.Z.Oehler *et al.* 1979). Mat-dwelling or 'guest' microbes (Golubic 1976) are diverse and modern in aspect. Similarly, in the approximately coeval Narssârssuk Formation of northwestern Greenland (Strother *et al.* 1983; and see Fig. 10.1 (a,b,d,e)) four distinct cyanobacterial assemblages are found in different facies of an ancient sabkha-lagoonal environment. Stromatolitic microstructures suggest that as many as eleven mat-building communities were present in the Narssârssuk lagoon, with such factors as water agitation, frequency and duration of subaerial exposure, and presence or absence of detrital particles determining the distribution of the biological elements. Different microbial assemblages are preserved in other stromatolites of late Precambrian age and, as in the case of the Early Proterozoic, the impressive variety of stromatolite types present (most of which do not contain microfossils) indicate the ecological diversity of microbial mat ecosystems in late Precambrian times. (The dramatic decline in global stromatolite diversity and abundance documented in Lower Palaeozoic rocks can be attributed to the evolution of multicellular animals (Awramik 1971). Metazoans restricted the environments available to mat-building microbes through grazing, burrowing and competition for space. This last factor appears to be particularly important in that the *Götterdämmerung* of carbonate shelf stromatolites coincides in time with the Ordovician radiation of bryozoans, rugose and tabulate corals, and articulate brachiopods.

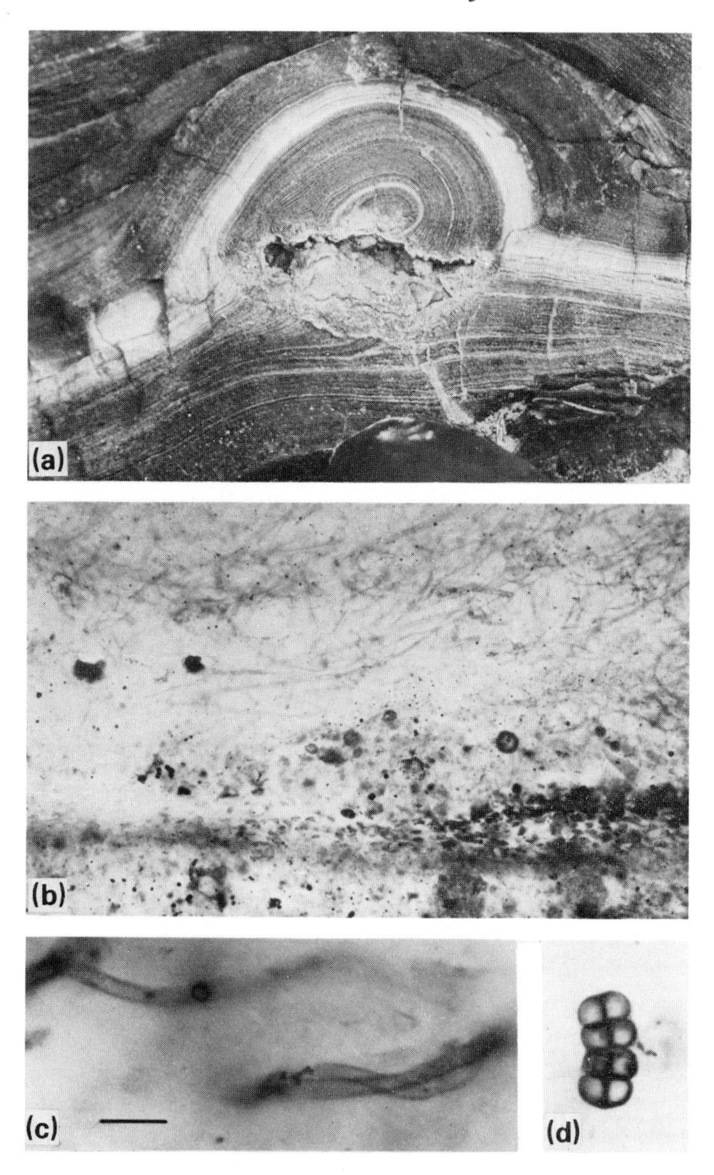

Fig. 10.3. (a) Domal stromatolite from the late Precambrian Narssârssuk Formation, Greenland (× 1—note end of hammer handle for scale). (b) Low-power microscopic view of a silicified flat laminate microbial mat from the late Precambrian Draken Conglomerate, Svalbard—note the abundant interwoven filamentous cyanobacteria and associated unicells (× 166). (c) Cyanobacterial sheaths (*Eomycetopsis* sp.) that weave the fabric shown in (b)—note that the trichomes internal to the sheaths are not preserved (× 1000). (d) Mat-dwelling or 'guest' cyanobacteria found within the Svalbard mats (× 1000). The bar in (c) = 10 μm for (c) and (d) and 60 μm in (b).

The lateral variation in stromatolite morphologies is matched by stratigraphic variation among Precambrian stromatolites. The practical consequence of this is that stromatolites can be used for the biostratigraphic subdivision of the Proterozoic Eon, or at least the latter half of it (Keller *et al.* 1960, Preiss 1976, Semikhatov 1976). The biological implication is that although the microbial associations comprising mat ecosystems have been modern in aspect for at least the last 2000 million years, evolution has continued in the mat-building cyanobacterial biota.

If the record of mat ecosystems can be traced forward from the Early Proterozoic with confidence, it is somewhat more difficult to trace that record backward in time. Archaean stromatolites are known from North America, Africa and Australia—microbially laminated sedimentary rocks in the Pongola Supergroup of South Africa are nearly 3000 million years old (Mason & von Brunn 1976), and recently discovered domal stromatolites from the Warrawoona Group, Western Australia, extend the record to 3500 million years (Lowe 1980, Walter *et al.* 1980). No microfossils are known from these stromatolites, but Awramik *et al.* (1983) have discovered filamentous microfossils in flat laminated cherts in the Warrawoona sequence.

Archaean stromatolites are found in carbonates (or cherts) and are similar morphologically to younger cyanobacterially built forms. The uniform biological organization of modern and, apparently, most fossil marine mat ecosystems suggests that Archaean stromatolites should be taken to indicate the great antiquity of this ecosystem organization. Certainly, the studies of Schopf *et al.* (1971) on 2800–2600 million-year-old stromatolites from near Bulawayo, Zimbabwe, demonstrate that the macrostructure, microstructure, and carbon isotope geochemistry of these stromatolites are consistent with such an interpretation. It should be noted, however, that in the absence of diagnostic microfossils, one cannot completely eliminate the possibility that the oldest stromatolites were built by anaerobic photoautotrophs or other bacteria (Walter 1978). Whatever the taxonomic nature of the constituent biota, Archaean stromatolites do demonstrate that complex microbial mat ecosystems are very nearly as ancient as the earth's oldest surviving rocks.

10.4 NON-STROMATOLITIC ECOSYSTEMS

Although stromatolites and stromatolitic microorganisms have received a great deal of attention from palaeontologists, non-stromatolitic microfossils, both benthic and planktonic, are also well represented in Precambrian rocks, and it is clear that these organisms played an important role in the early marine biota.

A planktonic mode of life can be inferred from a taxon's distribution in the rocks. Because benthic organisms are often specifically adapted to a single environmental zone, their fossil remains are likely to be confined to a restricted range of sedimentary facies. In contrast, the fossils of planktonic microbes are often widely distributed in different lithologies, a consequence of the dimensions of the planktonic niche. On a much smaller scale, planktonic and benthic microfossils can sometimes be distinguished by their distribution within a single petrographic thin-section. Benthic microbes often occur in large concentrations along bedding planes, while plankters are commonly scattered randomly throughout the section. This distinction is not absolute; indeed the distinction between the benthic and planktonic modes of life is not always well defined for shallow marine microorganisms. Very rare benthos and individuals transported to the site of burial from another environment may further complicate the picture.

As in the general case for the Precambrian microfossil record, preservational biases colour our perception of ancient non-stromatolitic assemblages. Some of these biases are geological. Currents frequently winnow small, light organic remains from relatively coarse-grained sediments and deposit them in finer-grained facies. Although presumably planktonic microfossils have been found in a wide variety of lithologies ranging from quartzitic sandstone to black shale, they are most often encountered in grey to green siltstone because it is in these rocks that the optimum conditions of current sorting, rapid burial, and a reducing diagenetic environment are met. A second geological bias is more restrictive; because plate tectonic processes continually destroy the sea floor, we have no record of Precambrian microbial life in the open ocean.

Biochemical biases also affect non-stromatolitic assemblages. Preservation favours those organisms having relatively decay-resistant walls, cysts, or spores, although the primary or early diagenetic deposition of silica may preserve extremely delicate forms.

Geological and geochemical evidence suggests that the Archaean biota must have been physiologically and, presumably, ecologically diverse; however, little is known of early planktonic communities. Walker (1978) has hypothesized that in the absence of an atmospheric ozone shield to filter out destructive ultraviolet radiation, planktonic photoautotrophs would have led a marginal existence. If true, this would mean that the initial radiation of marine prokaryotes into the near surface regions of the planktonic realm took place *after* the rise of atmospheric oxygen some 2200–2300 million years ago (see section 10.5). Unfortunately, present geobiological data do not allow critical evaluation of this hypothesis.

Early Proterozoic plankters are known with more certainty. Large (5–31 μm) spherical microbes of uncertain systematic position are found scattered

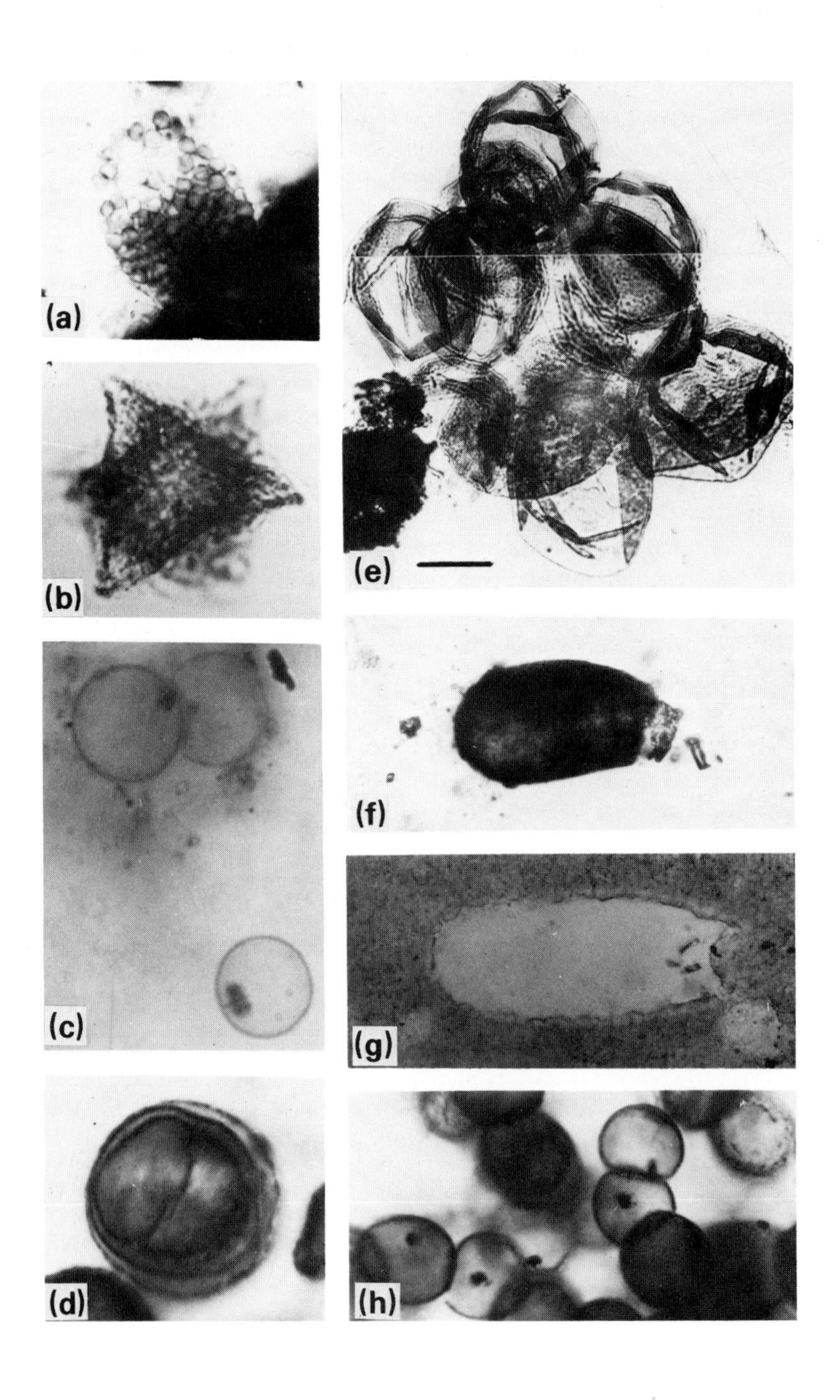
(a)
(b)
(c)
(d)
(e)
(f)
(g)
(h)

throughout distal cherty facies of the Gunflint Iron Formation (Knoll *et al.* 1978; see Fig. 10.4 (c)), and presumed planktonic unicells have also been reported from the Belcher Supergroup (Hofmann 1976). Trichospheric bacteria, morphologically indistinguishable from some modern bacteria that are capable of oxidizing iron and manganese, are also abundant in non-stromatolitic Gunflint cherts (Fig. 10.2). The occurrence of dense ($\sim 10^7$ colonies cm^{-3} rock), virtually monospecific assemblages of these bacteria in organic-rich cherts associated with siderite ($FeCO_3$) suggests that these Early Proterozoic microbes may well have been physiologically, as well as morphologically, similar to their extant counterparts. Considered as a whole, the Gunflint microbiota presents a picture of an ecologically complex marine platform in which stromatolitic communities became established in aerobic habitats, while heterotrophic bacteria proliferated in oxygen-starved waters. The co-occurrence of presumable microaerophilic Fe and Mn oxidizing bacterial microbenthos and the spheroidal plankters in the chert-siderite facies of the formation suggests that the Gunflint environment was zoned vertically as well as horizontally as a function of O_2 concentration gradients.

1400 million years ago appears to be a signal date in the history of planktonic microorganisms (Timofeev 1973, Schopf & Oehler 1976). In shales and siltstones deposited before this time, microfossils are small, rare, and generally rather poorly preserved. Younger clastic rocks, on the other hand, often contain abundant and diverse acritarchs (a term coined for organic-walled microfossils that cannot confidently be related to any single extant taxonomic group). This transition reflects the introduction into the

←

Fig. 10.4. Non-stromatolitic Precambrian microbes. (a) *Bavlinella faveolata* Schepeleva em. Vidal, the dominant planktonic photoautotroph in the glacially influenced late Precambrian Mineral Fork Formation, Utah (× 1000). (b), (e) and (g) Planktonic microfossils from the late Precambrian Visingsö Beds, Sweden—(b) = *Octoedryxium truncatum* Rudavskaja em. Vidal (× 1000); (e) = *Kildinella hyperborica* Timofeev (× 400); (g) = vase-shaped heterotrophic protist preserved in a phosphate nodule (× 400). (c) Planktonic microbes (*Leptoteichos golubicii* Knoll *et al.*) from non-stromatolitic cherts of the 2000 million-year-old Gunflint Iron Formation, Canada (× 1000). (d) and (h) Microfossils from the late Precambrian Bitter Springs Formation, Australia—(d) = Chroococcoid cyanobacterium (*Gloeodiniopsis lamellosa* Schopf em. Knoll and Golubic) interpreted as inhabiting or being trapped in evanescent pools associated with intertidal entophysalid mats; (h) = planktonic unicells (*Glenobotrydion aenigmatis* Schopf) that fell into and were preserved in flat laminated stromatolites (both × 1000). (f) Heterotrophic protist from the late Precambrian Backlundtoppen Formation, Svalbard—note distinct collar region of theca (× 400). The bar in (e) = 25 μm for (e)–(g) and 10 μm for (a)–(d) and (h). ((b) and (e) courtesy of G. Vidal, reprinted by permission from *Fossils and Strata* 9, Universitetsforlaget, Oslo.)

planktonic ecosystem of a new group of microorganisms capable of synthesizing decay-resistant wall structures. Microfossils preserved in 1300–1400 million-year-old shales of the Roper Group of northern Australia exemplify these newly evolved biotas (Peat *et al.* 1978). Roper shales contain at least 14 distinct types of spheroidal microfossils ranging in diameter from < 10 to > 500 μm and evidencing both endospory and simple binary division cycles. Among the filamentous algae also represented in the assemblage are oscillatorian-like sheaths and trichome fragments as large as 120 μm in cross-sectional diameter. Clearly, the Roper microbiota demonstrates that by 1400 million years ago, shallow marine planktonic ecosystems included diverse communities of structurally complex and decay-resistant organisms. Schopf and Oehler (1976) and others (including Peat *et al.* 1978) have voiced the reasonable hypothesis that the evolutionary event responsible for the observed fossil record was the radiation of nucleated algae. One fascinating, but unanswered, question concerning this radiation is whether eukaryotic phytoplankton diversified in an undersaturated environment or whether they displaced a pre-existing cyanobacterial biota.

Abundant acritarchs contained in the 800–600 million-year-old Visingsö Beds of southern Sweden (Vidal 1976) bear witness to the fact that throughout the late Precambrian the diversity of planktonic ecosystems grew. (Like the stromatolites found in contemporaneous carbonates, planktonic microfossils preserved in detrital rocks can be used in the biostratigraphic subdivision and correlation of late Precambrian sedimentary sequences (Volkova 1968, Timofeev 1969, 1973, Vidal 1976).) While the more than two dozen differentiable taxa of the Visingsö Beds (Fig. 10.4 (b),(e)) only hint at the true diversity of coastal seas in late Precambrian times, they include firm evidence for a new functional element in marine ecosystems, the heterotrophic protists. Vase-shaped, sturdy loricas 60–130 μm long and sometimes possessing distinct collar regions are preserved in phosphate nodules from the upper Visingsö Beds (Knoll & Vidal 1980; and see Fig. 10.4 (g)).

These remains are similar morphologically to several groups of loricate protists, especially the Tintinnida, a group of ciliate protozoans that in present-day marine environments are important predators on bacteria, algae, and other protists. Additionally, although these forms lack opercula and do not form chains of individuals, they do bear a close morphological resemblance to certain simple forms of chitinozoans (Bloeser *et al.* 1977). It is of interest to note that Reid and John (1980) believe that the Chitinozoa are polyphyletic and that some morphologically simple taxa are, indeed, the cysts of Palaeozoic tintinnids. Initially discovered in shales from the Grand Canyon, Arizona (Bloeser *et al.* 1977), these distinctive vase-shaped fossils are now also known from Brazil (Fairchild *et al.* 1978), East Greenland

(Vidal 1979), Saudi Arabia (Binda & Bokhari 1980), and Svalbard (Knoll 1981), not to mention southern Sweden. The addition of sophisticated heterotrophic protists to marine ecosystems in the late Precambrian undoubtedly contributed to the increasing complexity of planktonic communities.

Just as one can demonstrate ecologically controlled variation in the taxonomic composition of ancient mat-building cyanobacterial communities, one can document environmental differences among late Precambrian planktonic assemblages; for example, microfossils preserved in siltstones from the approximately 800 million-year-old Mineral Fork Formation, Utah, are quite unlike those of the Visingsö Beds in that they exhibit very low taxonomic diversity and are in fact dominated by a single species (Knoll *et al.* 1981). Such a palaeocommunity structure is suggestive of ecological stress. Sedimentary structures in the Mineral Fork rocks indicate that the fossiliferous sediments accumulated in a shallow marine embayment or fjord, marginal to a retreating glacier. High concentrations of nutrients introduced into the bay (a consequence of the leaching of exposed tills and outwash sediments), coupled with the rigorous salinity and temperature regime associated with the influx of glacial meltwater, produced the stressed conditions, allowing one organism (*Bavlinella faveolata*, normally a rare component of late Precambrian planktonic assemblages, Fig. 10.4) to proliferate and dominate the primary production level of the glacially influenced ecosystem.

Unicellular microfossils preserved in silicified microbial mats provide a final example of ecological variation among late Precambrian non-stromatolitic organisms. Flat laminate, possibly lagoonal, stromatolites from the Bitter Springs Formation contain abundant specimens of small (8–15 μm diameter) spheroidal unicells given the name *Glenobotrydion aenigmatis* (Fig. 10.4 (h)). The random distribution of *Glenobotrydion* specimens in Bitter Springs cherts led Schopf (1968) to conclude that these organisms were plankters that fell into and were preserved with the microbial mat assemblage. Confirming evidence for this view can be found in grits and fine conglomerates of the approximately coeval Draken Formation, Svalbard, where flakes of thin mats containing fossils of a mat-building community, very similar to that of the Bitter Springs, were deposited with grains and small pebbles of non-stromatolitic micritic mud, in part silicified, which contain abundant *Glenobotrydion* individuals (Knoll 1982). Clearly, these organisms were important photoautotrophs in the very shallow marine waters that bathed the flat laminate mats. Intertidal mats from the Bitter Springs Formation also contain large populations of chroococcoid cyanobacteria that inhabited or were trapped in very small evanescent pools of water occasionally left on uneven mat surfaces by retreating tides (Knoll &

Golubic 1979; and see Fig. 10.4 (d)). Many of the acritarchs present in late Precambrian siltstones and shales have not been identified as allochthonous elements in stromatolitic assemblages; conversely, some plankters preserved among microbial mat organisms are unknown from detrital facies. Such a distribution is at least consistent with the hypothesis of a partial ecological barrier between lagoonal environments and open continental shelf waters (Knoll 1982). Taken together, the Visingsö, Mineral Fork, and Bitter Springs reveal some measure of diversity and complexity of non-stromatolitic Precambrian microbial systems.

10.5 A BRIEF MODEL OF EARLY BIOLOGICAL, GEOLOGICAL AND ATMOSPHERIC EVOLUTION

In his famous *Principles of Geology* (1830–33), Charles Lyell forcefully advocated the Huttonian theory of a steady-state system in earth history. To Lyell, the opposing processes of uplift and subsidence, erosion and sedimentation, maintained a balance such that not only did the rock record of geological history evidence 'no vestige of a beginning,—no prospect of an end' (Hutton 1788), it showed no signs of directional change through time. Lyell's major stumbling block was, of course, the fossil record which clearly indicated the trajectory of time's arrow through sedimentary rocks. In the past two decades, intensive research on the Precambrian rock record has shown that it is not only the earth's biota that has changed through time; the atmosphere and the crust itself have evolved (e.g. Cloud 1976, Windley 1977).

From the preserved geological record, one can infer temporal patterns of biological, crustal, and atmospheric history. All three patterns show apparent transitions approximately 2300–2200 million years ago, during the Early Proterozoic Era (Fig. 10.5). The thesis of this brief summary is that these transitions are coincident in time because the courses of atmospheric and biological evolution have not been independent of shifts in the earth's crustal configuration (Knoll 1979a).

10.5.1 The Archaean

Direct evidence bearing on the nature of Archaean microbial ecosystems is meagre. Presumable spheroidal microfossils as old as 3400 million years are known from organic-rich cherts from South Africa (Muir & Grant 1976, Knoll & Barghoorn 1977), and as mentioned earlier, filamentous fossils have been discovered in approximately coeval rocks from Australia (Awramik

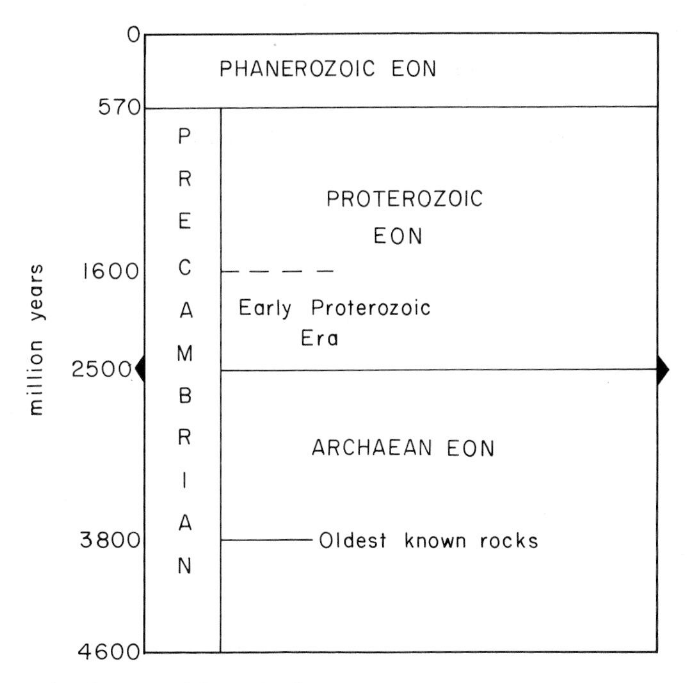

Fig. 10.5. The Geological Time Scale.

et al. 1983); these remains are morphologically simple and convey little physiological information. More enlightening in this respect are the stromatolites found in Archaean rocks. As discussed in section 10.3, the presence of stromatolitic structures in rocks as old as 3500 million years (Lowe 1980, Walter *et al.* 1980) prompts the hypothesis that a community structure and distribution similar to that characterizing modern mat-building microbial ecosystems existed in the Archaean. This would imply that aerobic and anaerobic photoautotrophs, chemolithotrophs and anaerobic heterotrophs were all present in early shallow marine habitats. Unfortunately, no Archaean stromatolites are known to contain preserved microfossils. Based on observations of flexibacteria which form part of the mat-building community in hot springs of Yellowstone National Park, Walter (1972, 1978) has raised the distinct possibility that early stromatolites could have accreted without cyanobacteria.

Ratios of stable carbon isotopes of Archaean organic matter (including kerogen removed from early stromatolites) strongly suggest photosynthetic isotope fractionation by ancient photoautotrophs (Eichmann & Schidlowski 1975). Indeed, on the basis of isotopic analyses of the world's oldest rocks, 3800 million-year-old metasediments from the Isua region of southwestern Greenland, Schidlowski *et al.* (1979) have inferred that, in the early

Archaean, the carbon cycle worked much as it does today. Although the caveats discussed in section 10.2.3 must be borne in mind, the clear implication of this study is that life began very early in earth history and that photosynthetic prokaryotes, must have appeared soon afterward.

It is attractive to hypothesize that aerobic photosynthesizers evolved 3800 million years ago or earlier, because of the ubiquity of the H_2O molecule used as an electron donor in oxidative photosynthesis; however, set against this is geological evidence suggesting that the Archaean atmosphere contained very little free O_2 (Walker 1977; see explanation of evidence in section 10.5.2). This apparent anomaly can be reconciled if one can envision a biosphere in which rates of O_2 consumption equal or exceed those of oxygen production (Margulis *et al.* 1976). A brief consideration of Archaean geology demonstrates that the early earth could well have been such a system (Knoll 1979).

A diverse array of stratigraphic, structural and geochemical evidence can be marshalled in support of the hypothesis that the early Archaean earth was characterized by abundant vulcanism and high levels of tectonic activity, but only limited areas of stable continental crust (e.g. Windley 1977; references cited in Knoll 1983). The effusion of volcanic gases, the weathering of minerals and buried organic matter exhumed by rapid rates of erosion associated with active tectonism, the oxidation of newly produced organic matter, and the oxidation of reduced mineral species released from crustal rifts would all contribute to high rates of oxygen consumption. At the same time, actualistic consideration of the sedimentary environments available to early photoautotrophs suggests that global rates of primary production were relatively low. Productivity data for modern benthic microfloras are few, but enough information does exist to indicate that microbes living in environments characterized by high rates of clastic influx have primary production rates one or two orders of magnitude lower than those of stromatolitic communities living in quieter water. (Compare the rates measured by Krumbein and co-workers (1977) for mat-building communities from Solar Lake and Sabkha Gavish, Sinai—1.5 to 12 and 1.8 to 9 g C m^{-1} d^{-1}—with those reported for benthic microfloras from shallow marine clastic bottom areas of the tropics, temperate zone North American estuaries, and Scotland—0.02–0.22, 0.08–0.53, and 0.01–0.03 g C m^{-2} d^{-1} (Bunt 1975).) The relatively low rates of primary production characterizing many Archaean environments, coupled with the inferred limited areal extent of early shallow marine habitats, would have resulted in relatively low total O_2 production by benthic microorganisms.

The foregoing arguments are not meant to suggest that no highly productive shallow platform or quiet water environments existed during the Archaean eon; such environments are represented in the sedimentary rock record, and it is in these strata that the oldest stromatolites are found. The point of the argument is that in the Archaean, the total extent of such environments was small relative to the succeeding Proterozoic eon.

Estimation of planktonic productivity is more difficult, but the absence of an effective ozone shield in the atmosphere, if not completely precluding the occupation of the planktonic niche by photoautotrophs, may well have barred cyanobacterial plankton from the upper few metres of ocean surface water (Walker 1978). In short, the sedimentary/tectonic nature of the Archaean provides a reasonable mechanism whereby O_2 could have been produced by physiologically modern cyanobacteria, but have been kept at low atmospheric concentration by high levels of O_2 consumption. Hematite in widely distributed Archaean iron formations makes it clear that *some* O_2 was available in the early ocean, but concentrations apparently remained low. (Although many authors contend that only a photosynthetic source could have produced the volumes of O_2 bound in Archaean iron formation, the possibility of abiological O_2 production via the photodissociation of H_2O molecules in the upper atmosphere should not be dismissed out of hand (e.g. Towe 1978). Possible modern analogues for Archaean prokaryotes may be those found in microbial mats of Sabkha Gavish in the Gulf of Elat (Krumbein *et al.* 1979). Here oxygenic cyanobacteria are able to coexist with anaerobic photosynthetic bacteria, sulphate reducers and anaerobic heterotrophs in a generally reducing environment by creating oxidizing micro-environments.

It is obvious that much in the above account is speculative, but as Charles Darwin wrote in 1857, 'without speculation there is no good original observation'. Continuing geological, geochemical and palaeobiological investigations of Archaean sedimentary sequences will test the model and clarify many uncertainties over the next decade. What does appear certain, however, is that by the end of the Archaean eon, defined as 2500 million years ago, prokaryotes had evolved most, if not all, of the manifold physiological pathways found in modern microorganisms. Aerobic respiration may not have been characteristic of Archaean microbes, but it certainly must have evolved by the Early Proterozoic. The various microbially mediated cycles discussed in previous chapters must have operated 2000 million years ago in a manner essentially similar, at least in qualitative terms, to the present-day biosphere.

10.5.2 The Proterozoic

The Archaean–Proterozoic transition marks a period major change in patterns of earth surface geology (Windley 1977, Salop 1977). Following a terminal Archaean episode of extensive continental growth and stabilization, a new tectonic/sedimentary regime characterized by large cratonic masses flanked by broad mio- and eugeo-synclines and flooded by shallow epicontinental seas became established in the Early Proterozoic era. Two ecological consequences of this geological transition are of particular import.

(1) The area available for colonization by benthic photoautotrophs must have increased substantially. Engel *et al.* (1974) have calculated that the ratio of quartzite plus carbonate to other sediments shifts from 1:100 in the Archaean to about 1:5 in the Proterozoic. Windley (1977, p. 335) regards this as an approximate estimate of the increase in stable shelf and platform environments at this time, suggesting a twentyfold increase in area.

(2) The average ecological rigour of these shallow-water environments decreased significantly—one can contrast the thick sequences of carbonates that accumulated over wide stretches of the Early Proterozoic sea floor to the highly clastic habitats of the Archaean.

The diverse forms of stromatolites that are so characteristic of these earliest widespread platform carbonates (see section 10.3) demonstrate that cyanobacterial communities quickly took advantage of the new environments. As highly productive blue-green mat communities colonized the shallow marine platforms, total benthic primary production must have risen substantially. At the same time the hypertensive rates of tectonic and volcanic activity that had predominated during the Archaean abated considerably. The terminal Archaean growth of continents have increased greatly the amount of phosphate entering the oceans via continental erosion, and this may have had important consequences for the productivity of oceanic plankton communities. Upwelling (Chamberlain & Marland 1977) and the establishment of a stratispheric ozone layer would have further increased primary production. Hence, one might predict that at last rates of photosynthetic O_2 production exceeded those of O_2 consumption, and indeed the geological record suggests that this is exactly what happened.

Earliest Proterozoic continental sedimentary rocks of four continents contain detrital uraninite (UO_2), an easily oxidizable mineral that is unlikely to survive even minimal exposure to an O_2-rich environment—Grandstaff (1973) has observed that O_2 in concentrations as low as 1% PAL quickly reacts with uraninite to form other uranium oxides. Thus, the contemporaneous deposition of uraninite-bearing continental sediments in widely scattered areas furnishes a strong argument that atmospheric O_2 concentrations were very low at the dawn of the Proterozoic.

Detrital uraninites are not found in any rocks deposited after the earliest

Proterozoic (Robertson *et al.* 1978).* Following the deposition of the initial Proterozoic sandstones and conglomerates, the first widespread stromatolitic carbonates accumulated, and among the clastic sediments associated with these platform and shelf deposits are red beds, siltstones and sandstones in which flecks of oxidized iron, especially hematite, occur among the clastic grains. Red beds appear to be a lithological species unknown before the Proterozoic (Salop 1977, Windley 1977, Robertson *et al.* 1978). From the time relations of uraninite-bearing alluvium and continental red beds, one can infer that the Early Proterozoic sedimentary record documents the initial rise of O_2 concentrations in the atmosphere (Cloud 1968, 1974, Frarey & Roscoe 1970) at exactly the time one would expect if photosynthetic productivity were linked to the tectonic control of sedimentary environments (Knoll 1979a).

According to the model discussed in this section, the coincidence in time of fundamental transitions in the nature of the Earth's tectonic framework, atmosphere and biota was not fortuitous—the events are related by microbial ecology. It is only fair to note that other scenarios have been proposed to explain the geological and palaeontological record of the early earth (Cloud 1976, Schopf 1978a), but these theories postulate a major evolutionary event, namely the evolution of physiologically modern cyanobacteria, to explain the Early Proterozoic atmospheric transition and require that the stromatolites, microfossils, carbon isotope ratios, reduced carbon production and distribution, and iron formations of the Archaean be interpreted in different terms from their Proterozoic and younger counterparts (see Knoll (1979) for a discussion of assumptions made in theories of early evolution and their consequences for the interpretation of Archaean palaeobiological data).

10.6 SUMMARY

(1) Any element cycle containing a biological link through a (non-intracellular) anaerobic environment requires a prokaryote. Only the silica cycle requires eukaryotes. Therefore, with the exception of the silica cycle, whose biological components do indeed appear to have Phanerozoic origins, the

* This is not entirely true. Although detrital uraninites are not found in any *lithified* rocks younger than about 2300 million years of age, Simpson and Bowles (1977) have described detrital uraninite, gold and pyrite in the sediment load of the modern Indus River. This illustrates the slenderness of this thread by which many arguments on early atmospheres hang. As summarized by Schopf (1979), most geological evidence still suggests an early Proterozoic transition to an oxygen-rich atmosphere. If this is incorrect, it is most likely to err in the direction of providing too young an estimate of atmospheric oxygenation (Dimroth & Kimberly 1976).

cycles discussed in previous chapters could have been, and probably were, operative early in earth history.

(2) The persistence of presumably primitive microbes in the Earth's present-day biota and the discovery of apparently modern prokaryotes in very ancient rocks suggests that, historically, microbial evolution is better viewed as a process of addition, rather than one of replacement (Knoll 1977).

(3) The coupling of statements (1) and (2) provides a conceptual framework within which the development of complex microbial ecosystems can be discussed. The string that ties together the various biological components of the ecosystem is the cycling of energy-yielding and structurally important compounds; the metabolic products and by-products of one organism constitute the raw materials for the metabolism of another. It is probable that the earliest microbial ecosystems were simple, that cycles had few steps; however, with time, new microbes evolved that were able to use various products formed by pre-existing organisms. The dependence on pre-existing microbes is important because it explains how new biological innovations can increase the complexity of an ecosystem without diminishing the fundamental importance of antecedent microbes. Thus, the evolution of aerobic photoautotrophy reduced the geographic dimensions of the niche occupied by anaerobic heterotrophs, but did not reduce the role of these bacteria in the recycling of reduced molecules. Physiological pathways that evolved early in earth history remain critical to the maintenance of biospheric equilibrium (Lovelock & Margulis 1974).

(4) Looking at microbial evolution as a whole, certain temporal themes are apparent. The Archaean can be viewed as the time during which many of the major metabolic pathways found among prokaryotes evolved and many microbially mediated element cycles became established. With good reason, then, we can refer to the Archaean as the *age of physiological evolution*.

During the succeeding Proterozoic eon, prokaryotic biochemical systems became increasingly fine tuned (as exemplified by the observed ecological heterogeneity of Proterozoic microfossils and stromatolites). Also, once O_2 concentrations reached sufficiently high levels, there began a period of increasingly obligate endosymbiotic associations and concomitant ultrastructural reorganization that culminated in the evolution of the eukaryotic cell. By analogy to the Archaean, the period 2500–1000 million years ago (these are *very* broadly painted boundary dates) can be called the *age of ultrastructural evolution*.

Finally, the late Precambrian and Phanerozoic have witnessed an impressive radiation of multicellular plants and animals in which increasing structural complexity has been superimposed on the major physiological mechanisms present among the microbes. This, then, has been the *age of structural evolution*. The evolutionary products of the latter two 'ages' have vastly increased the complexity of the biosphere, but basic microbial processes have remained at the core of functioning ecosystems. From a microbial perspective, the evolution of plants and animals has merely provided microorganisms with new kingdoms to conquer.

References

Awramik S.M. (1971) Precambrian columnar stromatolite diversity: reflection of metazoan appearance. *Science* **174,** 825–27.

Awramik S.M. & Barghoorn E.S. (1977) The Gunflint microbiota. *Precambrian Res.* **5,** 121–42.

Awramik S.M., Gebelein C.D. & Cloud P. (1978) Biogeologic relationships of ancient stromatolites and modern analogs. In *Environmental Biogeochemistry and Geomicrobiology* (Ed. W.Krumbein), pp. 165–78. Ann Arbor Science, Ann Arbor, Michigan.

Awramik S.M. & Semikhatov M.A. (1979) The relationship between morphology, microstructure, and microbiota in three vertically intergrading stromatolites from the Gunflint Iron Formation. *Can. J. Earth Sci.* **16,** 484–95.

Awramik S.M., Schopf J.W. & Walter M.R. (1983) Filamentous fossil bacteria from the Archean of western Australia. *Precambrian Res.*, in press.

Barghoorn E.S., Knoll A.H., Dembicki H. & Meinschein W.G. (1977) Variation in stable carbon isotopes in organic matter from the Gunflint Iron Formation. *Geochim. Cosmochim. Acta* **41,** 425–30.

Barghoorn E.S. & Tyler S.A. (1965) Microorganisms from the Gunflint Chert. *Science* **147,** 563–77.

Bertrand-Sarfati J. & Eriksson K.A. (1977) Columnar stromatolites from the Early Proterozoic Schmidtsdrift Formation, Northern Cape Province, South Africa—Part I: systematic and diagnostic features. *Palaeont. Afr.* **20,** 1–26.

Binda P.I. & Bokhari M.M. (1980) Chitinozoanlike microfossils in a late Precambrian dolostone from Saudi Arabia. *Geology* **8,** 70–1.

Bloeser B., Schopf J.W., Horodyski R.J. & Breed W.J. (1977) Chitinozoans from the late Precambrian Chuar Group of the Grand Canyon, Arizona. *Science* **195,** 676–9.

Bunt J.S. (1975) Primary productivity of marine ecosystems. In *Primary productivity of the Biosphere* (Eds H.Lieth & R.H.Whittaker), pp. 169–84. Springer-Verlag, Heidelberg.

Chamberlain W.M. & Marland G. (1977) Precambrian evolution in a stratified global sea. *Nature* **265,** 135–6.

Cloud P.E. (1965) Significance of the Gunflint (Precambrian) microflora. *Science* **148,** 27–45.

Cloud P. (1968) Atmospheric and hydrospheric evolution on the primitive earth. *Science* **160,** 729–36.

Cloud P.E. (1972) A working model of the primitive earth. *Am. J. Sci.* **272,** 537–48.

Cloud P.E. (1974) Evolution of ecosystems. *Am. Sci.* **62,** 54–66.

Cloud P.E. (1976) Beginnings of biospheric evolution and their biogeochemical consequences. *Paleobiology* **2,** 351–87.

Craig H. (1953) The geochemistry of stable carbon isotopes. *Geochim. Cosmochim. Acta* **3,** 53–92.

Darwin C. (1857) Letter to A.R.Wallace quoted in: George W.B. (1964) *Biologist Philosopher: a study of the life and writings of Alfred Russel Wallace*, p. 1. Abelard and Schuman, London.

Dimroth E. & Kimberley M.M. (1976) Precambrian atmospheric oxygen: evidence in the sedimentary distributions of carbon, sulfur, uranium, and iron. *Can. J. Earth Sci.* **13,** 1161–85.

Diver W.L. (1980) Some factors controlling cryptarch distribution in the late Precambrian Torridon Group. *Abstr. 5th Internat. Palynol. Conf. Cambridge* p. 113.

Doemel W.N. & Brock T.D. (1977) Structure, growth, and decomposition of laminated algal-bacterial mats in alkaline hot springs. *Appl. Environ. Microbiol.* **34,** 433–52.

Donaldson J.A. (1976) Aphebian stromatolites in Canada: implications for stromatolite zonation. In *Stromatolites* (Ed. M.R.Walter), pp. 371–80. Elsevier, Amsterdam.

Dunlop J.S.R., Muir M.D., Milne V.A. & Groves D.I. (1978) A new microfossil assemblage from the Archean of Western Australia. *Nature* **274,** 676–8.

Eichmann R. & Schidlowski M. (1975) Isotopic fractionation between coexisting organic carbon-carbonate pairs in Precambrian sediments. *Geochim. Cosmochim. Acta* **39,** 585–95.

Engel A.E.J., Itson S.P., Engel C.G., Stickney D.M. & Cray E.J. (1974) Crustal evolution and global tectonics: a petrogenic view. *Geol. Sci. Amer. Bull.* **85,** 843–58.

Fairchild T.R., Barbour A.P. & Haralyi N.L.E. (1978) Microfossils in the 'Eopaleozoic' Jacadigo Group at Urucum, Mato Grosso, southwest Brazil. *Bol. IG. Inst. Geociências, Univ. São Paulo* **9,** 74–9.

Frarey M.L. & Roscoe S.M. (1970) The Huronian Supergroup north of Lake Huron. *Geol. Surv. Can. Paper* 70–40, 143–58.

Gebelein C.D. (1974) Biologic control of stromatolite microstructure: implications for Precambrian time stratigraphy. *Am. J. Sci.* **274,** 575–98.

Golubic S. (1973) The relationship between blue-green algae and carbonate deposits. In *The Biology of Blue-Green Algae* (Eds N.G.Carr & B.A.Whitton), pp. 434–72. Blackwell Scientific Publications, Oxford.

Golubic S. (1976) Organisms that build stromatolites. In *Stromatolites* (Ed. M.R.Walter), pp. 113–26. Elsevier, Amsterdam.

Golubic S. & Barghoorn E.S. (1977) Interpretation of microbial fossils with special reference to the Precambrian. In *Fossil algae* (Ed. E.Flugel), pp. 1–14. Springer Verlag, Berlin.

Golubic S. & Hofmann H.J. (1976) Comparison of modern and mid-Precambrian Entophysalidaceae (Cyanophyta) in stromatolitic algal mats: cell division and degradation. *J. Paleontol.* **50,** 1074–82.

Goodwin A.M., Monster J. & Thode H.G. (1976) Carbon and sulfur isotope abundances in Archean iron-formations and early Precambrian life. *Econ. Geol.* **71,** 870–91.

Grandstaff D.E. (1973) Kinetics of uraninite oxidation: implications for the

Precambrian atmosphere. Ph.D. Dissertation, Princeton University, Princeton, New Jersey.
Hoering T.C. (1967) The organic geochemistry of Precambrian rocks. In *Researches in Geochemistry*, vol. 2 (Ed. P.H.Abelson), pp. 89–111. John Wiley & Sons, New York.
Hoering T.C. (1978) Molecular fossils from the Precambrian Nonesuch shale. In *Comparative Planetology* (Ed. C.Ponnamperuma), pp. 243–55. Academic Press, New York.
Hofmann H.J. (1976) Precambrian microflora, Belcher Islands, Canada: significance and systematics. *J. Paleontol.* **50,** 1040–73.
Hofmann H.J. (1977) On Aphebian stromatolites and Riphean stromatolite stratigraphy. *Precambrian Res.* **5,** 175–205.
Hutton J. (1788) Theory of the earth. *Trans. Roy. Soc. Edinburgh* **1**(2), 209–304.
Keller B.M., Kazakov G.A., Krylov I.N., Nuzhnov S.V. & Semikhatov M.A. (1960) New data on the stratigraphy of the Riphean Group (Upper Proterozoic). *Izv. Akad. Nauk SSSR, Ser. Geol.*, 1960(**12**), 26–41 (in Russian).
Knoll A.H. (1977) Paleomicrobiology. In *Handbook of Microbiology*, vol. I (Eds A.I.Laskin & H.A.Lechevalier), pp. 9–29. CRC Press, Cleveland.
Knoll A.H. (1979a) Archean photoautotrophy: some alternatives and limits. *Origins of Life*, **9,** 313–27.
Knoll A.H. (1981) Paleoecology of late Precambrian microbial assemblages. In *Paleobotany, Paleoecology and Evolution* (Ed. K. Niklas), vol. 1, pp. 17–54. Praeger, New York.
Knoll A.H. (1982) Microorganisms of the late Precambrian Draken Conglomerate, Ny Friesland, Spitzbergen. *J. Paleotol.* **56,** 755–790.
Knoll A.H. (1983) The Archaean/Proterozoic transition: a secondary and paleobiological perspective. In *Patterns of Change in Earth Evolution* (Eds H. D. Holland & A. F. Trendall). Dahlem Konferenzen. Springer Verlag, Heidelberg.
Knoll A.H. & Barghoorn E.S. (1977) Archaean microfossils showing evidence of cell division. *Science* **198,** 396–9.
Knoll A.H., Barghoorn E.S. & Awramik S.M. (1978) New microorganisms from the Aphebian Gunflint Iron Formation, Ontario. *J. Paleontol.* **52,** 976–92.
Knoll A.H., Blick N. & Awramik S.M. (1981) Stratigraphic and ecologic implications of late Precambrian microfossils from Utah. *Amer. J. Sci.* **281,** 247–63.
Knoll A.H. & Golubic S. (1979) Anatomy and taphonomy of a Precambrian algal stromatolite. *Precambrian Res.* **10,** 115–51.
Knoll A.H. & Vidal G. (1980) Late Proterozoic vase-shaped microfossils from the Visingsö Beds, Sweden. *Geol. Fören. Stockholm Förh.* **102,** 207–11.
Krumbein W.E., Bucholz H., Franke P., Giani D., Giel C. & Wonneberger K. (1979) A model for the origin of mineralogical lamination in stromatolites and banded iron formations. *Naturwiss.* **66,** 381–89.
Krumbein W.E., Cohen Y. & Shilo M. (1977) Solar Lake (Sinai). 4 Stromatolitic cyanobacterial mats. *Limnol. Oceanogr.* **22,** 635–56.
Lovelock J.E. & Margulis L. (1974) Homeostatic tendencies of the earth's atmosphere. *Origins of Life* **5,** 93–103.
Lowe D.R. (1980) Stromatolites 3,400-Myr old from the Archean of Western Australia. *Nature* **284,** 441–3.
Lyell Charles (1830–33) *Principles of Geology*, 3 vols, London.

Margulis L., Walker J.C.G. & Rambler M. (1976) Reassessment of roles of oxygen and ultraviolet light in Precambrian evolution. *Nature* **264,** 620–4.

Mason T.R. & von Brunn V. (1977) 3.0 G.y. old stromatolites from South Africa. *Nature* **266,** 47–9.

McKirdy D.M. (1974) Organic geochemistry in Precambrian research. *Precambrian Res.* **1,** 75–137.

Monster J., Appel P.W.U., Thode H.G., Schidlowski M., Carmichael C.M. & Bridgwater D. (1979) Sulfur isotope studies in early Archean sediments from Isua, West Greenland: implications for the antiquity of bacterial sulfate reduction. *Geochim. Cosmochim. Acta* **43,** 405–14.

Muir M.D. & Grant P.R. (1976) Micropaleontological evidence from the Onverwacht Group, South Africa. In *The Early History of the Earth* (Ed. B. Windley), pp. 595–604. John Wiley & Sons, New York.

Oehler D.Z. (1978) Microflora of the Middle Proterozoic Balbirini Dolomite (McArthur Group) of Australia. *Alcheringa* **2,** 269–309.

Oehler D.Z., Oehler J.H. & Stewart A.J. (1979) Algal fossils from a late Precambrian, hypersaline lagoon. *Science* **205,** 388–90.

Oehler D.Z., Schopf J.W. & Kvenvolden K.A. (1972) Carbon isotope studies of organic matter in Precambrian rocks. *Science* **175,** 1246–8.

Oehler J.H. (1977) Microflora of the H.Y.C. Pyritic Shale member of the Barney Creek Formation (McArthur Group) Middle Proterozoic of northern Australia. *Alcheringa* **1,** 315–49.

Park R. & Epstein S. (1960) Carbon isotopic fractionation during photosynthesis. *Geochim. Cosmochim. Acta* **21,** 110–26.

Peat C.J., Muir M.D., Plumb K.A., McKirdy D.M. & Norvick M.S. (1978) Proterozoic microfossils from the Roper Group, Northern Territory, Australia. *BMR J. Austral. Geol. Geophys.* **3,** 1–17.

Preiss W.V. (1976) Intercontinental correlations. In *Stromatolites* (Ed. M.R.Walter), pp. 359–70. Elsevier, Amsterdam.

Reid & John (1980) A possible relationship between chitinozoa and tintinnids. *Abstr. 5th Internat. Palynol. Conf., Cambridge*, p. 332.

Robertson D.S., Tilsley J.E. & Hogg G.M. (1978) The time-bound character of uranium deposits. *Econ. Geol.* **73,** 1409–19.

Salop L.J. (1977) *Precambrian of the Northern Hemisphere*. Elsevier, Amsterdam.

Schidlowski M., Appel P.W.U., Eichmann R. & Junge C.E. (1979) Carbon isotope geochemistry of 3.7×10^9-yr.-old Isua sediments, West Greenland: implications for the Archaean carbon and oxygen cycles. *Geochim. Cosmochim. Acta* **43,** 189–99.

Schopf J.W. (1968) Microflora of the Bitter Springs Formation, late Precambrian, central Australia. *J. Paleontol.* **42,** 651–88.

Schopf J.W. (1974) The development and diversification of Precambrian life. *Origins of Life* **5,** 119–35.

Schopf J.W. (1978a) The evolution of the earliest cells. *Sci. Amer.* **239(3),** 110–38.

Schopf J.W. (1978b) Biostratigraphic usefulness of stromatolitic Precambrian microbiotas: a preliminary analysis. *Precambrian Res.* **5,** 143–73.

Schopf J.W. & Blacic J.M. (1971) New microorganisms from the Bitter Springs Formation (late Precambrian) of the North-central Amadeus Basin, Australia. *J. Paleontol.* **45,** 925–60.

Schopf J.W. & Oehler D.Z. (1976) How old are the eukaryotes? *Science* **193,** 47–9.

Schopf J.W., Oehler D.Z., Horodyski R.J. & Kvenvolden K.A. (1971) Biogenicity and significance of the oldest known stromatolites. *J. Paleontol.* **45,** 477–85.

Schopf T.J.M. (1979) *Paleooceanography*. Harvard, Cambridge Mass.

Scrope G.P. (1858) *The Geology and Extinct Volcanoes of France*. London.

Semikhatov M.A. (1976) Experience in stromatolite studies in the U.S.S.R. In *Stromatolites* (Ed. M.R.Walter), pp. 337–58. Elsevier, Amsterdam.

Semikhatov M.A. (1978) Aphebian assemblage of stromatolites: general characteristics and comparison with the Riphean one. In *Lower Boundary of the Riphean and Stromatolites of the Aphebian*, pp. 148–58, Acad. Sci. USSR, Izd.-vo, Nauka, Trudy 312 (in Russian).

Semikhatov M.A., Gebelein C.D., Cloud P., Awramik S.M. & Benmore W.C. (1979) Stromatolite morphogenesis: progress and problems. *Can. J. Earth Sci.* **16,** 992–1015.

Simpson P.R. & Bowles J.F.W. (1977) Uranium mineralization of the Witwatersrand and Dominion Reef systems. *Phil. Trans. Royal Soc. London* **286A,** 527–48.

Sklarew D.S. & Nagy B. (1979) 2,5-dimethylfuran from 2.7×10^9-year-old Rupemba-Belingwe stromatolite, Rhodesia: potential evidence for remnants of carbohydrates. *Proc. Nat. Acad. Sci., U.S.A.* **76,** 10–14.

Strother P.K., Knoll A.H. & Barghoon E.S. (1983) Microfossils from the late Precambrian Narssârssuk formation, northwestern Greenland. *Palaentology* **26,** 1–32.

Timofeev B.V. (1969) *Sphaeromorphida of the Proterozoic*. Izd. Akad Nauk SSSR (in Russian).

Timofeev B.V. (1973) Plant microfossils from the Proterozoic and lower Paleozoic. In *Microfossils of the Oldest Deposits: Proceedings of the 3rd Intern. Palynol. Conf.* (Eds T.F.Vozzhennikova & B.V.Timofeev), pp. 7–12. Publ. House 'Nauka', Moscow (in Russian).

Towe K.M. (1978) Early Precambrian oxygen: a case against photosynthesis. *Nature* **274,** 651–61.

Tyler S.A. & Barghoorn E.S. (1954) Occurrence of structurally preserved plants in Precambrian rocks of the Canadian Shield. *Science* **119,** 606–8.

Vidal G. (1976) Late Precambrian microfossils from the Visingsö Beds in southern Sweden. *Fossils and Strata* **9,** 57.

Vidal G. (1979) Acritarchs from the Upper Proterozoic and Lower Cambrian of East Greenland. *Grönls. Geol. Undersögelse Bull.* **134,** 1–40.

Volkova N.L. (1968) Acritarcha of Precambrian and Lower Cambrian deposits of Estonia. *Akad. Nauk SSSR, Geol. Inst.* **188,** 8–36.

Walker J.C.G. (1977) *Evolution of the Atmosphere*. Macmillan, New York.

Walker J.C.G. (1978) The early history of oxygen and ozone in the atmosphere. *Pure Applied Geophys.* **116,** 222–31.

Walter M.R. (1972) A hot spring analog for the depositional environment of Precambrian iron formations. *Econ. Geol.* **67,** 965–72.

Walter M.R. (1978) Recognition and significance of Archaean stromatolites. *Publs. Geol. Dept. and Extension Service, Univ. Western Australia* **2,** 1–10.

Walter M.R., Buick J.R. & Dunlop J.S.R. (1980) Stromatolites 3400–3500 Myr old from the North Pole area, Western Australia. *Nature* **284,** 443–5.

Windley B.F. (1977) *The Evolving Continents*. John Wiley & Sons, New York.

INDEX